Lecture Notes in Bioinformatics 13447

Subseries of Lecture Notes in Computer Science

More information about this subseries at https://link.springer.com/bookseries/5381

Ion Petre · Andrei Păun (Eds.)

Computational Methods in Systems Biology

20th International Conference, CMSB 2022
Bucharest, Romania, September 14–16, 2022
Proceedings

 Springer

Editors
Ion Petre ⓘ
University of Turku
Turku, Finland

Andrei Păun ⓘ
University of Bucharest
Bucharest, Romania

ISSN 0302-9743   ISSN 1611-3349 (electronic)
Lecture Notes in Bioinformatics
ISBN 978-3-031-15033-3   ISBN 978-3-031-15034-0 (eBook)
https://doi.org/10.1007/978-3-031-15034-0

LNCS Sublibrary: SL8 – Bioinformatics

This Springer imprint is published by the registered company Springer Nature Switzerland AG
The registered company address is: Gewerbestrasse 11, 6330 Cham, Switzerland

Preface

This volume contains the paper presented at the 20th International Conference on Computational Methods in Systems Biology (CMSB 2022), held at the University of Bucharest, Romania, during September 14–16, 2022. The conference was organized as an on-site event, with support for online participation.

The CMSB annual conference series was initiated in 2003. (https://cmsb. sciencesconf.org/). The webpage for this year's edition can be found at https://fmi. unibuc.ro/en/cmsb-2022/. The previous editions of the conference were held in Rovereto, Italy (2003), Paris, France (2004, 2011), Edinburgh, UK (2005, 2007), Trento, Italy (2006, 2010), Rostock, Germany (2008), Bologna, Italy (2009), London, UK (2012), Klosterneuburg, Austria (2013), Manchester, UK (2014), Nantes, France (2015), Cambridge, UK (2016), Darmstadt, Germany, (2017), Brno, Czech Republic (2018), Trieste, Italy (2019), Konstanz, Germany (2020, online), and Bordeaux, France (2021, hybrid).

The conference brings together researchers from across biological, mathematical, computational, and physical sciences who are interested in the modeling, simulation, analysis, inference, design, and control of biological systems. It covers the broad field of computational methods and tools in systems and synthetic biology and their applications. Topics of interest to the conference include, but are not limited to, methods and tools for biological system analysis, modeling, and simulation; high-performance methods for computational systems biology; identification of biological systems; applications of machine learning; network modeling, analysis, and inference; automated parameter and model synthesis; model integration and biological databases; multi-scale modeling and analysis methods; design, analysis, and verification methods for synthetic biology; methods for biomolecular computing and engineered molecular devices; data-based approaches for systems and synthetic biology; optimality and control of biological systems; modeling, analysis, and control of microbial communities. The conference welcomes new theoretical results with potential applications to systems and synthetic biology, as well as novel applications and case studies of existing methods, tools, or frameworks.

CMSB 2022 received 44 submissions: 20 regular papers (13 of them accepted), 4 tool papers (all accepted), 17 extended abstracts for poster presentations (13 accepted, but not included in this volume), and 3 extended abstracts for highlight talks (2 accepted, but not included in this volume). The conference followed a single blind review process. Each regular paper was reviewed by three program committee members. Each tool paper had four reviewers: two member of the Tool Evaluation Committee for the quality of the tool itself, its usability and documentation, and two members of the Program Committee for the scientific contribution of the paper. The extended abstracts were subject to a lighter review process, coordinated by the PC chairs. The submissions accepted to CMSB 2022 span a broad range of topics in chemical reaction networks, Boolean networks, continuous and hybrid models, machine learning, and software tools.

We had five invited talks at CMSB 2022 given by Alessandra Carbone (Sorbonne Université, France), Gabriela Chiosis (Memorial Sloan Kettering Cancer Center, USA), Loïc Paulevé (CNRS, Laboratoire Bordelais de Recherche en Informatique, France), Alfonso Rodríguez-Patón (Universidad Politécnica de Madrid, Spain), and Erik Sonnhammer (Stockholm University, Sweden). The abstracts of their talks are included in the front matter of this volume.

We are very grateful to the members of the Program Committee, to the members of the Tool Evaluation Committee, and to all our reviewers for their thoughtful and diligent work. We thank Springer for the generous sponsorship of the best paper awards and for the support in producing this volume. We also want to thank the CMSB steering committee, especially François Fages, for their support in organizing this edition of the conference.

July 2022

Ion Petre
Andrei Păun

Organization

Program Committee Chairs

Ion Petre	University of Turku, Finland, and National Institute for R&D in Biological Sciences, Romania
Andrei Păun	University of Bucharest and National Institute of R&D for Biological Sciences, Romania

Steering Committee

Alessandro Abate (Guest Member)	University of Oxford, UK
Luca Cardelli	University of Oxford, UK
Eugenio Cinquemani (Guest Member)	Inria Grenoble Rhône-Alpes, France
François Fages	Inria Saclay, France
Monika Heiner	Brandenburg University of Technology Cottbus-Senftenberg, Germany
Tommaso Mazza	Istituto CSS-Mendel, Italy
Satoru Miyano	University of Tokyo, Japan
Loïc Paulevé (Guest Member)	CNRS, LaBRI, France
Ion Petre	University of Turku, Finland, and National Institute of R&D for Biological Sciences, Romania
Tatjana Petrov (Guest Member)	University of Konstanz, Germany
Gordon Plotkin	University of Edinburgh, UK
Corrado Priami	University of Pisa, Italy
Andrei Păun (Guest Member)	University of Bucharest and National Institute for R&D in Biological Sciences, Romania
Carolyn Talcott	SRI International, USA
Adelinde Uhrmacher	University of Rostock, Germany
Verena Wolf	Saarland University, Germany

Program Committee

Alessandro Abate	University of Oxford, UK
Paolo Ballarini	Centrale Supélec, France
Daniela Besozzi	University of Milano-Bicocca, Italy
Luca Cardelli	University of Oxford, UK

Milan Češka	Brno University of Technology, Czech Republic
Eugenio Cinquemani	Inria Grenoble Rhône-Alpes, France
Eugen Czeizler	University of Helsinki, Finland
Franck Delaplace	Université Paris-Saclay, France
François Fages	Inria Saclay, France
Jérôme Feret	Inria Paris, France
Clémence Frioux	Inria Bordeaux, France
Ashutosh Gupta	IIT Bombay, India
Monika Heiner	Brandenburg University of Technology Cottbus-Senftenberg, Germany
Jane Hillston	University of Edinburgh, UK
Tommaso Mazza	Istituto CSS-Mendel, Italy
Andrzej Mizera	University of Luxembourg, Luxembourg
Jun Pang	University of Luxembourg, Luxembourg
Nicola Paoletti	Royal Holloway, University of London, UK
Loïc Paulevé	CNRS, LaBRI, France
Tatjana Petrov	University of Konstanz, Germany
Ovidiu Radulescu	University of Montpellier, France
Andre Ribeiro	University of Tampere, Finland
Maria Rodriguez Martinez	IBM Zürich Research Laboratory, Switzerland
Olivier Roux	École Centrale de Nantes, France
David Šafránek	Masaryk University, Czech Republic
Carolyn Talcott	SRI International, USA
Jing Tang	University of Helsinki, Finland
Adelinde Uhrmacher	University of Rostock, Germany
Andrea Vandin	Sant'Anna School for Advanced Studies, Italy
Verena Wolf	Saarland University, Germany
Christoph Zechner	Max Planck Institute of Molecular Cell Biology and Genetics, Germany

Tool Evaluation Committee

Georgios Argyris	Technical University of Denmark, Denmark
Candan Çelik	Comenius University, Slovakia
Anastasis Georgoulas	University College London, UK
Lukrécia Mertová	Heidelberg Institute for Theoretical Studies, Germany
Samuel Pastva	IST Austria, Austria
Misbah Razzaq	INRAE Tours, France
David Šafránek (Chair)	Masaryk University, Czech Republic
Jakub Šalagovič	KU Leuven, Belgium
Matej Trojak	Masaryk University, Czech Republic

Additional Reviewers

Georgios Argyris
Salvatore Daniele Bianco
Gerrit Grossmann
Till Köster
Lukrécia Mertová
Gareth Molyneaux

Samuel Pastva
Sylvain Soliman
Mirco Tribastone
Max Tschaikowski
Hanna Wiederanders

Local Organization

Florin Bîlbîe
Daniela Cheptea
Marian Guşatu
David Păcioianu
Andrei Păun
Ion Petre
Nicoleta Siminea

Invited Talks

Targeting and Controlling Protein-Protein Interaction Networks in Disease

Gabriela Chiosis ⓘ

Program in Chemical Biology, Memorial Sloan Kettering Cancer Center
New York, NY10021, USA
chiosisg@mskcc.org

Disease states represent perturbations of the underlying tissue- and cell-specific molecular networks arising from both internal perturbations (e.g., genetic mutations, epigenetic alterations, proteotoxic stress) and external stressors (chemical or other environmental exposures, and/or lifestyle choices). The disease interactome is therefore a map of how individual stressors or a combination thereof alter interaction networks, including protein-protein interaction networks, and perturb the system as a whole. Analysis of cell- and tissue-specific networks may therefore shed light on organization of biological systems and subsequently, on disease vulnerabilities. However, deriving human interactomes across different cell and disease contexts remains a challenge. To this end, a solution is provided by discoveries in disease biology that link stressors-induced protein-protein interaction networks perturbations to the formation of epichaperomes, pathologic scaffolds composed of tightly bound chaperones, co-chaperones, and other factors. Epichaperomes mediate how thousands of proteins anomalously interact and organize inside cells, which aberrantly affects the function of protein networks, and in turn, cellular phenotypes. Therefore, capturing epichaperomes and the proteome at large negatively impacted by these critical scaffolds provides informative clues for direct access to protein-protein interaction networks perturbations in diseases, and to the functional outcome of such changes in native biological systems, providing previously unattainable systems level insights into disease-specific stressor adaptation mechanisms. We introduced the term epichaperomics to describe the affinity-purification method that uses epichaperomes as baits to analyze context-specific alterations in protein connectivity and study disease specific protein-protein interaction networks. We here present an overview of the platform, from epichaperomics probes to bioinformatics pipelines, and provide proof-of-principle applications of the method in investigating the biology of Alzheimer's disease and in deriving new treatment paradigms for cancer.

Keywords: Epichaperomics · Cell- and tissue-specific protein-protein interaction networks · Alzheimer's disease and cancer

References

1. Rodina, A., et al.: The epichaperome is an integrated chaperome network that facilitates tumour survival. Nature. **538**(7625), 397–401 (2016). https://doi.org/10.1038/nature19807
2. Joshi, S., et al.: Pharmacologically controlling protein-protein interactions through epichaperomes for therapeutic vulnerability in cancer. Commun Biol. **4**(1), 1333 (2021). https://doi.org/10.1038/s42003-021-02842-3
3. Ginsberg, S.D., et al.: Disease-specific interactome alterations via epichaperomics: the case for Alzheimer's disease. FEBS J. **289**(8), 2047–2066 (2022). doi: https://doi.org/10.1111/febs.16031
4. Inda, M.C., et al.: The epichaperome is a mediator of toxic hippocampal stress and leads to protein connectivity-based dysfunction. Nat Commun. **11**(1), 319 (2020). https://doi.org/10.1038/s41467-019-14082-5
5. Ginsberg, S.D., et al.: The penalty of stress - epichaperomes negatively reshaping the brain in neurodegenerative disorders. J Neurochem. **159**(6), 958–979 (2021). doi:https://doi.org/10.1111/jnc.15525
6. Joshi, S., Wang, T., Araujo, T.L.S., Sharma, S., Brodsky, J.L., Chiosis, G.: Adapting to stress - chaperome networks in cancer. Nat Rev Cancer. **18**(9), 562–575 (2018). https://doi.org/10.1038/s41568-018-0020-9

Boolean Networks as a Link Between Knowledge, Data, and Quantitative Models

Loïc Paulevé

University of Bordeaux, CNRS, Bordeaux INP, LaBRI, UMR 5800, F-33400
Talence, France
loic.pauleve@labri.fr
https://loicpauleve.name

By bridging the gap between dynamical systems and their partial observation, computational models of biological processes aim at uncovering key mechanisms driving cellular dynamics, and ultimately predict their behavior under unobserved conditions. With this perspective, Boolean Networks (BNs) are widely adopted for modeling signaling pathways and gene and transcription factor networks, as their specification requires little parameters on top of the known molecular interactions.

The "Most Permissive" (MP) [3] dynamics of BNs allow a formal reasoning on behaviors of underlying quantitative models, without additional parameters. Moreover, the complexity for characterizing dynamical properties related to trajectories and attractors makes them considerably more tractable than with traditional synchronous and asynchronous modes and enables genome-scale analysis, as demonstrated with the mpbn tool[1].

By leveraging the properties of MPBNs, we are developing BoNesis[2], a logic programming interface for the automatic synthesis of BNs from complex dynamical properties [1, 2]. This approach allows integrating knowledge on pairwise gene and transcription factor interactions, as available in databases or by inference methods, with observation data such as single-cell RNA sequencing, to infer BNs reproducing cellular differentiation processes, and predict their reprogramming.

References

1. Chevalier, S., Froidevaux, C., Paulevé, L., Zinovyev, A.: Synthesis of boolean networks from biological dynamical constraints using answer-set programming. In: 2019 IEEE 31st International Conference on Tools with Artificial Intelligence (ICTAI), pp. 34–41. IEEE (2019)
2. Chevalier, S., Noël, V., Calzone, L., Zinovyev, A., Paulevé, L.: Synthesis and simulation of ensembles of boolean networks for cell fate decision. In: Abate, A., Petrov, T., Wolf, V. (eds.) Computational Methods in Systems Biology. CMSB 2020. LNCS. vol. 12314. Springer, Cham (2020). https://doi.org/10.1007/978-3-030-60327-4_11
3. Paulevé, L., Kolcák, J., Chatain, T., Haar, S.: Reconciling qualitative, abstract, and scalable modeling of biological networks. Nat. Commun. 11(1) (2020)

---

[1] https://github.com/bnediction/mpbn
[2] https://github.com/bnediction/bonesis

How to Infer Accurate Gene Regulatory Networks From Gene Expression

Erik Sonnhammer

Stockholm University, Sweden
erik.sonnhammer@scilifelab.se

System-wide measurements of gene expression responses to perturbations such as knock-down experiments offer the possibility to reconstruct the underlying gene regulatory network (GRN). With no or little noise this is feasible, but at the high noise levels typically found in experimental data it is very challenging to infer accurate GRNs.

I will present some approaches that we have developed to counteract issues arising from noise. We show that explicit usage of the perturbation design adds important information and strongly boosts GRN inference accuracy. Some improvements to classical inference algorithms that can further boost accuracy for noisy data will be discussed, such as selection of informative genes or inference of the perturbation design [3, 4]. To control the false discovery rate of inferred GRNs we developed a general bootstrap-based method called NestBoot [1], and to assess the predictiveness of inferred GRNs despite the lack of a gold standard we developed a cross-validation approach BFECV [2].

Another challenge in GRN inference is selecting the optimal sparsity, and I will present a new approach for this based on the "GRN Information Criterion" [5]. Finally, to stimulate the GRN inference field to develop more accurate algorithms, we have created a web server https://GRNbenchmark.org/ [6] where anybody can benchmark their method online on a range of datasets with varying noise levels.

References

1. Morgan, D., Tjärnberg, A., Nordling, T.E.M., Sonnhammer, E.L.L.: A generalized framework for controlling FDR in gene regulatory network inference. Bioinformatics, **35,** 1026–1032 (2019)
2. Morgan, D., et al.: Perturbation-based gene regulatory network inference to unravel oncogenic mechanisms, Sci. Rep. **10** ,14149 (2020)
3. Seçilmiş, D. et al: Uncovering cancer gene regulation by accurate regulatory network inference from uninformative data. NPJ Syst. Biol. Appl. **6,** 37 (2020)
4. Seçilmiş, D., Hillerton, T., Nelander, S., Sonnhammer, E.L.L.: Inferring the experimental design for accurate gene regulatory network inference, Bioinformatics, **37,** 3553 (2021)
5. Seçilmiş, D., Nelander, S., Sonnhammer, E.L.L.: Optimal sparsity selection based on an information criterion for accurate gene regulatory network inference, Frontiers in Genetics, in press (2022)

6. Seçilmiş, D., Hillerton, T., Sonnhammer, E.L.L.: GRNbenchmark - a web server for benchmarking directed gene regulatory network inference methods, Nucleic Acids Res., in press (2022). https://doi.org/10.1093/nar/gkac377

Sequence Space and Deep Learning to Better Understand Proteins

Alessandra Carbone

Department of Computational and Quantitative Biology, Sorbonne Université
and CNRS, France
alessandra.carbone@lip6.fr

I will explain how rethinking the sequence space with multiple probabilistic models leads to the functional classification of proteins and the reconstruction of protein-protein interaction networks with deep learning, two important problems in computational biology.

GRO: A Multicell Bacterial Simulator

Alfonso Rodríguez-Patón

Universidad Politécnica de Madrid, Spain
arpaton@fi.upm.es

GRO is an agent-based model for simulating the growth of programmed multicell bacterial populations. The goal of GRO is to predict the dynamics of complex microbial consortia. I will present some of the most relevant features of GRO.

Contents

Machine Learning

Software

Chemical Reaction Networks

Algebraic Biochemistry: A Framework for Analog Online Computation in Cells

Mathieu Hemery and François Fages[(✉)]

Inria Saclay, Palaiseau, France
`Francois.Fages@inria.fr`

Abstract. The Turing completeness of continuous chemical reaction networks (CRNs) states that any computable real function can be computed by a continuous CRN on a finite set of molecular species, possibly restricted to elementary reactions, i.e. with at most two reactants and mass action law kinetics. In this paper, we introduce a notion of online analog computation for the CRNs that stabilize the concentration of their output species to the result of some function of the concentration values of their input species, whatever changes are operated on the inputs during the computation. We prove that the set of real functions stabilized by a CRN with mass action law kinetics is precisely the set of real algebraic functions.

Keywords: Chemical reaction networks · stabilization · analog computation · online computation · algebraic functions

1 Introduction

Chemical Reaction Networks (CRNs) are a standard formalism used in chemistry and biology to describe complex molecular interaction systems. In the perspective of systems biology, they are a central tool to analyze the high-level functions of the cell in terms of their low-level molecular interactions. In that perspective, the Systems Biology Markup Language (SBML) [19] is a common format to exchange CRN models and build CRN model repositories, such as Biomodels.net [5] which contains thousands of CRN models of a large variety of cell biochemical processes. In the perspective of synthetic biology, they constitute a target programming language to implement in chemistry new functions either *in vitro*, e.g. using DNA polymers [20], or in living cells using plasmids [10] or in artificial vesicles using proteins [8].

The mathematical theory of CRNs was introduced in the late 70's, on the one hand by Feinberg in [15], by focusing on perfect adaptation properties and multistability analyses [9], and on the other hand, by Érdi and Tóth by characterizing the set of Polynomial Ordinary Differential Equation systems (PODEs) that can be defined by CRNs with mass action law kinetics, using dual-rail encoding for negative variables [11].

I. Petre and A. Păun (Eds.): CMSB 2022, LNBI 13447, pp. 3–20, 2022.
https://doi.org/10.1007/978-3-031-15034-0_1

More recently, a computational theory of CRNs was investigated by formally relating their Boolean, discrete, stochastic and differential semantics in the framework of abstract interpretation [14], and by studying the computational power of CRNs under those different interpretations [6,7,12].

In particular, under the continuous semantics of CRNs interpreted by ODEs, the Turing-completeness result established in [12] states that any computable real function, i.e. computable by a Turing machine with an arbitrary precision given in input, can be computed by a continuous CRN on a finite set of molecular species, using elementary reactions with at most two reactants and mass action law kinetics. This result uses the following notion of analog computation of a non-negative real function computed by a CRN, where the result is given by the concentration of one species, y_1, and the error is controlled by the concentration of one second species, y_2:

Definition 1 [12]. *A function $f : \mathbb{R}_+ \to \mathbb{R}_+$ is CRN-computable if there exist a CRN over some molecular species $\{y_1, ..., y_n\}$, and a polynomial $q \in \mathbb{R}_+^n[\mathbb{R}_+]$ defining their initial concentration values, such that for all $x \in \mathbb{R}_+$ there exists some (necessarily unique) function $y : \mathbb{R} \to \mathbb{R}^n$ such that $y(0) = q(x)$, $y'(t) = p(y(t))$ and for all $t > 1$*

$$|y_1(t) - f(x)| \leq y_2(t),$$

$y_2(t) \geq 0$, $y_2(t)$ is decreasing and $\lim_{t \to \infty} y_2(t) = 0$.

From the theoretical point of view of computability, the control of the error which is explicitly represented in the above definition by the auxiliary variable y_2, is necessary to decide when the result is ready for a requested precision, and to mathematically define the function computed by a CRN if any.

From a practical point of view however, precision is of course an irrelevant issue since chemical reactions are stochastic in nature and the stability of the CRN and robustness with respect to the concentration species variations is a more important criterion than the precision of the result. With this provision to omit error control, the Turing-completeness result of continuous CRNs was used in [12] to design a compilation pipeline to implement any mathematical elementary function in abstract chemistry. This compiler, implemented in our CRN modeling, analysis and synthesis software Biocham [2] as the one presented here[1], generates a CRN over a finite set of abstract molecular species, through several symbolic computation steps, mainly composed of polynomialization [17], quadratization [16] and lazy dual-rail encoding of negative variables. A similar approach is undertaken in the CRN++ system [22], also related to [3].

Now, it is worth remarking that in the definition above, and in our implementation in Biocham, the input is defined by the initial concentration of the input species which may be consumed by the synthesized CRN to compute the result. This marks a fundamental difference with many natural CRNs which perform a kind of online computation by adapting the response to the evolution of the

---

[1] All the computational results presented in this paper are available in an executable Biocham notebook at https://lifeware.inria.fr/wiki/Main/Software#CMSB22.

input. This is the case for instance of the ubiquitous MAPK signaling network which computes an input/output function well approximated by a Hill function of order 4.9 [18], while our synthesized CRNs to compute the same function consume their input and do not correctly adapt to change of the input value during computation [17].

In this paper, we are interested in a notion of online computation for continuous CRNs, by opposition to our previous notion of static computation of the result of a function for any initially given input. Our main theorem shows that the set of input/output functions stabilized online by a CRN with mass action law kinetics, is precisely the set of real algebraic functions.

Example 1. We can illustrate this result with a simple example. Let us consider a cell that produces a receptor, I, which is transformed in an active form, A, when bound to an external ligand L, and which stays active even after unbinding:

$$L + I \to L + A$$
$$\emptyset \leftrightarrow I \tag{1}$$
$$A \to \emptyset$$

The differential semantics with mass action law of unitary rate constant is the PODE:

$$\frac{dI}{dt} = 1 - I - LI$$
$$\frac{dA}{dt} = LI - A \tag{2}$$
$$\frac{dL}{dt} = 0$$

At steady state, all the derivatives are null and by eliminating I, we immediately obtain the polynomial equation: $L - LA - A = 0$. Thinking of this simple CRN as a kind of signal processing with the ligand as input and the active receptor as output, it is possible to find a polynomial relation between the input and the output. In this case, this relation entirely defines the function computed by the CRN:

$$A(L) = \frac{L}{1+L}.$$

For a given concentration of ligand, this is the only stable state of the system, independently of the initial concentrations of A or I. This is why we say that the CRN stabilizes the function.

Such functions, for which there exists a polynomial relation between the inputs and output, are called algebraic functions in mathematics. We show here that the set of real algebraic functions is precisely the set of input/output functions stabilized by CRNs with mass action law kinetics. Furthermore, our constructive proof provides a compilation method to generate a stabilizing CRN for any real algebraic curve, i.e. any curve defined by the zeros of some polynomial.

1.1   Related Work

Our CRN synthesis results can be compared to the ones of Buisman et al. who present in [1] a method to implement any algebraic expression by an abstract CRN[2]. They rely on a direct expression of the function and a compilation process that mimics the composition of the elementary algebraic operations. We improve their results in three directions. First, our compilation pipeline is able to generate stabilizing CRNs for any algebraic function, including those algebraic functions that cannot be defined by algebraic expressions, such as the Bring radical (see Example 6). Second, our theoretical framework shows that the general set of algebraic functions precisely characterizes the set of functions that can be stably computed online by a CRN. Third, the quadratization and lazy-negative optimization algorithms presented in this paper allow us to generate more concise CRNs. On the example given in Sect. 3.4 of [1] for the quadratic expression

$$y = \frac{b - \sqrt{b^2 - 4ac}}{2a}$$

used to find the root of a polynomial of second order, our compiler generates a CRN of 7 species (including the 3 inputs) and 11 reactions, while their CRN following the syntax of the expression uses 10 species and 14 reactions. Moreover, our dual-rail encoding allows us to give correct answers for negative values of y.

2   Definitions and Main Theorem

For this article, we denote single chemical species with lower case letters and set of species with upper case letters, e.g. $X = \{x_1, x_2, \ldots\}$. By abuse of notation, we will use the same symbol for the variables of the ODEs, the chemical species and their concentrations, the context being sufficient to remove any ambiguity.

2.1   Chemical Reaction Networks

A chemical reaction with mass action law kinetics is composed of a multiset of reactants, a multiset of products and one rate constant. Such a reaction can be written as follows:

$$a + b \xrightarrow{k} 2a \tag{3}$$

where k is the rate constant, and the multisets are represented by linear expressions in which the (stoichiometric) coefficients give the multiplicity of the species in the multisets, here 2 for the product a, the coefficients equal to 1 being omitted. In this example, the velocity of the reaction is the product kab, i.e. the rate constant k times the concentration of the reactants, a and b.

In this paper, we consider CRNs with mass action law kinetics only. It is well known that the other kinetics, such as Michaelis-Menten or Hill kinetics, can be

---

[2] The terminology of "algebraic functions" used in the title of [1] refers in fact to its restriction to algebraic expressions.

derived by quasi-steady state or quasi-equilibrium reductions of more complex CRNs with mass action kinetics [21]. Furthermore, the Turing-completeness of this setting [12] shows that there is no loss of generality with that restriction.

We also study the case where one or several species, called *pinned (input) species*, are present in such a way that their concentrations remain constant, independently of the activity of the CRN under study.

Definition 2. *The differential semantics of a CRN with a distinguished set of pinned species* S^p, *is composed of the usual ODEs for the non pinned species* $s \notin S^p$, *and null differential functions for the pinned species:*

$$\forall s \in S^p, \quad \frac{ds}{dt} = 0. \tag{4}$$

This pinning process may be due to a scale separation between the different concentrations (one of the species is so abundant that the CRN essentially do not modify its concentration), to a scale separation of volume (e.g. a compartment within a cell and a freely diffusive small molecule) or to an active mechanism ensuring perfect adaptation (e.g. the input is produced and consumed by some external reactions faster than the CRN itself, thus locking its concentration).

2.2 Stabilization

We are interested in the case where one particular species of the CRN, called the output, is such that, whatever moves the inputs may do, once the inputs are fixed, the concentration of the output species stabilizes on the result of some function of the fixed inputs. Furthermore, we want this value to be robust to small perturbations of both the auxiliary variables and the output. Of course, if the inputs are modified, the output has to be modified. The output thus encodes a particular kind of robust computation of a function which we shall call stabilization.

Definition 3. *We say that a CRN over a set of* $m + 1 + n$ *species* $\{X, y, Z\}$ *with pinned inputs* X *of cardinality* m *and distinguished outputs* y, *stabilizes the function* $f : I \mapsto \mathbb{R}_+$, *with* $I \subset \mathbb{R}_+^m$, *over the domain* $\mathcal{D} \subset \mathbb{R}_+^{m+1+n}$ *if:*

1. $\forall X^0 \in I$ *the restriction of the domain* $\mathcal{D}$ *to the slice* $X = X^0$ *is of plain dimension* $n + 1$, *and*
2. $\forall (X^0, y^0, Z^0) \in \mathcal{D}$ *the Polynomial Initial Value Problem (PIVP) given by the differential semantic with pinned input species* X *and the initial conditions* X^0, y^0, Z^0 *is such that:* $\lim_{t \to \infty} y(t) = f(X)$.

This definition is extended to functions of $\mathbb{R}^n$ *in* $\mathbb{R}$ *by dual-rail encoding* [11]: *for a CRN over the species* $\{X^+, X^-, y^+, y^-, Z\}$ *we ask that* $\lim_{t \to \infty} (y^+ - y^-)(t) = f(X^+ - X^-)$, *for all initial conditions in the validity domain* $\mathcal{D}$.

Let $\mathcal{F}_S$ *be the set of functions stabilized by a CRN.*

Several remarks are in order. A first remark concerns the fact that we ask for a domain $\mathcal{D}$ of plain dimension $n + 1$, i.e. non-null measure in $\mathbb{R}^{n+1}$. That constraint is imposed in order to benefit from a strong form of robustness: there exists an open volume containing the desired fixed point such that it is the unique attractor in this space. Hence in this setting, minor perturbations are always corrected. This requirement of an isolated fixed point also impedes from hiding information in the initial conditions. The following example illustrates the crucial importance of that condition on the dimension of the domain $\mathcal{D}$

Example 2. The following PODE is constructed in such a way that z_2 goes exponentially to x while y and z_1 remain equal to $\cos(z_2)$ and $\sin(z_2)$ respectively.

$$
\begin{aligned}
\frac{dx}{dt} &= 0, & x(t = 0) &= \text{input} \\
\frac{dy}{dt} &= -z_1(x - z_2) & y(t = 0) &= 1 \\
\frac{dz_1}{dt} &= y(x - z_2) & z_1(t = 0) &= 0 \\
\frac{dz_2}{dt} &= (x - z_2) & z_2(t = 0) &= 0
\end{aligned}
\tag{5}
$$

One might think that this PODE stabilizes the cosine as we have $\lim y(t) = \cos(x)$ for any value of x. But cosine is not an algebraic function, and indeed, the only requirement for this PODE to be at steady state is: $x = z_2$, meaning that there exist fixed points for any value of z_1 and x. So this PODE does not stabilize the cosine function. The reason is that the cosine computation is encoded in the initial state. It is only for the domain where $y = \cos(z_2)$ and $z_1 = \sin(z_2)$ that the computation works, but this domain is of null measure in $\mathbb{R}^3$ which breaks the first condition of Definition 3.

A second remark is that since the inputs are fixed in our semantics (they are by definition pinned species), the target of the output species which is the result $f(X)$ of some function f is not a fluctuating goal: it is fixed by the initial conditions. In practice, what we ask is that the dynamics of the ODE for the slice of the domain $\mathcal{D}$ defined by imposing the inputs have a unique attractor satisfying $y = f(X^0)$. But as we do not impose any constraint on the other variables (Z), we cannot speak of a fixed point since the dynamics on the other variables may not stabilize (e.g. oscillations, divergence, etc.). We will nevertheless speak of these object as pseudo-fixed point. If we start from a point on this pseudo fixed point, we will have: $\forall t, y(t) = f(X^0)$.

A third remark is that our definition implies that apart from a transient behaviour of characteristic time τ, the whole system is constrained to live in, or nearby, the subspace defined by $y = f(X)$. What is interesting is that if the inputs are themselves varying with a characteristic time that is slower than τ, the output will follow those variations, hence preserving our desired property up to an error coming from the delay as long as the system stays in the domain $\mathcal{D}$. In a synthetic biology perspective, it is in principle possible to use a time-rescaling to modify the value of τ. While a small τ allows for a faster adaptation, this usually comes at the expense of a greater energetic cost as the proteins turn-over tends to increase.

2.3 Algebraic Curves and Algebraic Functions

In mathematics, an algebraic curve (or surface) is the zero set of a polynomial in two (or more) variables. It is a usual convention in mathematics to speak indifferently of the polynomial and the curve it defines, seen as the same object. For instance, $x^2 + y^2 - 1$ is seen as the unit circle.

Any polynomial P can be expressed as a product of irreducible polynomials, i.e. polynomials that cannot be further factorized, up to a constant k:

$$P = k. \prod_{i=1...n} P_i^{a_i}.$$

The P_i's are called the components of P, and a_i the multiplicity of P_i. We say that P is in reduced form if all the components have multiplicity one, $\forall i \ a_i = 1$. This is justified by one important result of algebraic geometry: in an algebraic closed field, such as the complex numbers $\mathbb{C}$, the set of points of an algebraic curve given with their multiplicity, suffices to define the polynomial in reduced form. This makes algebraic geometry an elegant and powerful theory.

In a non-algebraically closed field such as $\mathbb{R}$, a polynomial may have no real root. This difficulty is however irrelevant to us in this paper since we start with an algebraic real function, thus assuming the existence of real roots. For the purpose of this article, this fundamental result provides a canonical correspondence between an algebraic real function and its polynomial of minimal degree, i.e. a polynomial in reduced form, up to a multiplicative factor.

Definition 4. *A function $f : I \subset \mathbb{R}^m \mapsto \mathbb{R}$ is algebraic if there exists a polynomial P of $m + 1$ variables such that:*

$$\forall X \in I, P(X, f(X)) = 0. \tag{6}$$

We denote $\mathcal{F}_A$ the set of real algebraic functions.

We shall prove the following central theorem:

Theorem 1. *The set of functions stabilized by a CRN with mass action law kinetics is the set of algebraic real functions: $\mathcal{F}_S = \mathcal{F}_A$.*

One technical difficulty comes from the fact that it is not immediate to determine the function f from the polynomial. Indeed for a given polynomial pinning the value of the inputs results in one, several or no possible value for the output. Hence, a given polynomial actually defines several functions on the domain of its input. This is for instance the case of the unit circle curve defined by $x^2 + y^2 - 1$. If we see it as an equation to solve upon y, it admits two solutions when $x \in]-1, 1[$, exactly one for $x = -1$ or $x = 1$, and no solution for other values of x. Hence, that curve defines two continuous functions $y(x)$, each of them with support $]-1,1[$.

To overcome that difficulty, let us call branch point (or branch curve), the set of points where the number of real roots of an algebraic function changes $(-1,1$ in the previous example). Now for a polynomial $P(X,y)$ and a given root X,y that is not a branch point, the implicit function theorem ensures the existence and uniqueness of the implicit function up to the next branch point/curve.

Example 3. The branch points of the unit circle polynomial $x^2 + y^2 - 1$ are $(-1,0),(1,0)$. If we provide an additional point on the curve, e.g. $(0,1)$, one can define the function that contains it and that goes from one branch point to the other one, here:

$$]-1,1[\to \mathbb{R}$$
$$f : x \mapsto \sqrt{x^2 - 1}$$

Figure 1 in a latter section illustrates the flow diagram used in this example by our CRN compiler to approximate that function.

Similarly in the case of the sphere defined by the polynomial $x_1^2 + x_2^2 + y^2 - 1$, the branch curve is the whole unit circle contained in the plane $y = 0$. And giving the point $0,0,-1$ is enough to define the whole surface corresponding to the down part of the sphere inside the branch-curve circle.

3   Proof

Lemma 1. $\mathcal{F}_A \subset \mathcal{F}_S$.

Proof. Suppose that $f : I \mapsto \mathbb{R}$ is an algebraic real function and let P_f denote the canonical polynomial such that $\forall X \in I, P_f(X, f(X)) = 0$. Let us choose a vector X_0 in the domain of f.

Then, the PODE

$$\frac{dy}{dt} = \pm P(X,y),$$
$$\frac{dX}{dt} = 0, \tag{7}$$

is such that $Y = f(X)$ is a fixed point. By choosing the sign such that, locally $\pm P(X,y)$ is negative above $Y = f(X)$ and positive below, we ensure that this point is stable.

The fact that the polynomial has to change the sign across the fixed point is dut to the fact that we choose the polynomial of minimal degree, hence it has to be in reduced form and the multiplicity of every branch of the curve is one: the sign cannot be the same on both sides of the curve.

It is worth remarking that any ODE system made of elementary mathematical functions can be transformed in a polynomial ODE system [17], hence one can wonder why we restrict here to polynomial expressions. This comes from the condition that asks that the domain $\mathcal{D}$ be of plain dimension in Definition 3. The polynomialization of an ODE system may indeed introduce constraints between the initial concentrations which is precisely what is forbidden by the requirement upon $\mathcal{D}$.

Now, let us note $Y^+ = \inf(Y \mid P(X,Y) = 0, Y > f(X))$ and $Y^- = \sup(Y \mid P(X,Y) = 0, Y < f(X))$, with $\pm\infty$ values if the set is empty. We know that for all y in $]Y^-, Y^+[$, the only attractor is $f(X)$ and as a polynomial can only have a finite number of zeros, those sets are non empty.

For all variables that are not bound to be positive, the dual-rail encoding consists in splitting the variable into two positive variables corresponding to the positive and negative parts. Then, all positive monomials can be dispatched to the positive part and all negative ones to the negative part (with a positive sign), with the addition of an mutual annihilation reaction between the variables as described in [12]. It is worth noting that dual-rail encoding is necessary for positive functions whenever the auxiliary variables may take negative values.

Lemma 2. $\mathcal{F}_S \subset \mathcal{F}_A$.

Proof. Let us suppose that f is a function stabilized by a mass action CRN. The idea is to use the characterization of functions that are projectively polynomial, as defined in [4]. By using the higher-order derivatives of the stabilized variable, it is shown in [4] that one can eliminate all the auxiliary variables and obtain a single equation of the form:

$$P(X, y, y^{(1)}, \ldots, y^{(n)}) = 0.$$

Using the fact that for all X, $y = f(X)$ is a pseudo fixed point by definition, if we use it as initial condition we immediately get:

$$X = X,$$
$$y = f(X),$$
$$y^{(k)} = 0 \quad \forall k \in [1, n].$$

Injecting this in the characterization of the function y, we obtain:

$$\forall X, P^\star(X, f(X)) = 0. \tag{8}$$

There are now two cases. Either $P^\star$ is not trivial and effectively defines the surface of fixed points: this gives a polynomial for f, hence f is algebraic. Either $P^\star$ is the uniformly null polynomial. But in this case, every points in the X, y plane may be a fixed point and the domain $\mathcal{D}$ of the definition of stabilization is reduced to a single point, yet we asked it to be of non-null measure. Therefore, $P^\star$ is not trivial and f is algebraic.

4   Compilation Pipeline for Generating Stabilizing CRNs

The proof of Lemma 1 is constructive and provides a method to transform any algebraic function defined by a polynomial and one point, in an abstract CRN that stabilizes it. This is implemented with a command

    stabilize_expression(Expression, Output, Point)

with three arguments:

Expression: For a more user friendly interface, we accept in input more general mathematical expressions than polynomials; the non polynomial parts are detected and transformed by introducing new variable/species to compute their values;

Output: a name of the Output species different from the input;

Point: a point on the algebraic curve that is used to determine the branch of the curve to stabilize if several exist.

Similarly to our previous pipeline for compiling any elementary function in an abstract CRN that computes it [12,16,17], our compilation pipeline for generating stabilizing CRNs follows the same sequence of symbolic transformations:

1. polynomialization
2. stabilization
3. quadratization
4. dual-rail encoding
5. CRN generation

yet with some important differences.

4.1   Polynomialization

This optional step has been added just to obtain a more user friendly interface, since polynomials may sometimes be cumbersome to manipulate. The first argument thus admits algebraic expressions instead of being limited to polynomials.

The rewriting simply consists in detecting all the non-polynomial terms of the form $\sqrt[a]{b}$ or $\frac{a}{b}$ in the initial expression and replace them by new variables, hence obtaining a polynomial.

Then to compute the variables that just have been introduced, we perform the following basic operations on each of them to recover polynomiality:

$$n = \sqrt[a]{b} \rightarrow n^a - b$$

$$n = \frac{a}{b} \rightarrow nb - a$$

and recursively call `stabilize_expression` on these new expressions with the introduced variable (here n) as desired output.

4.2   Stabilization

To select the branch of the curve to stabilize, it is sufficient to choose the sign in front of the polynomial in Eq. 7. such that at the designated point, the second derivative of y is positive. For this, we use a formal derivation to compute the sign of the polynomial, and reverse it if necessary.

4.3   Quadratization

The quadratization of the PODE is an optional transformation which aims at generating elementary reactions, i.e. reactions having at most two reactants each, that are fully decomposed and more amenable to concrete implementations with real biochemical molecular species. It is worth noting that the quadratization problem to solve here is a bit different from the one of our original pipeline studied in [16] since we want to preserve a different property of the PODE. It is necessary here that the introduced variables stabilize on their target value independently of their initial concentrations. While it was possible in our previous framework to initialize the different species with a precise value given by a polynomial of the input, this is no more the case here as the domain $\mathcal{D}$ has to be of plain dimension.

The variables introduced by quadratization correspond to monomials of order higher than 2 that can thus be separated as the product of two variables corresponding to monomials of lower order: A and B. Those variables can be either present in the original polynomial or introduced variables. The following set of reactions:

$$A + B \rightarrow A + B + M$$

$$M \rightarrow \emptyset,$$

gives the associated ODE:

$$\frac{dM}{dt} = AB - M, \tag{9}$$

for which the only stable point satisfies: $M = AB$.

Furthermore as before, we are interested in computing a quadratic PODE of minimal dimension. In [16], we gave an algorithm in which the introduced variables were always equal to the monomial they compute, whereas in our online stabilization setting, this is true only when $t \rightarrow \infty$. For instance, if we replaced AB by M in Eq. 9, the system would no longer adapt to changes of the input. To circumvent this difficulty, it is possible to modify the PIVP and use it as input of our previous algorithm to take this constraint and still obtain the minimal set of variables. In our previous computation setting, the derivatives of the different variables where simply the derivatives of the associated monomials computed in the flow generated by the initial ODE. In Algorithm 1, we construct a pseudo-ODE containing twice as many variables, the derivatives of which being built to ensure that the solution is correct. The idea is that the actual variables are of the form Mb and the Mb^2 variables exist only to construct the solution. To compute quadratic monomials with a b^2 term present in the derivatives of the

Mb variables (the "true" variables), one can either add two Mb variables to the solution set or add a single Mb^2 variable. As can be seen on the lines 5 and 9 of Algorithm 1, Mb^2 variables require that the corresponding Mb is in the solution set.

Algorithm 1. Quadratization algorithm for a PODE stabilizing a function. The $minimal\_quadratic\_set(PODE, y)$ returns the minimal set of variables containing y sufficient to express all its derivatives in quadratic form [16].

Input: A PODE of the form $\frac{dx_i}{dt} = P_i(X)$, with $i \in [1, n]$ to compute x_n
Output: A set S of monomials to quadratize the input.

   $ODE_{\text{aux}} \leftarrow \emptyset$
   find an unused variable name: b
   for all $i \in [1, n]$ **do**
       add $\frac{dx_i b}{dt} = P_i(X) \times b^2$ to ODE_{aux}
5:    add $\frac{dx_i b^2}{dt} = x_i b$ to ODE_{aux}
   $AllMonomials \leftarrow$ the set of monomials that are less or equal to a monomial present in one of the P_i and not in X.
   for all $M \in AllMonomials$ **do**
       add $\frac{dMb}{dt} = Mb^2$ to ODE_{aux}
       add $\frac{dMb^2}{dt} = Mb$ to ODE_{aux}
10: $S_{\text{aux}} \leftarrow minimal\_quadratic\_set(ODE_{\text{aux}}, x_n b)$
   $S \leftarrow \emptyset$
   for all $Mb \in S_{\text{aux}}$ **do**
       add M to S
   return S

This variant of the quadratization problem studied in [16] has the same theoretical complexity, as shown by the following proposition:

Proposition 1. *The quadratization problem of a PODE for stabilizing a function and minimizing the number of variables is NP-hard.*

Proof. The proof proceeds by reduction of the vertex covering of a graph as in [16]. Let us consider the graph $G = (V, E)$ with vertex set $v_i, i \in [1, n]$ and edge set $E \in V \times V$. And let us study the quadratization of the PODE with input variables $V \cup \{a\}$ and output variable y such that the y computes $\sum_{v_i v_j \in E} v_i v_j a$. The derivative is:

$$\frac{dy}{dt} = \sum_{v_i v_j \in E} v_i v_j a - y. \tag{10}$$

An optimal quadratization contains variable corresponding either to $v_i a$ or $v_i v_j$ indicating that an optimal covering of the graph G contains either the node v_i either indifferently v_i or v_j. Hence en optimal quadratization gives us an optimal covering which concludes the proof.

Our previous MAXSAT algorithm [16] and heuristics [17] can again be used here with the slight modification mentioned above concerning the introduction of new variables.

Algorithm 1, when invoked with an optimal search for *minimal_quadratic_set*, is nevertheless not guaranteed to generate optimal solutions, because of the "pseudo" variables noted Mb^2. Despite those theoretical limitations, the CRNs generated by Algorithm 1 are particularly concise, as shown in the example section below and already mentioned above for the compilation of algebraic expressions compared to [1].

4.4 Lazy Dual-Rail Encoding

As in our original compilation pipeline [12], it is necessary to modify our PODE in order to impose that no variable may become negative. This is possible through a lazy version of dual-rail encoding. First by detecting the variable that are or may become negative and then by splitting them between a positive and negative part, thus implementing a dual-rail encoding of the variable: $y = y^+ - y^-$. Positive terms of the original derivative are associated to the derivative of y^+ and negative terms to the one of y^- and a fast mutual degradation term is finally associated to both derivative in order to impose that one of them stays close to zero [12].

4.5 CRN Generation

The same back-end compiler as in our original pipeline is used, i.e. introducing one reaction for each monomial. It is worth remarking that this may have for effect to aggregate in one reaction several occurrences of a same monomial in the ODE system [13].

5 Examples

Example 4. As a first example, we can study the unit circle: $x^2 + y^2 - 1$. Our pipeline gives us for the upper part of the circle, the following CRN.

$$
\begin{aligned}
& \emptyset \rightarrow y^+ && 2y^- \rightarrow 3y^- \\
& 2x \rightarrow y^- + 2x && 2y^+ \rightarrow y^- + 2y^+ \\
& y^+ + y^- \xrightarrow{\text{fast}} \emptyset
\end{aligned}
\tag{11}
$$

the flow of the PODE associated to this model can be seen in Fig. 1**A** and the steady state is depicted in Fig. 1**B** as a function of x in the positive quadrant.

Example 5. Even rather involved algebraic curves need surprisingly few species and reactions. This is the case of the serpentine curve, or anguinea, defined by the polynomial $(y - 2)\left((x - 10)^2 + 1\right) = 4(x - 10)$ for which we choose the point

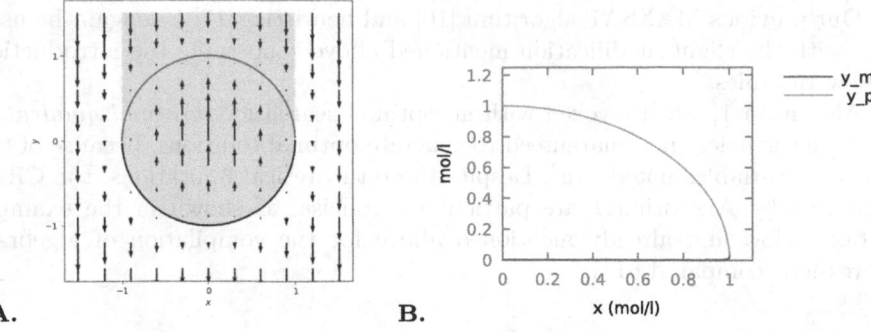

Fig. 1. A. Flow diagram in the x, y plane before the dual rail encoding for the stabilization of the unit circle. The arrows indicate the direction and strength of the flow. The upper part of the curve (in red) indicates the stable branch of the system and we colored in light red the domain $\mathcal{D}$ in which the system will reach the desired steady state. Outside of $\mathcal{D}$, the system is driven to the divergent state $\lim y = -\infty$. **B.** Dose response diagram of the generated CRN where the input concentration (x) is gradually increased from 0 to 1 while recording the steady state value of the output species y^+, y^-. (Color figure online)

$x = 10, y = 2$ to enforce stability. The compilation process takes less than 100 ms on a typical laptop[3]. The generated CRN reproduces the anguinea curve on the y variable, as shown in Fig. 2. It is composed of the following 4 species and 12 reactions:

$$
\begin{aligned}
y_m + y_p &\xrightarrow{\text{fast}} \emptyset, & 2x &\rightarrow z + 2x, \\
z &\rightarrow \emptyset, & \emptyset &\xrightarrow{162} y_p, \\
y_m &\xrightarrow{101} \emptyset, & x + y_p &\xrightarrow{20} x + 2y_p, \\
z + y_m &\rightarrow z, & z &\xrightarrow{2} z + y_p, \\
x &\xrightarrow{36} x + y_m, & y_p &\xrightarrow{101} \emptyset, \\
x + y_m &\xrightarrow{20} x + 2y_m, & z + y_p &\rightarrow z.
\end{aligned}
\tag{12}
$$

Example 6. In the field of real analysis, the Bring radical of a real number x is defined as the unique real root of the polynomial: $y^5 + y + x$. The Bring radical is an algebraic function of x that cannot be expressed by any algebraic expression.

The stabilizing CRN generated by our compilation pipeline is composed of 7 species $(y_m, y_p, y2_m, y2_p, y3_m, y3_p, x)$ and 20 reactions presented in model 13. A dose-response diagram of that CRN is shown in Fig. 3.

---

[3] An Ubuntu 20.04, with an Intel Core i6, 2.4 GHz x 4 cores and 15.5 GB of memory.

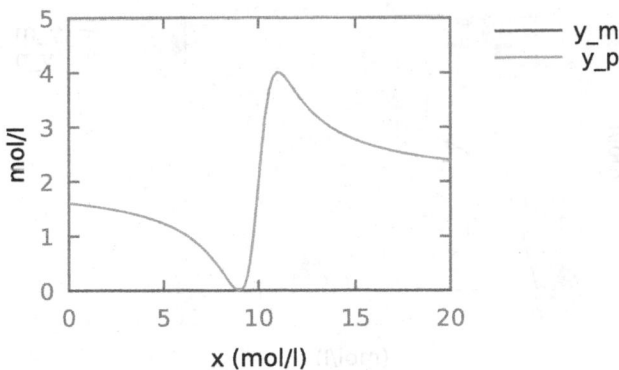

Fig. 2. Dose-response diagram of the CRN generated by compilation of the serpentine algebraic curve, $(y - 2)\left((x - 10)^2 + 1\right) = 4(x - 10)$, with x as input and y as output.

$$
\begin{aligned}
y_m + y_p &\xrightarrow{\text{fast}} \emptyset, & y2_m + y2_p &\xrightarrow{\text{fast}} \emptyset, \\
y3_m + y3_p &\xrightarrow{\text{fast}} \emptyset, & y_p &\to \emptyset, \\
y2_m + y3_p &\to y2_m + y3_p + y_p, & y2_p + y3_m &\to y2_p + y3_m + y_p, \\
x &\to x + y_m, & y_m &\to \emptyset, \\
y2_p + y3_p &\to y2_p + y3_p + y_m, & y2_m + y3_m &\to y2_m + y3_m + y_m, \quad (13) \\
2 \cdot y_p &\to y2_p + 2 \cdot y_p, & 2 \cdot y_m &\to y2_p + 2 \cdot y_m, \\
y2_p &\to \emptyset, & y2_m &\to \emptyset, \\
y2_p + y_p &\to y2_p + y3_p + y_p, & y2_m + y_m &\to y2_m + y3_p + y_m, \\
y3_p &\to \emptyset, & y2_p + y_m &\to y2_p + y3_m + y_m, \\
y2_m + y_p &\to y2_m + y3_m + y_p, & y3_m &\to \emptyset.
\end{aligned}
$$

Example 7. To generate the CRN that stabilize the Hill function of order 5, we can use the expression $y - \frac{x^5}{1+x^5}$ along with the point $x = 1, y = \frac{1}{2}$. Our compilation pipeline generates the following model with 6 species and 10 reactions:
$2x \to z_1 + 2x, 2z_1 \to z_2 + 2z_1, x + z_2 \to x + z_2 + z_3, \emptyset \to z_4, z_4 + z_3 \to z_3 + y,$
$z_1 \to \emptyset, \quad\quad z_2 \to \emptyset, \quad\quad z_3 \to \emptyset, \quad\quad\quad\quad z_4 \to \emptyset, y \to \emptyset,$
all kinetics being mass action law with unit rate. The $z's$ are auxiliary variables corresponding to the following expressions:

$$
z_1 = x^2, \quad z_2 = x^4, \quad z_3 = x^5, \quad z_4 = \frac{1}{1 + x^5}.
$$

The production and degradation of z_4 may be surprising, but looking at all the reactions implying both z_4 and y, we can see that their sum follow the equation $\frac{d(z_4 + y)}{dt} = 1 - (z_4 + y)$ hence ensuring that the sum of the two is fixed independently of their initial concentrations. It is worth remarking that another way of reaching the same result would be to directly build-in conservation laws

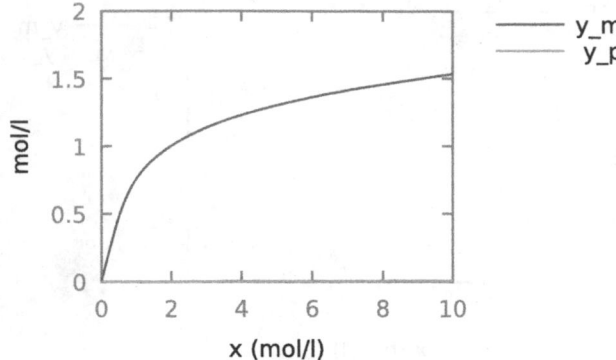

Fig. 3. The bring radical is the real root of the polynomial equation $y^5 + y + x = 0$. As this quantity is negative, the output is read on the negative part of the output y_m. It is the simplest equation for which there is no expression for y as a function of x.

into our CRN, hence using both steady state and invariant laws to define our steady state which however would make us sensitive to the initial concentrations.

6   Conclusion and Perspectives

We have introduced a notion of on-line analog computation for CRNs in which the concentration of one output species stabilizes to the result of some function of the concentrations of the input species, whatever perturbations are applied to the species concentrations during computation before the inputs stabilize. We have shown that the real functions that can be stably computed by a CRN in that way is precisely the set of real algebraic functions, defined by a polynomial and one point. Furthermore, we have derived from the constructive proof of this result a compilation pipeline to transform any algebraic function in a stabilizing CRN which computes it online.

These results open a whole research avenue for both the understanding of the structure of natural CRNs that allow cells to adapt to their environment, and for the design of artificial CRNs to implement high-level functions in chemistry. In the latter perspective of synthetic biology, our compilation pipeline makes it possible to automatically generate an abstract CRN which remains to be implemented with real enzymes, as in [8]. Taking into account a catalog of concrete enzymatic reactions earlier on in our compilation pipeline, in the polynomialization, quadratization and dual-rail encoding steps, is a particularly interesting, yet hard, challenge in order to guide search towards both concrete and economical solutions.

Our main theorem describes only CRNs at steady state only, while important aspects of signal processing are linked to the temporal evolution of the signals. Since our computation framework relies on ratios between production and degradation, a multiplication of both terms by some factor might be the matter of future work to control the characteristic time τ of equilibration, with high value of τ filtering out the high frequency noise of the inputs, and small τ values resulting in a more accurate output, yet at the expense of a higher protein turnover.

Acknowledgments. We are grateful to Amaury Pouly and Sylvain Soliman for interesting discussions on this work, and to ANR-20-CE48-0002 and Inria AEx GRAM grants for partial support.

References

1. Buisman, H.J., ten Eikelder, H.M.M., Hilbers, P.A.J., Liekens, A.M.L.: Computing algebraic functions with biochemical reaction networks. Artif. Life **15**(1), 5–19 (2009)
2. Calzone, L., Fages, F., Soliman, S.: BIOCHAM: an environment for modeling biological systems and formalizing experimental knowledge. Bioinformatics **22**(14), 1805–1807 (2006)
3. Cardelli, L., Tribastone, M., Tschaikowski, M.: From electric circuits to chemical networks. Nat. Comput. **19**(1), 237–248 (2019). https://doi.org/10.1007/s11047-019-09761-7
4. Carothers, D.C., Parker, G.E., Sochacki, J.S., Warne, P.G.: Some properties of solutions to polynomial systems of differential equations. Electron. J. Differ. Equ. **2005**(40), 1–17 (2005)
5. Chelliah, V., Laibe, C., Novére, N.L.: Biomodels database: a repository of mathematical models of biological processes. In: Schneider, M. (ed.) Silico Systems Biology. Methods in Molecular Biology, vol. 1021, pp. 189–199. Humana Press, Totowa (2013). https://doi.org/10.1007/978-1-62703-450-0_10
6. Chen, H.-L., Doty, D., Soloveichik, D.: Deterministic function computation with chemical reaction networks. Nat. Comput. **7433**, 25–42 (2013)
7. Cook, M., Soloveichik, D., Winfree, E., Bruck, J.: Programmability of chemical reaction networks. In: Condon, A., Harel, D., Kok, J., Salomaa, A., Winfree, E. (eds.) Algorithmic Bioprocesses, pp. 543–584. Springer, Heidelberg (2009). https://doi.org/10.1007/978-3-540-88869-7_27
8. Courbet, A., Amar, P., Fages, F., Renard, E., Molina, F.: Computer-aided biochemical programming of synthetic microreactors as diagnostic devices. Mol. Syst. Biol. **14**(4), e7845 (2018)
9. Craciun, G., Feinberg, M.: Multiple equilibria in complex chemical reaction networks: II. The species-reaction graph. SIAM J. Appl. Math. **66**(4), 1321–1338 (2006)
10. Duportet, X., et al.: A platform for rapid prototyping of synthetic gene networks in mammalian cells. Nucleic Acids Res. **42**(21), 13440–13451 (2014)
11. Érdi, P., Tóth, J.: Mathematical Models of Chemical Reactions: Theory and Applications of Deterministic and Stochastic Models. Nonlinear Science: Theory and Applications. Manchester University Press, Manchester (1989)
12. Fages, F., Le Guludec, G., Bournez, O., Pouly, A.: Strong Turing completeness of continuous chemical reaction networks and compilation of mixed analog-digital programs. In: Feret, J., Koeppl, H. (eds.) CMSB 2017. LNCS, vol. 10545, pp. 108–127. Springer, Cham (2017). https://doi.org/10.1007/978-3-319-67471-1_7

13. Fages, F., Gay, S., Soliman, S.: Inferring reaction systems from ordinary differential equations. Theor. Comput. Sci. **599**, 64–78 (2015)
14. Fages, F., Soliman, S.: Abstract interpretation and types for systems biology. Theor. Compu. Sci. **403**(1), 52–70 (2008)
15. Feinberg, M.: Mathematical aspects of mass action kinetics. In: Lapidus, L., Amundson, N.R. (eds.) Chemical Reactor Theory: A Review, chap. 1, pp. 1–78. Prentice-Hall (1977)
16. Hemery, M., Fages, F., Soliman, S.: On the complexity of quadratization for polynomial differential equations. In: Abate, A., Petrov, T., Wolf, V. (eds.) CMSB 2020. LNCS, vol. 12314, pp. 120–140. Springer, Cham (2020). https://doi.org/10.1007/978-3-030-60327-4_7
17. Hemery, M., Fages, F., Soliman, S.: Compiling elementary mathematical functions into finite chemical reaction networks via a polynomialization algorithm for ODEs. In: Cinquemani, E., Paulevé, L. (eds.) CMSB 2021. LNCS, vol. 12881, pp. 74–90. Springer, Cham (2021). https://doi.org/10.1007/978-3-030-85633-5_5
18. Huang, C.-Y., Ferrell, J.E.: Ultrasensitivity in the mitogen-activated protein kinase cascade. PNAS **93**(19), 10078–10083 (1996)
19. Hucka, M., et al.: The systems biology markup language (SBML): a medium for representation and exchange of biochemical network models. Bioinformatics **19**(4), 524–531 (2003)
20. Qian, L., Soloveichik, D., Winfree, E.: Efficient turing-universal computation with DNA polymers. In: Sakakibara, Y., Mi, Y. (eds.) DNA 2010. LNCS, vol. 6518, pp. 123–140. Springer, Heidelberg (2011). https://doi.org/10.1007/978-3-642-18305-8_12
21. Segel, L.A.: Modeling Dynamic Phenomena in Molecular and Cellular Biology. Cambridge University Press, Cambridge (1984)
22. Vasic, M., Soloveichik, D., Khurshid, S.: CRN++: molecular programming language. In: Doty, D., Dietz, H. (eds.) DNA 2018. LNCS, vol. 11145, pp. 1–18. Springer, Cham (2018). https://doi.org/10.1007/978-3-030-00030-1_1

Abstract Simulation of Reaction Networks via Boolean Networks

Joachim Niehren[1,2] (ID), Athénaïs Vaginay[3,4](✉) (ID), and Cristian Versari[1]

[1] BioComputing Team of CRIStAL Lab, Université de Lille, Lille, France
{joachim.niehren,cristian.versari}@univ-lille.fr
[2] Inria Lille, Lille, France
[3] Université de Lorraine, CNRS, CRAN, 54000 Nancy, France
athenais.vaginay@loria.fr
[4] Université de Lorraine, CNRS, Inria, LORIA, 54000 Nancy, France

Abstract. We propose to simulate chemical reaction networks with the deterministic semantics abstractly, without any precise knowledge on the initial concentrations. For this, the concentrations of species are abstracted to Booleans stating whether the species is present or absent, and the derivatives of the concentrations are abstracted to signs saying whether the concentration is increasing, decreasing, or unchanged. We use abstract interpretation over the structure of signs for mapping the ODEs of a reaction network to a Boolean network with nondeterministic updates. The abstract state transition graph of such Boolean networks can be computed by finite domain constraint programming over the finite structure of signs. Constraints on the abstraction of the initial concentrations can be added naturally, leading to an abstract simulation algorithm that produces only the part of the abstract state transition graph that is reachable from the abstraction of the initial state. We prove the soundness of our abstract simulation algorithm, and show its applicability to reaction networks in the SBML format from the BioModels database.

Keywords: Systems Biology · Reaction networks · SBML · Boolean networks · Abstract interpretation · Logic · Constraint programming

1 Introduction

Reaction networks [6,10,12,14] are the most prominent formalism for modeling the continuous dynamics of biological systems. Boolean networks [13,21,26,28] are the most prominent formalism for modeling discrete abstractions of the continuous dynamics. Hybrid automata [3] offer a framework for mixed continuous and discrete modeling. How to discretize the continuous semantics of reaction networks or of hybrid automata into Boolean networks is a long standing general question [4,5,7,15,16,27]:

Reaction networks can be given different semantics. The continuous dynamics of a reaction network is given by its deterministic semantics: a system of ordinary differential equations (ODEs), one per species, that is composed from

I. Petre and A. Păun (Eds.): CMSB 2022, LNBI 13447, pp. 21–40, 2022.
https://doi.org/10.1007/978-3-031-15034-0_2

the kinetic expressions of the reactions producing or consuming the species. The non-deterministic rewrite semantics [19], in contrast, ignores the kinetic expressions, while the stochastic semantics uses them differently, for computing the probability of a reaction to happen. Reaction networks also have a Boolean semantics which abstracts from the rewrite semantics [11]. In the present paper, we abstract the continuous semantics, which is notoriously difficult to compare to the rewrite semantics and thus to the Boolean semantics.

Any solution of the ODEs from the deterministic semantics provides a derivable function of type $\mathbb{R}_+ \to \mathbb{R}$ per species, that is called the trajectory. At any time point, the value of a trajectory must be positive, since it stands for the concentration of the species. In contrast, the value of its derivative may be negative, meaning that the concentration of the species is decreasing. The value of the trajectory at time point 0 is called the initial concentration of the species. It is well-known that for any fixed collection of initial concentrations per species, there exists at most one solution of the ODEs. This solution can be approximated numerically by using Euler's deterministic simulation algorithm [9].

A concrete continuous state of a reaction network at a given time point is a vector of positive real numbers, one for the concentration of each species. Any concrete state can be abstracted to a vector of Booleans, stating for each species whether its concentration is zero or not. The possible trajectories of a reaction network can thus be abstracted to a state transition graph whose states are bit vectors. The graph can be enriched, when not only considering the trajectories but also their derivatives. Since these may become negative, the concrete states now become vectors of real numbers that can be abstracted to vectors of signs: increasing $\nearrow = 1$, decreasing $\searrow = -1$, and no-change $\to = 0$. In this way, we obtain an enriched abstract state transition graph between sign vectors.

In the present paper, we study an instance of the general discretization problem, which is whether one can compute the abstract state transition graph from a given reaction network. Clearly, abstract state transition graphs are finite, but may be large, since having $\Theta(2^{|\mathcal{S}|})$ many states where $|\mathcal{S}|$ is the number of species. Computing the abstract state transition graph completely quickly becomes impossible, given that the size grows exponentially in the number of species. Therefore, we propose to study the problem of abstract simulation, which is to compute the part of the abstract state transition graph that is accessible from the abstraction of the initial concentrations. This also has the advantage that the concrete initial concentrations do not need to be known precisely. Nevertheless, the problem remains non-trivial, given that trajectories are infinite objects, and that there are infinitely many trajectories depending on the choice of the initial concentrations.

Our idea for abstract simulation is based on the abstract interpretation of the system of ODEs of the reaction network over the structure of signs $\mathbb{S} = \{\nearrow, \to, \searrow\}$. This abstraction introduces non-determinism, since $\nearrow +^{\mathbb{S}} \searrow$ may be evaluated to any sign. It can be proven to provide a sound over-approximation based on John's soundness theorem for the abstract interpretation of logic formulas [1,17,24]. We show that the sign abstraction of the ODEs of a reaction network can be used to define a Boolean network with non-deterministic updates [25].

It will have rules stating that a species A is present in the next step, if A was already present at the previous step, or if the derivative of A was positive at the previous step. Since such rules can be defined by first-order (FO) formulas, we propose the notion of first-order Boolean networks with non-deterministic semantics (FO-BNNs).

We provide a soundness theorem for abstraction of reaction networks to FO-BNNs. It will rely on a causal next transition relation rather than on a temporal next transition relation inferred from the trajectories, given that concrete simulation algorithms too are based on causality. This may lead to approximation errors for the concrete numerical simulation, so we have to take care of approximation errors for abstract simulation too.

Given that FO-BNNs are first-order formulas that are to be interpreted over the finite structure of signs, we use finite domain constraint programming to compute the abstract state transition graphs of FO-BNNs. Constraints on the abstraction of the initial concentrations can be added naturally, leading to an ' algorithm for abstract simulation based on constraint programming. We have implemented this algorithm based on the Minizinc constraint solver [23].

While abstract interpretation enables qualitative reasoning, we can still support exact quantitative reasoning about thresholds. We show that whether $A \leq \epsilon$ for some threshold $\epsilon > 0$ can be tested by introducing an artificial species B so that $\dot{B} = A - \epsilon$. In this way, the sign of the derivative of B indicates, whether the concentration of A is above, below or equal ϵ. One can then use exact reasoning with linear equation systems [1] to improve the quality of our abstract simulation algorithm, while taking thresholds into account. For instance, we can show for the usual enzymatic reaction network that if the initial concentration of the substrate is above of a given threshold ϵ, then (1) the concentration of the product may eventually become bigger than ϵ, and (2) once this happens, it can never become smaller than ϵ again. Most interestingly, the precise initial concentration of the substrate does not matter for this argument, as long as it is above ϵ. In this way, abstract simulation can sometimes show properties of infinitely many concrete simulations.

We apply our abstract simulation algorithm to a reaction network represented in the Systems Biology Markup Language (SBML) [18] from the BioModels database [20]. We consider the model https://www.ebi.ac.uk/biomodels/ BIOMD0000000448 that we will call B448 for short. This network describes the insulin signalling in human adipocytes in normal conditions [8]. It has 27 species and 34 reactions, and its graph covers one full page (see Fig. 13). The abstract simulation algorithm successfully yields a very small subgraph of the abstract state transition graph with more than 2^{27} states.

Related Work on Exact Computation of State Transition Graphs.
Mover et al. [22] develop an efficient implicit method for the exact abstraction of dynamical systems, whose abstract state space description and ODE dynamics are restricted to be systems of polynomial equations. While polynomial equations allow the exact description of abstract state spaces that are more general and fine-grained than the ones used for Boolean networks, they do not provide

enough expressiveness to describe the kinetic expressions frequently used for the modeling of chemical reaction networks.

Outline. After a few preliminaries in Sect. 2, we discuss arithmetic expressions with three different interpretations in Sect. 3. We recall the notion of reaction networks and their deterministic semantics via ODEs in Sect. 4. In Sect. 5, we recall the first-order logic, which permits us to formally capture ODEs in Sect. 6, enables their abstract interpretation in Sect. 7, and lays the foundation of FO-BNNs in Sect. 8. We present our compiler from reaction networks to FO-BNNs and prove its soundness in Sect. 9. The treatment of thresholds is discussed in Sect. 10. It illustrates exact reasoning at the example of the enzymatic reaction network. The application of abstract simulation to reaction network B448 of the Biomodels database is shown in Sect. 11. The conclusion and future work are given in Sect. 12.

2   Preliminaries

The Cartesian product of sets $A_1, \ldots, A_n$ is denoted by $A_1 \times \ldots \times A_n$. The domain of a partial function $f \subseteq A \times B$ is denoted by $dom(f)$. The restriction of f to a subset $A' \subseteq dom(f)$ is written as $f_{|A'}$. We write $[a_1/b_1, \ldots, a_n/b_n]$ for the finite function f with $dom(f) = \{a_1, \ldots, a_n\}$ and $f(a_i) = b_i$ for all $1 \leq i \leq n$. For any two sets A, B, the power set $B^A = A \to B = \{f \mid f : A \to B\}$ is the set of total functions from A to B. A multiset M with elements in A is an element $M \in \mathbb{N}^A$. For any $a \in A$, the multiplicity of a in M is $M(a)$.

Let $\mathbb{B} = \{0, 1\}$ be the set of Booleans, $\mathbb{S} = \{-1, 0, 1\}$ the set of signs, $\mathbb{N}$ the set of natural numbers including 0, $\mathbb{Z}$ the set of integers, $\mathbb{R}$ the set of real numbers, and $\mathbb{R}_+$ the set of positive real numbers including 0. Note that $\mathbb{B} \subseteq \mathbb{N} \subseteq \mathbb{R}_+ \subseteq \mathbb{R}$ and that $\mathbb{S} \subseteq \mathbb{Z} \subseteq \mathbb{R}$. For signs, we use the symbols $\nearrow = 1$ for increase, $\to = 0$ for no-change and $\searrow = -1$ for decrease.

3   Arithmetic Expressions

We recall the syntax and semantics of arithmetic expressions while pointing out three different interpretations and usages. Let $\mathcal{V}$ be a set of variables and $\Sigma_{arith}^{(2)}$ be the set of binary operators $\{+, *, -, /, exp\}$. The set of arithmetic expressions $e \in \mathcal{E}_{arith}(\mathcal{V})$ is the least set of terms that can be build from variables $x \in \mathcal{V}$, real numbers $\rho \in \mathbb{R}$, and binary operators $\odot \in \Sigma_{arith}^{(2)}$:

$$e_1, e_2 \in \mathcal{E}_{arith}(\mathcal{V}) ::= \rho \mid x \mid e_1 \odot e_2$$

Definition 1. *A relational structure with binary operators in $\Sigma_{arith}^{(2)}$ and constants in $\mathbb{R}$ is a tuple $S = (dom(S), (\odot^S)_{\odot \in \Sigma_{arith}^{(2)}}, (\rho^S)_{\rho \in \mathbb{R}})$ where $D = dom(S)$ is a set called the domain, $\odot^S : D^2 \times D$ is the interpretation of $\odot$, and $\rho^S \in D$ the interpretation of ρ.*

$$[\![\rho]\!]^{\alpha,S} = \{\rho^S\}, \quad [\![x]\!]^{\alpha,S} = \{\alpha(x)\},$$
$$[\![e_1 \odot e_2]\!]^{\alpha,S} = \{s \mid s_1 \in [\![e_1]\!]^{\alpha,S},\ s_2 \in [\![e_2]\!]^{\alpha,S},\ (s_1, s_2, s) \in \odot^S\}$$

Fig. 1. Interpretation of arithmetic expressions over a relational structure S and a variable assignment $\alpha : \mathcal{V} \to dom(S)$, where $e_1, e_2 \in \mathcal{E}_{arith}(\mathcal{V})$, $\odot \in \Sigma^{(2)}_{arith}$, $\rho \in \mathbb{R}$, and $x \in \mathcal{V}$.

For any relational structure S, variable assignment $\alpha : \mathcal{V} \to dom(S)$ and arithmetic expression $e \in \mathcal{E}_{arith}(\mathcal{V})$, Fig. 1 defines the interpretation $[\![e]\!]^{S,\alpha} \subseteq dom(S)$. We consider interpretations over three different relational structures $\mathbb{R}$, $\mathbb{R}_+ \to \mathbb{R}$, and $\mathbb{S}$, that we freely confuse with their domain.

First, arithmetic expressions are used as kinetic expressions. They are then interpreted in the relational structure of the reals $\mathbb{R}$. The binary operators are interpreted as the binary partial functions $\odot^{\mathbb{R}}$ for the addition, multiplication, subtraction, division, and exponentiation of real numbers respectively. Note that $/^{\mathbb{R}}$ is not a total function, since division by zero is not defined. Therefore, $[\![x/0]\!]^{\mathbb{R},\alpha} = \emptyset$ for any $\alpha : \mathcal{V} \to \mathbb{R}$.

Second, we use arithmetic expressions in ODEs. Then they are interpreted in the structure of real-valued functions $\mathbb{R}_+ \to \mathbb{R}$. The binary operators now denote the partial functions $\odot^{\mathbb{R}_+ \to \mathbb{R}}$ for the addition, multiplication, subtraction, division, and exponentiation of real-valued functions respectively. Note that any constant $\rho \in \mathbb{R}$ is interpreted as the constant function with $\rho^{\mathbb{R}_+ \to \mathbb{R}}(\rho') = \rho$ for all $\rho' \in \mathbb{R}_+$.

Third, we will interpret arithmetic expressions over the structure of signs $\mathbb{S} = \{\nearrow, \to, \searrow\}$. The binary operators need to be interpreted as ternary relations $\odot^{\mathbb{S}} \subseteq \mathbb{S}^2 \times \mathbb{S}$. Let $h_{\mathbb{S}} : \mathbb{R} \to \mathbb{S}$ be the unique homomorphism between the structures of reals and of signs. It satisfies for all $\rho \in \mathbb{R}$ that $h_{\mathbb{S}}(\rho) = 1$ if $\rho > 0$, $h_{\mathbb{S}}(\rho) = 0$ if $\rho = 0$ and $h_{\mathbb{S}}(\rho) = -1$ if $\rho < 0$. Then we define for any constant $\rho \in \mathbb{R}$ that $\rho^{\mathbb{S}} = h_{\mathbb{S}}(\rho)$ and for any operator $\odot \in \Sigma^{(2)}_{arith}$ that $\odot^{\mathbb{S}} = \{(h_{\mathbb{S}}(\rho), h_{\mathbb{S}}(\rho'), h_{\mathbb{S}}(\rho'')) \mid (\rho, \rho', \rho'') \in \odot^{\mathbb{R}}\}$. We note that the addition $+^{\mathbb{S}}$ is not even a partial function, since $(\nearrow, \searrow, \sigma) \in +^{\mathbb{S}}$ for all three signs $\sigma \in \mathbb{S}$. For this reason, $[\![\nearrow + \searrow]\!]^{\mathbb{S},\gamma} = \mathbb{S}$ for any $\gamma : \mathcal{V} \to \mathbb{S}$. The intuition is that an arithmetic expression is evaluated non-deterministically to the set of all possible signs.

4   Chemical Reaction Networks

Let $\mathcal{S}$ be a finite set. A chemical solution with species in $\mathcal{S}$ is a multiset $M : \mathcal{S} \to \mathbb{N}$. The multiset $[A/3, B/2]$ for instance is often written as $3A + 2B$.

Definition 2. *A reaction with species in $\mathcal{S}$ is an element of $\mathbb{N}^{\mathcal{S}} \times \mathcal{E}_{arith}(\mathcal{S}) \times \mathbb{N}^{\mathcal{S}}$. A reaction network with species in $\mathcal{S}$ is a subset of reactions with species in $\mathcal{S}$.*

For instance, if $e = 5.1 * exp(A, 2) * B$ then $r = (3A + B, e, A + 2C)$ is a chemical reaction, that we denote as usual as $r : 3A + B \xrightarrow{e} A + 2C$. A state α of a reaction network R assigns each species of R a concentration, which is a positive

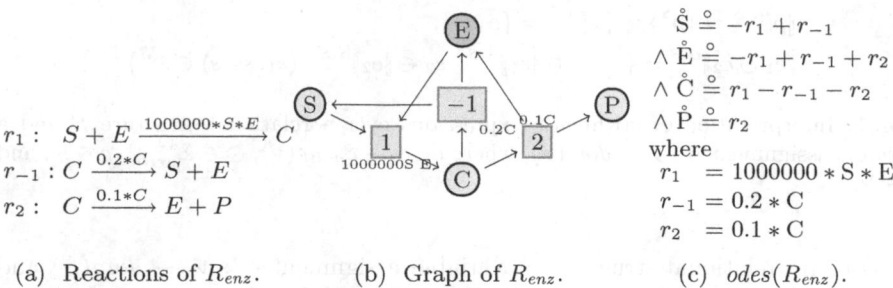

(a) Reactions of R_{enz}.   (b) Graph of R_{enz}.   (c) $odes(R_{enz})$.

Fig. 2. The enzymatic reaction network R_{enz}.

real number, so $\alpha : \mathcal{S} \to \mathbb{R}_+$. Let $[\![e]\!]^{\mathbb{R},\alpha} = 5.1 *^{\mathbb{R}} exp^{\mathbb{R}}([\![A]\!]^{\mathbb{R},\alpha}, 2) *^{\mathbb{R}} [\![B]\!]^{\mathbb{R},\alpha}$. The above reaction states that the concentration of A changes at any time point with state α with speed $-2 *^{\mathbb{R}} [\![e]\!]^{\mathbb{R},\alpha}$, the concentration of B with speed $-1 *^{\mathbb{R}} [\![e]\!]^{\mathbb{R},\alpha}$, and the concentration of C with speed $2 *^{\mathbb{R}} [\![e]\!]^{\mathbb{R},\alpha}$. Negative speeds mean that the species is consumed, while positive speeds mean that the species is produced.

Let $r = (M, e, M')$ be a chemical reaction with species in $\mathcal{S}$. We denote the kinetic expression of r by $kin_r = e$. For any species $A \in \mathcal{S}$, the stoichiometry of A in r is defined by $stoic_r(A) = M'(A) - M(A)$. The ODE system of a reaction network R is the following equation system:

$$odes(R) =_{def} \bigwedge_{A \in \mathcal{S}} \dot{A} \stackrel{\circ}{=} \sum_{r \in R} stoic_r(A) * kin_r \wedge A \geq 0$$

A formal definition of the syntax and semantics of ODEs will be given in Sect. 6 based on notions from the first-order logic in Sect. 5. For now, we just state that all species occurring in an arithmetic expression denote some real valued function of type $\mathbb{R}_+ \to \mathbb{R}$, that must be positive in addition. An expression $\dot{A}$ denotes the derivative of the denotation of A if its derivative exists, and is undefined otherwise. Note that derivatives may become negative. The arithmetic operators are interpreted as arithmetic operations in the structure of real-valued functions $\mathbb{R}_+ \to \mathbb{R}$.

As an example of a CRN, we show the network of enzymatic reactions in Fig. 2. It has species $\mathcal{S} = \{S, E, C, P\}$ and the three reactions in Fig. 2a, all with mass action kinetics. Reaction r_1 transforms a pair of a substrate S and an enzyme E to a complex C, reaction r_{-1} does the inverse, and reaction r_2 transforms the complex C into the free enzyme E and the product P. The graph of the CRN R_{enz} is given Fig. 2b and its ODEs in Fig. 2c.

5 First-Order Logic

We recall the syntax and semantics of first-order logic formulas. Rather than opting for maximal generality, we restrict ourselves to signatures able to capture ODEs and FO-BNNs.

$$\llbracket e \stackrel{\circ}{=} e' \rrbracket^{\alpha,S} = \begin{cases} 1 \text{ if } \llbracket e \rrbracket^{\alpha,S} \cap \llbracket e' \rrbracket^{\alpha,S} \neq \emptyset \\ 0 \text{ else} \end{cases} \qquad \llbracket \phi \wedge \phi' \rrbracket^{\alpha,S} = \llbracket \phi \rrbracket^{\alpha,S} \wedge^{\mathbb{B}} \llbracket \phi' \rrbracket^{\alpha,S}$$

$$\llbracket \neg \phi \rrbracket^{\alpha,S} = \neg^{\mathbb{B}}(\llbracket \phi \rrbracket^{\alpha,S}) \qquad \llbracket \exists x.\phi \rrbracket^{\alpha,S} = \begin{cases} 1 \text{ if exists } s \in dom(S). \\ \quad \llbracket \phi \rrbracket^{\alpha[x/s],S} = 1 \\ 0 \text{ else} \end{cases}$$

Fig. 3. Interpretation of formulas $\phi \in \mathcal{F}_\Sigma$ over a relational structure S with signature Σ with respect to a variable assignment $\alpha : \mathcal{V} \to dom(S)$.

We consider signatures $\Sigma = \Sigma^{(2)} \cup \Sigma^{(1)} \cup \mathbb{R}$ with a set of binary operators in $\Sigma^{(2)}$, unary operators in $\Sigma^{(1)}$, and constants in $\mathbb{R}$. Arithmetic expression have the signature $\Sigma_{arith} = \Sigma_{arith}^{(2)} \cup \Sigma_{arith}^{(1)} \cup \mathbb{R}$ with $\Sigma_{arith}^{(1)} = \emptyset$. An expression in $\mathcal{E}_\Sigma(\mathcal{V})$ then has the following form where $x \in \mathcal{V}$, $\rho \in \mathbb{R}$, $\odot \in \Sigma^{(2)}$ and $o \in \Sigma^{(1)}$.

$$e, e_1, e_2 \in \mathcal{E}_\Sigma(\mathcal{V}) \quad ::= x \mid \rho \mid e_1 \odot e_2 \mid o(e)$$

Clearly, these expressions generalize on arithmetic expressions since $\mathcal{E}_{arith}(\mathcal{V}) = \mathcal{E}_{\Sigma_{arith}}(\mathcal{V})$. The notion of relational structures can be lifted from the signature of arithmetic expressions to expressions with general signatures in the obvious manner, and also the semantics from arithmetic expressions to expressions in $\mathcal{E}_\Sigma(\mathcal{V})$.

We are particularly interested in the signature $\dot{\Sigma}_{arith}$ with $\dot{\Sigma}_{arith}^{(2)} = \Sigma_{arith}^{(2)}$ and $\dot{\Sigma}_{arith}^{(1)} = \{\dot{}\}$. The structure of real-valued functions can be extended to give an interpretation of the unary operator $\dot{}$, so that $^{\mathbb{R}_+ \to \mathbb{R}}(f)$ is the derivative of f for any derivable real-valued function $f : \mathbb{R}_+ \to \mathbb{R}$, and undefined otherwise.

The set of first-order formulas $\mathcal{F}_\Sigma(\mathcal{V})$ is constructed from equations between expressions in $e, e' \in \mathcal{E}_\Sigma(\mathcal{V})$ and the usual first-order connectives:

$$\phi \in \mathcal{F}_\Sigma(\mathcal{V}) ::= e \stackrel{\circ}{=} e' \mid \exists x.\phi \mid \phi \wedge \phi \mid \neg \phi \qquad \text{where } x \in \mathcal{V}$$

We sometimes use shortcuts $e \geq 0$ for the formula $\exists x.e \stackrel{\circ}{=} x * x$ and $e \leq e'$ for $e' - e \geq 0$. The set of free variables $fv(\phi)$ contains all those variables of ϕ that occur outside the scope of any occurrence of the existential quantifier.

The semantics of a first-order formula $\phi \in \mathcal{F}_\Sigma(\mathcal{V})$ is the truth value $\llbracket \phi \rrbracket^{\alpha,S} \in \mathbb{B}$ defined in Fig. 3. It depends on some relational structure S with signature Σ and variable assignment $\alpha : \mathcal{V} \to dom(S)$. An equation $e \stackrel{\circ}{=} e'$ is true if the intersection of the possible values for e and the possible values for e' is non-empty, that is, if $\llbracket e \rrbracket^{\alpha,S} \cap \llbracket e' \rrbracket^{\alpha,S} \neq \emptyset$. The set of solutions of a formula $\phi \in \mathcal{F}_\Sigma(\mathcal{V})$ over a relational structure S with the same signature Σ is $sol^S(\phi)=\{\alpha_{|fv(\phi)} \mid \alpha : \mathcal{V} \to dom(S), \llbracket \phi \rrbracket^{\alpha,S} = 1\}$.

6   Sign Abstraction of ODE Trajectories

We define systems of ODEs as formulas of first-order logic in order to formalize their syntax and semantics in a framework suitable for abstract interpretation.

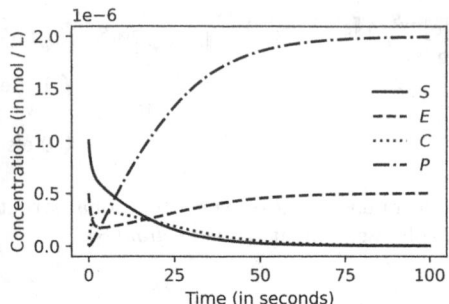

Fig. 4. The deterministic numerical simulation of $odes(R_{enz})$ with initial concentrations $S(0) = 1.0 * 10^{-5}$, $E(0) = 0.5 * 10^{-5}$ and $P(0) = C(0) = 0$ mol/L.

We then introduce a temporal and a causal transition relation on abstract states, by applying the sign abstraction to the trajectories of the ODEs.

Definition 3. *An ODE system is a first-order logic formula $\phi \in \mathcal{F}_{\dot{\Sigma}_{arith}}(\mathcal{V})$ such that all subexpressions of ϕ that are rooted by the dot-operator have the form $\dot{x}$ for some $x \in \mathcal{V}$ and do not occur below any existential quantifier.*

Note in particular that higher-order derivations $\ddot{x}$ are not permitted. The semantics of an ODE system ϕ is the set of its solutions over the relational structure of real-valued functions, i.e., $sol^{\mathbb{R}_+ \to \mathbb{R}}(\phi)$. For each such solution β and variable $x \in fv(\phi)$, we call $\beta(x) : \mathbb{R}_+ \to \mathbb{R}$ a trajectory of x.

For any choice of initial concentrations $\alpha : fv(\phi) \to \mathbb{R}^+$, there exists at most one solution $\beta \in sol^{\mathbb{R}_+ \to \mathbb{R}}(\phi)$, such that $\beta(x)(0) = \alpha(x)$ for all $x \in fv(\phi)$.

This solution can be computed numerically by the usual integration methods for ODEs starting with the initial concentrations. If some operations of the ODEs are undefined during the integration, no solution exists. For illustration, we show in Fig. 4 the solution of the ODEs of the reaction network R_{enz} with initial concentrations $S(0) = 1.0 * 10^{-5}$, $E(0) = 0.5 * 10^{-5}$ and $P(0) = C(0) = 0$ mol/L.

A (concrete) state with variables $V \subseteq \mathcal{V}$ is a function $\alpha : V \to \mathbb{R}$, and an abstract state a function $\gamma : V \to \mathbb{S}$. We next show how to define a successor relation on the abstract states of ODEs.

Definition 4 (Temporal next transitions). *Let $\gamma_1, \gamma_2 : S \to \mathbb{S}$ be two abstract states and ϕ an ODE system. We call γ_2 a (temporally) next state of γ_1 with respect to ϕ and write $(\gamma_1, \gamma_2) \in next_\phi$ if there exists a real-valued function $\beta \in sol^{\mathbb{R}_+ \to \mathbb{R}}(\phi)$ and two time points $0 \leq t_1 < t_2$ such that, for all species $A \in S$ and time points $t_2' \in]t_1, t_2]$: $\gamma_1(A) = h_{\mathbb{S}}(\beta(A)(t_1))$ and $\gamma_2(A) = h_{\mathbb{S}}(\beta(A)(t_2'))$.*

For instance, for $odes(R_{enz})$, the next state of $[S/1, E/1, C/0, P/0]$ is $[S/1, E/1, C/1, P/1]$, which has itself as next state. For this example, the next states are always unique, but in general this may not be the case.

Interestingly, $[S/1, E/1, C/0, P/0]$ does not have $[S/1, E/1, C/1, P/0]$ as next state. The reason is that instantaneously when C is produced, reaction r_2 starts producing P, so that both C and P will appear at the same time point. Nevertheless, the creation of C causes the creation of P, but this is not observable in the temporal order and thus not in the relation $next_{ode(R_{enz})}$.

So far, the states do not contain any information about the values of the derivatives (since $\dot{x}$ is not a variable but an expression). In order to change this, let $^{\circ}: \mathcal{V} \to \mathcal{V}$ be a bijection such that $\mathring{V}$ is disjoint from V for any $V \subseteq \mathcal{V}$. For any ODE system ϕ we define a formula $\mathring{\phi} \in \mathcal{F}_{\Sigma_{arith}}$ without dot operator, by replacing any subexpression of the form $\dot{x}$ in ϕ by $\mathring{x}$:

$$\mathring{\phi} =_{\text{def}} \phi[\dot{x}/\mathring{x} \mid x \in \mathit{fv}(\phi)]$$

Note that no dot operators remain in $\mathring{\phi}$, since all subexpressions of ϕ that are rooted by the dot operator must be of the form $\dot{x}$ by the definition of ODEs. For any assignment $\beta : V \to (\mathbb{R}_+ \to \mathbb{R})$ and time point $t \geq 0$ we define an assignment $\beta_t : \{x, \mathring{x} \mid x \in V\} \to \mathbb{R}$ such that: $\beta_t(x) = \beta(x)(t)$ and $\beta_t(\mathring{x}) = {}^{\mathbb{R}_+ \to \mathbb{R}}(\beta(x))(t)$.

Lemma 5. *If $\beta \in \mathit{sol}^{\mathbb{R}_+ \to \mathbb{R}}(\phi)$ then $\beta_t \in \mathit{sol}^{\mathbb{R}}(\mathring{\phi})$.*

When interested in derivatives, we consider the successor relation of the formula $next_{\mathring{\phi}}$. For instance, with respect to $next_{odes(R_{enz})}$, the abstract state $\gamma_1 = [S/1, E/1, C/0, P/0, \mathring{S}/\searrow, \mathring{E}/\searrow, \mathring{C}/\nearrow, \mathring{P}/\to]$ has the successor $\gamma_2 = [S/1, E/1, C/1, P/1, \mathring{S}/\searrow, \mathring{E}/\searrow, \mathring{C}/\nearrow, \mathring{P}/\nearrow]$. Furthermore, causality can be observed in the signs of the derivatives: we have $\gamma_1(\mathring{C}) = \nearrow$ since $\gamma_1(E) = \gamma_1(S) = 1$. In contrast, we have $\gamma_1(\mathring{P}) = \to$ since $\gamma_1(C) = 0$. As a consequence, for any solution $\beta \in \mathit{sol}^{\mathbb{R}_+ \to \mathbb{R}}(odes(R_{enz}))$, the value of $\lim_{t \to 0} \beta(C)(t)/t \neq 0$ so the change of $C(t)$ at $t = 0$ can be observed in the limit, while $\lim_{t \to 0} \beta(P)(t)/t = 0$, so the change of $P(t)$ at $t = 0$ cannot be observed in the limit. Nevertheless $\gamma_2(C) = \gamma_2(P) = 1$, since the successor time point of 0 is not in the limit.

Definition 6 (Causal next transitions). *Let ϕ be an ODE system, $V = fv(\phi)$, and $\gamma_1, \gamma_2 : (V \cup \mathring{V}) \to \mathbb{S}$ abstract states. We call γ_2 a causally-next dotted state of γ_1 and write $(\gamma_1, \gamma_2) \in cnext_{\phi}$ if $\gamma_2 \in \mathit{sol}^{\mathbb{S}}(\mathring{\phi})$ and there exists an abstract state γ_2' such that $(\gamma_1, \gamma_2') \in next_{\mathring{\phi}}$ and for all $x \in V$: $\gamma_2(x) = \gamma_1(x)$ if $\gamma_2'(\mathring{x}) = 0$ and $\gamma_2(x) = \gamma_2'(x)$ otherwise. We say that $\gamma_{2|V}$ is a causally-next state of $\gamma_{1|V}$ and write $(\gamma_{1|V}, \gamma_{2|V}) \in cnext_{\phi}$ if $(\gamma_1, \gamma_2) \in cnext_{\phi}$.*

For $odes(R_{enz})$, the causally next state of $[S/1, E/1, C/0, P/0]$ is $[S/1, E/1, C/1, P/0]$, of which the causally next state is $[S/1, E/1, C/1, P/1]$. As here we often have $next_{odes(R)} \subseteq (cnext_{odes(R)})^*$. We believe that this holds more generally for any reaction network R for which the numerical simulation by Euler's algorithm is sound when using exact arithmetics. Euler's algorithm performs simulation steps, which may introduce approximation errors. These errors may lead to arbitrarily false traces in some case, but may also be ignored in many others. For R_{enz}, this is the error of setting the value of P at the next step to 0 if the value of P was 0 at the previous step and $\dot{P} = 0$ (since C was absent). In reality, P should be set to a small non-zero value at the next step.

7   Abstract Interpretation of ODEs

We recall John's soundness theorem [17] for the abstract interpretation of first-order logic formulas, and apply it for interpreting ODEs abstractly over signs.

Theorem 7 (John's Soundness Theorem [1,17,24]**).** *For any homomorphism $h : S \to \Delta$ between relational structures with signature Σ and any negation-free formula $\phi \in \mathcal{F}_\Sigma(\mathcal{V})$:* $\quad h \circ sol^S(\phi) \subseteq sol^\Delta(\phi).$

Proof We only give a sketch of the proof. Let $\alpha : \mathcal{V} \to dom(S)$. For any expression $e \in \mathcal{E}_\Sigma(\mathcal{V})$ we can show that $h(\llbracket e \rrbracket^{\alpha,S}) = \llbracket e \rrbracket^{h \circ \alpha, \Delta}$ by induction on the structure of e. It then follows for any positive formula $\phi \in \mathcal{F}_\Sigma(\mathcal{V})$ that $\llbracket \phi \rrbracket^{\alpha,S} \le \llbracket \phi \rrbracket^{h \circ \alpha, \Delta}$. This is equivalent to that: $\{h \circ \alpha \mid \alpha \in sol^S(\phi)\} \subseteq sol^\Delta(\phi)$ and thus $h \circ sol^S(\phi) \subseteq sol^\Delta(\phi)$.   □

Recall that $h_\mathbb{S} : \mathbb{R} \to \mathbb{S}$ is a homomorphism between relational structures with signature Σ_{arith} and that the formula $\mathring{\phi}$ has the same signature Σ_{arith} for any ODE ϕ. John's theorem thus shows:

Corollary 8. *For any ODE ϕ:* $\quad h_\mathbb{S} \circ sol^\mathbb{R}(\mathring{\phi}) \subseteq sol^\mathbb{S}(\mathring{\phi}).$

This corollary states that the set of abstract dotted states of an ODE ϕ can be overapproximated by interpreting $\mathring{\phi}$ abstractely in the structure of signs. It can be used to reason about the temporal and causal next transition relations of ODEs as follows:

Lemma 9. *For any ODE ϕ, if $(\gamma_1, \gamma_2) \in next_{\mathring{\phi}} \cup cnext_\phi$, then $\{\gamma_1, \gamma_2\} \subseteq sol^\mathbb{S}(\mathring{\phi})$.*

Proof. 1. Definition 4 of the temporal next relation and Lemma 5 show that for any pair $(\gamma_1, \gamma_2) \in next_{\mathring{\phi}}$ that it satisfies $\gamma_1 = h_\mathbb{S} \circ \alpha_1$ and $\gamma_2 = h_\mathbb{S} \circ \alpha_2$ for some $\alpha_1, \alpha_2 \in sol^\mathbb{R}(\mathring{\phi})$. Corollary 8 of John's soundness theorem for abstract interpretation of logic formulas applied to ODEs shows that $h_\mathbb{S} \circ sol^\mathbb{R}(\mathring{\phi}) \subseteq sol^\mathbb{S}(\mathring{\phi})$ and thus $\{\gamma_1, \gamma_2\} \subseteq sol^\mathbb{S}(\mathring{\phi})$.
2. If $(\gamma_1, \gamma_2) \in cnext_\phi$ then $\gamma_2 \in sol^\mathbb{S}(\mathring{\phi})$ by Definition 6. Furthermore, there exists γ_2' such that $(\gamma_1, \gamma_2') \in next_{\mathring{\phi}}$. The first property shows that $\gamma_1 \in sol^\mathbb{S}(\mathring{\phi})$ too.   □

8   Boolean Networks with Non-deterministic Updates

Any (abstract) state in $\mathbb{B}^V$ is a function $\beta : V \to \mathbb{B}$ that we call a bit vector. For instance, the state $[S/1, E/1, C/0, P/1]$ can be identified with the bit vector 1101 when ordering the species as in above. In the pictures of state transition graphs, the states are drawn as bit vectors in oval nodes and the state transitions as arrows linking these nodes. The legends in blue boxes specify the species order.

Following [25], a Boolean network B with non-deterministic updates (BNN) and species in $\mathcal{S}$ is generally some kind of definition of an abstract state transition

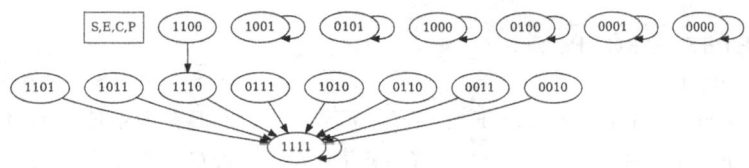

Fig. 5. The state transition graph of the FO-BNN $bnn(R_{enz})$ in Fig. 6.

graph $B \subseteq \mathbb{B}^{\mathcal{S}} \times \mathbb{B}^{\mathcal{S}}$, as for instance in Fig. 5. Definitions of state transition graphs can be expressed in various manners. Here we propose a novel alternative that is based on formulas of first-order logic interpreted over the structure of signs $\mathbb{S}$. We assume that $\mathcal{S} \subseteq \mathcal{V}$ and fix two bijections $\rightarrow : \mathcal{V} \to \mathcal{V}$ and $\circ : \mathcal{V} \to \mathcal{V}$ such that $\mathcal{S}$, $\mathring{\mathcal{S}}$, $\overrightarrow{\mathcal{S}}$, and $\overrightarrow{\mathring{\mathcal{S}}}$ are disjoint subsets of $\mathcal{V}$. Furthermore, we assume that $\overrightarrow{\mathring{x}} = \overrightarrow{\mathring{x}}$ for any $x \in \mathcal{V}$.

From the perspective of the sign abstraction of a reaction network, variable $A \in \mathcal{S}$ states whether species A is present at the current time point and $\mathring{A}$ is the sign of $\dot{A}$. Variable $\overrightarrow{A}$ stands for the presence of A at the next time point, and similarly, $\overrightarrow{\mathring{A}}$ for the sign of the derivative at the next time point $\overrightarrow{\dot{A}}$.

Definition 10. *A first-order Boolean network with nondeterministic updates (FO-BNN) and variables in $V \subseteq \mathcal{V}$ is a first-order formula $\phi \in \mathcal{F}_{\Sigma_{arith}}$ with free variables $fv(\phi) = V \cup \overrightarrow{V}$.*

Notice that the dot operator cannot occur in FO-BNNs ϕ. Any variable assignment $\gamma : V \cup \overrightarrow{V} \to \mathbb{S}$ yields an abstract state transition:

$$trans(\gamma) = (\gamma_{|V}, \gamma_{|\overrightarrow{V}} \circ \rightarrow) \in \mathbb{S}^V \times \mathbb{S}^V$$

For illustration, an FO-BNN for R_{enz} with the (free) variables in $\mathcal{S} = \{S, E, C, P\}$ is shown in Fig. 6. It can be inferred from the ODEs of the reaction network as follows: First the formula $ode(R_{enz})$ is added. Second, a copy of $ode(R_{enz})$ is added, in which all variables x are replaced by $\overrightarrow{x}$. All variables in $\mathring{\mathcal{S}} \cup \overrightarrow{\mathring{\mathcal{S}}}$ are existentially quantified. Furthermore, for any species $A \in \mathcal{S}$, we relate the variable $\overrightarrow{A}$ to the variables A and $\mathring{A}$ by the equation $\overrightarrow{A} \doteq A + \mathring{A}$. This states that $\overrightarrow{A}$ can be true only if either A or $\mathring{A}$ are true, i.e. if the concentration of A is either present or increasing at the previous time point. Finally, we impose $A \leq \overrightarrow{A}$ for stating that species A can never become absent when it was present before. Because the kinetics in R_{enz} are all mass-action laws, this invariant holds here for all the species. It may be false for other reaction networks though.

Furthermore, the solution sets of FO formulas over finite relational structures such as $\mathbb{S}$ can be computed by finite set constraint programming. We have implemented a constraint solver for the relational structure $\mathbb{S}$ in Minizinc [23] which allows us to compute the transition graph of FO-BNNs, i.e., the relation of bit

$\exists \mathring{S} \exists \mathring{E} \exists \mathring{C} \exists \mathring{P} \exists \vec{\mathring{S}} \exists \vec{\mathring{E}} \exists \vec{\mathring{C}} \exists \vec{\mathring{P}}.$

$$\mathring{S} \overset{\circ}{=} -r_1 + r_{-1} \qquad \wedge \vec{\mathring{S}} \overset{\circ}{=} -\vec{r_1} + \vec{r_{-1}} \qquad \wedge \vec{S} \overset{\circ}{=} S + \mathring{S} \qquad \wedge S \le \vec{S}$$

$$\wedge \mathring{E} \overset{\circ}{=} -r_1 + r_{-1} + r_2 \qquad \wedge \vec{\mathring{E}} \overset{\circ}{=} -\vec{r_1} + \vec{r_{-1}} + \vec{r_2} \qquad \wedge \vec{E} \overset{\circ}{=} E + \mathring{E} \qquad \wedge E \le \vec{E}$$

$$\wedge \mathring{C} \overset{\circ}{=} r_1 - r_{-1} - r_2 \qquad \wedge \vec{\mathring{C}} \overset{\circ}{=} \vec{r_1} - \vec{r_{-1}} - \vec{r_2} \qquad \wedge \vec{C} \overset{\circ}{=} C + \mathring{C} \qquad \wedge C \le \vec{C}$$

$$\wedge \mathring{P} \overset{\circ}{=} r_2 \qquad \wedge \vec{\mathring{P}} \overset{\circ}{=} \vec{r_2} \qquad \wedge \vec{P} \overset{\circ}{=} P + \mathring{P} \qquad \wedge P \le \vec{P}$$

where

$$r_1 = 1000000 * S * E \qquad r_{-1} = 0.2 * C \qquad r_2 = 0.1 * C$$
$$\vec{r_1} = 1000000 * \vec{S} * \vec{E} \qquad \vec{r_{-1}} = 0.2 * \vec{C} \qquad \vec{r_2} = 0.1 * \vec{C}$$

Fig. 6. An FO-BNN for the reaction network R_{enz}.

vectors that it defines. The transition graph in Fig. 5 is defined by the FO-BNN for the reaction network R_{enz} in Fig. 6.

The set of species of reaction network R_{enz} has cardinality 4. Therefore, the state transition graph of the FO-BNN of R_{enz} has $2^4 = 16$ states. So clearly, the number of states of a FO-BNN may be exponential in the number of its species. Therefore, it is generally advantageous if one does not have to compute the whole state transition graph, but only the needed part of it. Suppose that we know the sign abstraction of the initial state, we can then generate the subgraph of the state transition graph that is accessible from the abstraction of the initial state, without computing any further states or transitions. In this way, much smaller subgraphs can be observed. For R_{enz}, for instance, for any Boolean state the accessible subgraph contains at most 3 Boolean states.

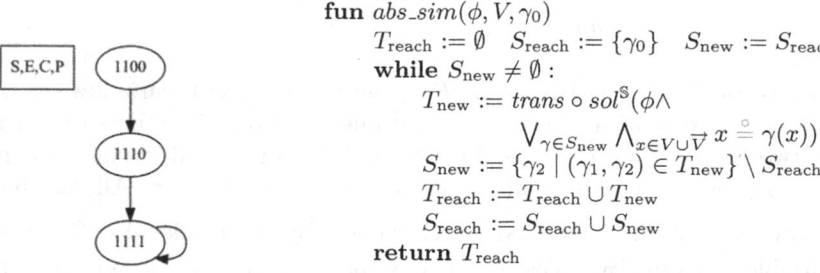

fun $abs\_sim(\phi, V, \gamma_0)$
$\quad T_{\text{reach}} := \emptyset \quad S_{\text{reach}} := \{\gamma_0\} \quad S_{\text{new}} := S_{\text{reach}}$
$\quad$**while** $S_{\text{new}} \ne \emptyset$:
$\qquad T_{\text{new}} := trans \circ sol^{\mathbb{S}}(\phi \wedge$
$$\bigvee_{\gamma \in S_{\text{new}}} \bigwedge_{x \in V \cup \vec{V}} \vec{x} \overset{\circ}{=} \gamma(x))$$
$\qquad S_{\text{new}} := \{\gamma_2 \mid (\gamma_1, \gamma_2) \in T_{\text{new}}\} \setminus S_{\text{reach}}$
$\qquad T_{\text{reach}} := T_{\text{reach}} \cup T_{\text{new}}$
$\qquad S_{\text{reach}} := S_{\text{reach}} \cup S_{\text{new}}$
$\quad$**return** T_{reach}

Fig. 7. Abstract simulation of the FO-BNN in Fig. 6 for the reaction network R_{enz}.

Fig. 8. The abstract simulation of an FO-BNN ϕ with variables in V from an abstract initial state γ_0 computes the set of abstract state transitions T_{reach}.

The abstract simulation of the FO-BNN of R_{enz}, starting with the abstract state $[S/1, E/1, C/0, P/0]$ is given in Fig. 7. This example illustrates, that abstract simulation is related to the causality rather than the temporality of species production. For instance, the temporal transition $(1100, 1111) \in next_{ode(R_{enz})}$ is represented by two causal edges $1100 \rightarrow 1110 \rightarrow 1111$ in the abstract simulation.

These show the causality of the production: C is produced if S and E are present, and P is produced if C is present. But when C is produced then instantaneously P is produced too, so even though C causally precedes P (as shown by the abstract simulation), they are both produced at the same time (in any solution of the ODEs). We notice that causality also plays for concrete numerical simulation with Euler's method: P will be produced shortly after C, depending on the step size that is admitted. The reachable subgraph can be computed by repeated constraint solving. In each step, the values of the variables in $\mathcal{S}$ are constrained to the Boolean states from which the subgraph is to be explored.

In general, for any FO-BNN ϕ with species in $\mathcal{S}$ and abstract initial state $\gamma_0 : V \cup \overrightarrow{V} \to \mathbb{S}$, the abstract simulation represented by the set S_{reach} of all reachable states and the corresponding transition relation T_{reach} can be obtained by iteratively computing the sets of new transitions T_{new} and new reachable states S_{new} starting from the initial state γ_0, as in Fig. 8. The algorithm ends when no new reachable states are obtained from the available transitions.

9   Abstract Simulation of Reaction Networks

To simulate reaction networks abstractly, we propose to translate them to FO-BNNs based on the abstract interpretation of logic formulas over the structure of signs. For any variable assignment $\alpha : V \to \mathbb{R}$, we define $\overrightarrow{\alpha} : \overrightarrow{V} \to \mathbb{R}$ such that $\alpha(x) = \overrightarrow{\alpha}(\overrightarrow{x})$ for all $x \in V$. For any reaction network R, let $\overrightarrow{R}$ be the reaction network with species in $\overrightarrow{S}$ obtained from R by replacing any species $A \in \mathcal{S}$ by $\overrightarrow{A}$. For any $A \in \mathcal{S}$, let $vars_A$ be the sequence of the four variables $A, \mathring{A}, \overrightarrow{A}, \overset{\circ}{\overrightarrow{A}}$.

Our objective is to approximate the relation $cnext_{ode(R)}$ on abstract states by a Boolean network. For any $A \in \mathcal{S}$, we consider formulas $next\_spec(vars_A)$ with the following property. For all $\gamma_1, \gamma_2 : \mathcal{S} \cup \mathring{\mathcal{S}} \to \mathbb{S}$ and all reaction networks R with species in $\mathcal{S}$:

$$(\gamma_1, \gamma_2) \in cnext_{ode(R)} \Rightarrow (\gamma_1 \cup \overrightarrow{\gamma_2})_{|\{vars_A\}} \in sol^{\mathbb{S}}(next\_spec(vars_A))$$

There are several possibilities to define $next\_spec(vars_A)$, of which we propose:

- $next\_spec_1(vars_A) =_{\text{def}} (\overrightarrow{A} \overset{\circ}{=} 1 \to (A \overset{\circ}{=} 0 \wedge \mathring{A} \overset{\circ}{=} 1) \vee A \overset{\circ}{=} 1)) \wedge (\overrightarrow{A} \overset{\circ}{=} 0 \to (A \overset{\circ}{=} 0 \wedge \mathring{A} \overset{\circ}{=} 0) \vee \mathring{A} \overset{\circ}{=} -1))$
- $next\_spec_2(vars_A) =_{\text{def}} next\_spec_1(vars_A) \wedge (\overrightarrow{A} \overset{\circ}{=} 0 \to ((A \overset{\circ}{=} 0 \wedge \mathring{A} \overset{\circ}{=} 0) \vee (A \overset{\circ}{=} 1 \wedge \mathring{A} \overset{\circ}{=} -1 \wedge \overset{\circ}{\overrightarrow{A}} \overset{\circ}{=} 1))))$
- $next\_spec_3(vars_A) =_{\text{def}} next\_spec_1(vars_A) \wedge (\overrightarrow{A} \overset{\circ}{=} 0 \to A \overset{\circ}{=} 0)$

It is not difficult to see that $next\_spec_1(vars_A)$ satisfies the above requirement since using causally-next relation (but this would not hold for the temporal-next relation). If all kinetic expressions are infinitely derivable, then, when a concentration becomes 0, the derivative requires an increase immediately after, in order to not become negative. If all reactions follow the mass action law then non-zero concentrations can never become zero later on, so $next\_spec_3(vars_A)$ should satisfy the requirement too.

Let $next\_spec(vars_A)$ be one of the three formulas $next\_spec_i(vars_A)$ above or any other formula satisfying the above property. Which of these choices is applicable or best depends on properties of the reaction network.

Definition 11. *For any reaction network R with species $\mathcal{S} = \{A_1, \ldots, A_n\}$, we define the FO-BNN $bnn(R)$ depending on the choice of $next\_spec$ as follows:*

$$\exists \mathring{A}_1 \ldots \exists \mathring{A}_n. \ odes(R) \wedge \exists \overrightarrow{\mathring{A}_1} \ldots \exists \overrightarrow{\mathring{A}_n}. \ odes(\overrightarrow{R}) \wedge \bigwedge_{i=1}^n next\_spec(A_i, \mathring{A}_i, \overrightarrow{A}_i, \overrightarrow{\mathring{A}_i})$$

Figure 6 gives the FO-BNN of the reaction network R_{enz} with $next\_spec_3$.

Theorem 12 (Soundness of $bnn(R)$). $cnext_{odes(R)} \subseteq trans \circ sol^{\mathbb{S}}(bnn(R))$.

Proof Let $(\gamma_1, \gamma_2) \in cnext_{odes(R)}$. Then there exists $(\gamma_1', \gamma_2') \in \mathring{cnext}_{odes(R)}$ such that $\gamma_1' = \gamma_{1|\mathcal{S}}$ and $\gamma_2' = \gamma_{2|\mathcal{S}}$. By assumption on $next\_spec(vars_A)$, this implies for all $A \in \mathcal{S}$ that $\gamma_1' \cup \overrightarrow{\gamma_2'} \in sol^{\mathbb{S}}(\bigwedge_{A \in \mathcal{S}} next\_spec(vars_A))$. Lemma 9 shows that $\gamma_1', \gamma_2' \in sol^{\mathbb{S}}(odes(R))$ so that $\overrightarrow{\gamma_2'} \in sol^{\mathbb{S}}(odes(\overrightarrow{R}))$. By definition of $bnn(R)$, we obtain $\gamma_{1|\mathcal{S}}' \cup \overrightarrow{\gamma_{2|\mathcal{S}}'} \in sol^{\mathbb{S}}(bnn(R))$. Hence, $(\gamma_1, \gamma_2) = (\gamma_{1|\mathcal{S}}', \gamma_{2|\mathcal{S}}') \in trans \circ sol^{\mathbb{S}}(bnn(R))$ as stated by the theorem. $\square$

Based on the Soundness Theorem 12 we can simulate any reaction network R by abstractly simulate the Boolean network $bnn(R)$. The abstract simulation of $bnn(R_{enz})$ with $next\_spec_3$ for $next\_spec$, for instance, was shown earlier.

10   Thresholds

We use Booleans to distinguish whether the concentration of a species is zero or not. It often happens, thought, that we would like to know whether the concentration of a species is above or below a given threshold. We now show that this can be treated with the above techniques.

Suppose we are given a species $S \in \mathcal{S}$ and a threshold $\epsilon > 0$, say $\epsilon = 0.3$, and we want to know whether the concentration of S is above, equal, or below ϵ, so whether $S - \epsilon < 0$. The idea is to add an artificial species S_ϵ to the network, such that $\dot{S}_\epsilon = S - \epsilon$. This can be done by adding the following two reactions:

$$S_\epsilon^{cons} : \quad S_\epsilon \xrightarrow{0.3}$$
$$S_\epsilon^{prod} : \quad \xrightarrow{S} S_\epsilon$$

The ODEs of the so extended reaction network contain the expected equation. We can thus run the abstract simulation algorithm on the extended reaction network. When applied to the reaction network R_{enz} with the same initial concentrations than above, this yields the following accessible transition graph:

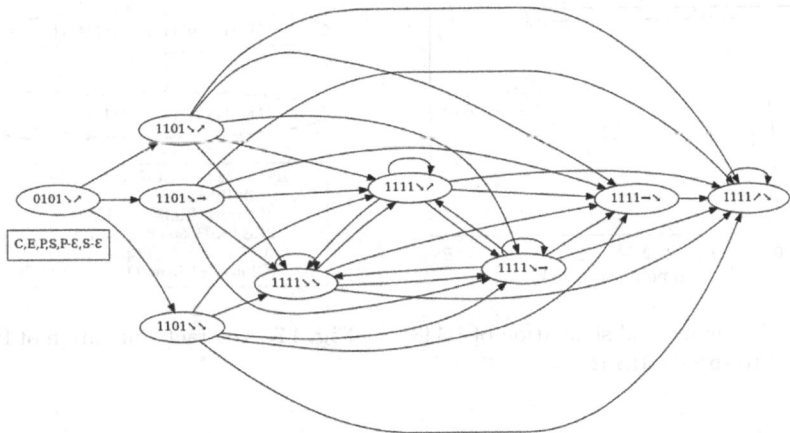

Fig. 9. Abstract simulation for the R_{enz} network with two thresholds.

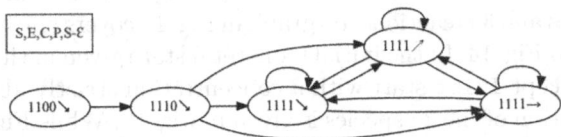

In this picture, we write $S-\epsilon$ instead of $\mathring{S}_\epsilon$. The negative sign $S-\epsilon = \searrow$ means $S < \epsilon$, and a positive sign $S-\epsilon = \nearrow$ means $S > \epsilon$, and $S-\epsilon = \rightarrow$ means $S = \epsilon$.

The proper addition of thresholds, combined with the utilisation of the exact Boolean abstraction algorithm [2] for the set of linear equations of the extended reaction network, provides a considerably more fine-grained abstract simulation of the network. For example, with the addition of a further threshold for P in R_{enz} and an upper bound on the sum of the initial concentrations of S, C, P, the abstract simulation allows us to conclude that only one final state may be reachable during the abstract simulation, where the concentration of S is below the given threshold, as shown in Fig. 9. The automatic application of the exact Boolean abstraction algorithm to the simulation of Boolean networks with thresholds requires however an extension of the algorithm to the inhomogeneous case, which is under implementation. For the simulation shown in Fig. 9, a subset of the inhomogeneous equations was reduced to the homogeneous case by manual rewriting, so that the original algorithm could be applied.

11   Application to Biomodel's Reaction Networks

Biomodels [20] is an online repository which contains a curated collection of over a thousand published models about various biological systems [20]. Most of these were previously published as ODE, but are now provided as reaction networks in the SBML format [18]. We applied abstract simulation to reaction

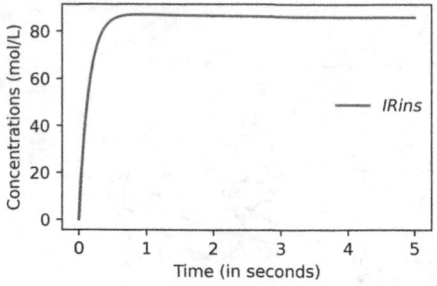

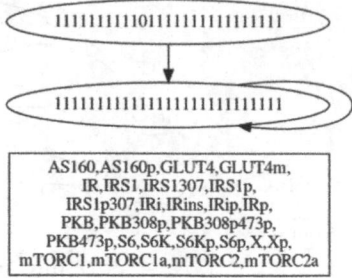

Fig. 10. The numerical simulation of B448 projected to species IRins.

Fig. 11. Abstract simulation of B448

network B448 of the Biomodels database at https://www.ebi.ac.uk/biomodels/ BIOMD0000000448. It models insulin signalling in human adipocytes in normal conditions [8]. Once converted to the BioComputing's XML format, the network involves 27 species and 34 reactions. Its graph in Fig. 13 covers one full page, and its ODEs are given in Fig. 14. In the initial (concrete) state given in the SBML model, all the species except IRins start with a concentration strictly above 0. The full numerical simulation of the 27 species is given in Fig. 12. While Fig. 10 focuses on the concentration of IRins over time.

The total number of abstract states is 2^{27}. However, by starting from the abstraction of the initial concentrations in the SBML model (all species are present except IRins), the state transition graph is reduced to 2 edges between 2 states (Fig. 11). One of these states being the initial abstract state mentioned above, and the other is the 1-only bit vector. The latter is an attractor, as its exiting edge is making a self-loop. This is consistent with the concrete simulation of the model (Fig. 10) and the steady-state computed numerically, but independent of the precise initial concentrations chosen in the SBML model.

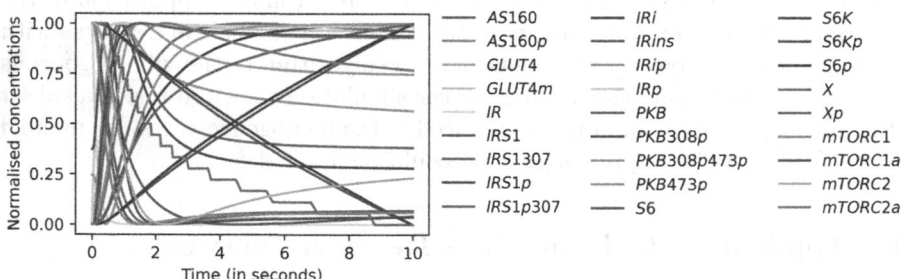

Fig. 12. The numeric simulation of B448 with all species.

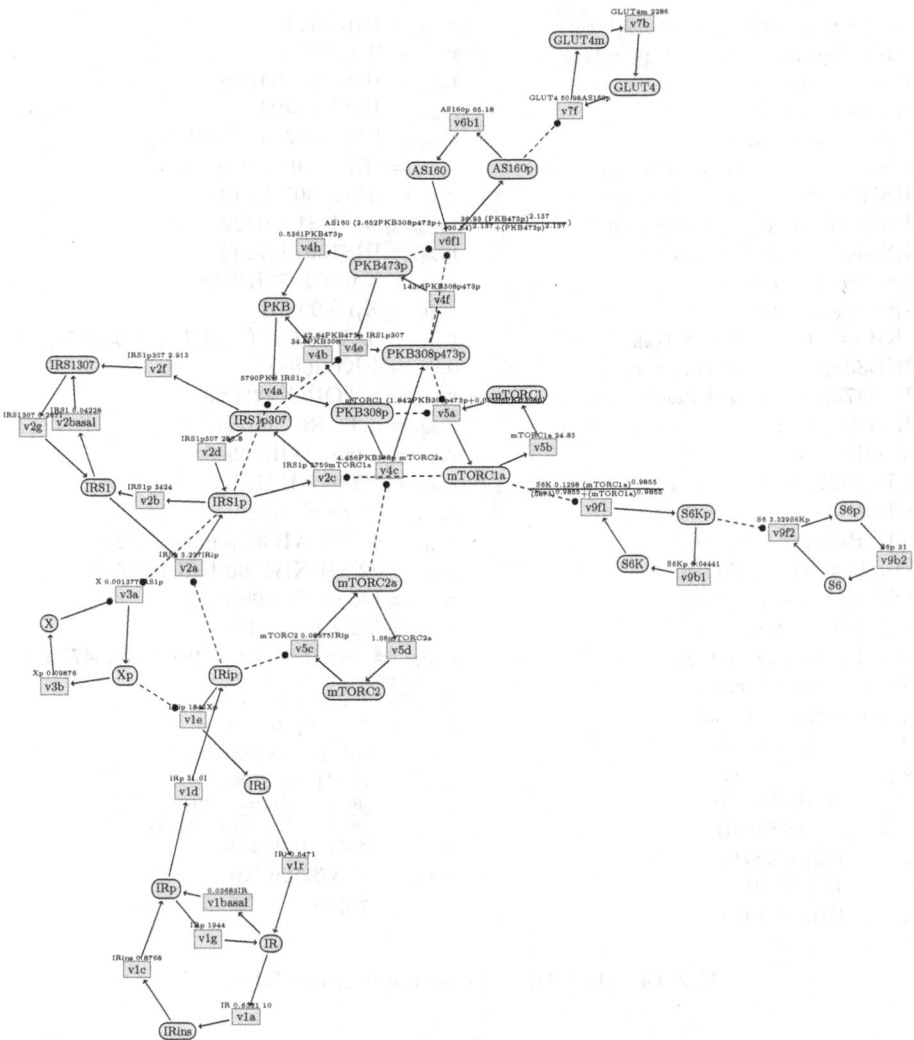

Fig. 13. The graph of reaction network B448

$$\mathring{IR} = -r_{v1a} - r_{v1basal} + r_{v1g} + r_{v1r}$$
$$\mathring{IRp} = r_{v1basal} + r_{v1c} - r_{v1d} - r_{v1g}$$
$$\mathring{IRins} = r_{v1a} - r_{v1c}$$
$$\mathring{IRip} = r_{v1d} - r_{v1e}$$
$$\mathring{IRi} = r_{v1e} - r_{v1r}$$
$$\mathring{IRS1} = -r_{v2a} + r_{v2b} - r_{v2basal} + r_{v2g}$$
$$\mathring{IRS1p} = r_{v2a} - r_{v2b} - r_{v2c} + r_{v2d}$$
$$\mathring{IRS1p307} = r_{v2c} - r_{v2d} - r_{v2f}$$
$$\mathring{IRS1307} = r_{v2f} + r_{v2basal} - r_{v2g}$$
$$\mathring{X} = -r_{v3a} + r_{v3b}$$
$$\mathring{Xp} = r_{v3a} - r_{v3b}$$
$$\mathring{PKB} = -r_{v4a} + r_{v4b} + r_{v4h}$$
$$\mathring{PKB308p} = r_{v4a} - r_{v4b} - r_{v4c}$$
$$\mathring{PKB473p} = -r_{v4e} + r_{v4f} - r_{v4h}$$
$$\mathring{PKB308p473p} = r_{v4c} + r_{v4e} - r_{v4f}$$
$$\mathring{mTORC1} = -r_{v5a} + r_{v5b}$$
$$\mathring{mTORC1a} = r_{v5a} - r_{v5b}$$
$$\mathring{mTORC2} = -r_{v5c} + r_{v5d}$$
$$\mathring{mTORC2a} = r_{v5c} - r_{v5d}$$
$$\mathring{AS160} = -r_{v6f1} + r_{v6b1}$$
$$\mathring{AS160p} = r_{v6f1} - r_{v6b1}$$
$$\mathring{GLUT4m} = r_{v7f} - r_{v7b}$$
$$\mathring{GLUT4} = -r_{v7f} + r_{v7b}$$
$$\mathring{S6K} = -r_{v9f1} + r_{v9b1}$$
$$\mathring{S6Kp} = r_{v9f1} - r_{v9b1}$$
$$\mathring{S6} = -r_{v9f2} + r_{v9b2}$$
$$\mathring{S6p} = r_{v9f2} - r_{v9b2}$$
$$r_{v1a} = \text{IR } 0.6331\ 10$$
$$r_{v1basal} = 0.03683 \text{IR}$$
$$r_{v1c} = \text{IRins } 0.8768$$
$$r_{v1d} = \text{IRp } 31.01$$
$$r_{v1e} = \text{IRip } 1840 \text{Xp}$$

$$r_{v1g} = \text{IRp } 1944$$
$$r_{v1r} = \text{IRi } 0.5471$$
$$r_{v2a} = \text{IRS1 } 3.227 \text{IRip}$$
$$r_{v2b} = \text{IRS1p } 3424$$
$$r_{v2c} = \text{IRS1p } 5759 \text{mTORC1a}$$
$$r_{v2d} = \text{IRS1p307 } 280.8$$
$$r_{v2f} = \text{IRS1p307 } 2.913$$
$$r_{v2basal} = \text{IRS1 } 0.04228$$
$$r_{v2g} = \text{IRS1307 } 0.2671$$
$$r_{v3a} = \text{X } 0.001377 \text{IRS1p}$$
$$r_{v3b} = \text{Xp } 0.09876$$
$$r_{v5a} = \text{mTORC1 } (1.842 \text{PKB308p473p} + 0.05506 \text{PKB308p})$$
$$r_{v5b} = \text{mTORC1a } 24.83$$
$$r_{v5c} = \text{mTORC2 } 0.08575 \text{IRip}$$
$$r_{v5d} = 1.06 \text{mTORC2a}$$
$$r_{v4a} = 5790 \text{PKB IRS1p}$$
$$r_{v4b} = 34.8 \text{PKB308p}$$
$$r_{v4c} = 4.456 \text{PKB308p mTORC2a}$$
$$r_{v4e} = 42.84 \text{PKB473p IRS1p307}$$
$$r_{v4f} = 143.6 \text{PKB308p473p}$$
$$r_{v4h} = 0.5361 \text{PKB473p}$$
$$r_{v6f1} = \text{AS160 } (2.652 \text{PKB308p473p} + \frac{36.93\ (\text{PKB473p})^{2.137}}{(30.54)^{2.137} + (\text{PKB473p})^{2.137}})$$
$$r_{v6b1} = \text{AS160p } 65.18$$
$$r_{v7f} = \text{GLUT4 } 50.98 \text{AS160p}$$
$$r_{v7b} = \text{GLUT4m } 2286$$
$$r_{v9f1} = \frac{\text{S6K } 0.1298\ (\text{mTORC1a})^{0.9855}}{(5873)^{0.9855} + (\text{mTORC1a})^{0.9855}}$$
$$r_{v9b1} = \text{S6Kp } 0.04441$$
$$r_{v9f2} = \text{S6 } 3.329 \text{S6Kp}$$
$$r_{v9b2} = \text{S6p } 31$$

Fig. 14. The ODEs of reaction network B448.

12   Conclusion and Future Work

We presented an algorithm that simulates a reaction network abstractely without any exact knowledge on the initial concentrations. The soundness Theorem 12 relies on the causal next relation of the reaction network, rather than on its temporal next relation. The precise relationship between these two relations seems to be related to approximation errors of Euler's numeric simulation.

One open question is whether one can compute the causal next relation exactly, similarly to the exact computation of the Boolean abstraction of linear equation systems from [2]. Another question is whether more accurate approximations may be possible than those presented here.

The next step will be to lift our abstract simulation algorithm to reaction networks for which the kinetic expressions are only partially known. The most frequent case is that some parameters of the kinetic expressions are unknown. Alternatively, the form of the kinetic expressions may be known only up to similarity [1,24]. Such networks cannot even be simulated concretely without estimating the missing kinetic information from data, so abstract simulation may provide an interesting alternative for the qualitative analysis of such networks.

We hope that the present work could be of interest to the research community of Boolean networks. We believe that FO-BNNs offer an interesting alternative to classical Boolean networks with deterministic updates. So the classical questions for Boolean networks should be reconsidered for FO-BNNs.

Acknowledgements. We thank Jun Pang and Loïc Paulevé for the helpful discussions and references on the relation of reaction and Boolean networks as well as our colleagues from BioComputing Emilie Allart, Maxime Folschette and Cédric Lhoussaine.

References

1. Allart, E., Niehren, J., Versari, C.: Computing difference abstractions of metabolic networks under kinetic constraints. In: Bortolussi, L., Sanguinetti, G. (eds.) CMSB 2019. LNCS, vol. 11773, pp. 266–285. Springer, Cham (2019). https://doi.org/10.1007/978-3-030-31304-3_14
2. Allart, E., Niehren, J., Versari, C.: Exact Boolean abstraction of linear equation systems. Computation **9**(11), 32 (2021)
3. Alur, R., Courcoubetis, C., Henzinger, T.A., Ho, P.-H.: Hybrid automata: an algorithmic approach to the specification and verification of hybrid systems. In: Grossman, R.L., Nerode, A., Ravn, A.P., Rischel, H. (eds.) HS 1991-1992. LNCS, vol. 736, pp. 209–229. Springer, Heidelberg (1993). https://doi.org/10.1007/3-540-57318-6_30
4. Batt, G., de Jong, H., Page, M., Geiselmann, J.: Symbolic reachability analysis of genetic regulatory networks using discrete abstractions. Automatica **44**(4), 982–989 (2008)
5. Benes, N., Brim, L., Drazanova, J., Pastva, S., Safranek, D.: Facetal abstraction for non-linear dynamical systems based on δ-decidable SMT. In: Proceedings of the 22nd ACM International Conference on Hybrid Systems: Computation and Control, Montreal Quebec Canada, pp. 99–108. ACM (2019)
6. Calzone, L., Fages, F., Soliman, S.: BIOCHAM: an environment for modeling biological systems and formalizing experimental knowledge. Bioinformatics **22**(14), 1805–1807 (2006)
7. Collins, P.J., Habets, L., van Schuppen, J.H., Cerná, I., Fabriková, J., Safránek, D.: Abstraction of biochemical reaction systems on polytopes. IFAC Proc. Volumes **44**(1), 14869–14875 (2011)
8. Di Guglielmo, G.M., Drake, P.G., Baass, P.C., Authier, F., Posner, B.I., Bergeron, J.J.: Insulin receptor internalization and signalling. Mol. Cell. Biochem. **182**(1–2), 59–63 (1998). https://doi.org/10.1023/A:1006883311233
9. Euler, L.: Institutionum Calculi Integralis. Number, vol. 1 in Institutionum Calculi Integralis. imp. Acad. imp. Saènt (1768)
10. Fages, F., Gay, S., Soliman, S.: Inferring reaction systems from ordinary differential equations. Theor. Comput. Sci. **599**, 64–78 (2015)
11. Fages, F., Soliman, S.: Abstract interpretation and types for systems biology. Theor. Comput. Sci. **403**(1), 52–70 (2008)

12. Feinberg, M.: Chemical reaction network structure and the stability of complex isothermal reactors–I. The deficiency zero and deficiency one theorems. Chem. Eng. Sci. **42**(10), 2229–2268 (1987)
13. Glass, L., Kauffman, S.A.: The logical analysis of continuous, non-linear biochemical control networks. J. Theor. Biol. **39**(1), 103–129 (1973)
14. Hoops, S., et al.: Copasi—a complex pathway simulator. Bioinformatics **22**(24), 3067–3074 (2006)
15. Jamshidi, S., Siebert, H., Bockmayr, A.: Comparing discrete and piecewise affine differential equation models of gene regulatory networks. In: Lones, M.A., Smith, S.L., Teichmann, S., Naef, F., Walker, J.A., Trefzer, M.A. (eds.) IPCAT 2012. LNCS, vol. 7223, pp. 17–24. Springer, Heidelberg (2012). https://doi.org/10.1007/978-3-642-28792-3_3
16. Jamshidi, S., Siebert, H., Bockmayr, A.: Preservation of dynamic properties in qualitative modeling frameworks for gene regulatory networks. Biosystems **112**(2), 171–179 (2013)
17. John, M., Nebut, M., Niehren, J.: Knockout prediction for reaction networks with partial kinetic information. In: Giacobazzi, R., Berdine, J., Mastroeni, I. (eds.) VMCAI 2013. LNCS, vol. 7737, pp. 355–374. Springer, Heidelberg (2013). https://doi.org/10.1007/978-3-642-35873-9_22
18. Keating, S.M., et al.: SBML level 3: an extensible format for the exchange and reuse of biological models. Mol. Syst. Biol. **16**(8), e9110 (2020)
19. Madelaine, G., Lhoussaine, C., Niehren, J.: Attractor equivalence: an observational semantics for reaction networks. In: Fages, F., Piazza, C. (eds.) FMMB 2014. LNCS, vol. 8738, pp. 82–101. Springer, Cham (2014). https://doi.org/10.1007/978-3-319-10398-3_7
20. Malik-Sheriff, R.S., et al.: BioModels—15 years of sharing computational models in life science. Nucleic Acids Res. **48**(D1), D407–D415 (2020)
21. Mizera, A., Pang, J., Qu, H., Yuan, Q.: Taming asynchrony for attractor detection in large Boolean networks. IEEE/ACM Trans. Comput. Biol. Bioinf. **16**(1), 31–42 (2019)
22. Mover, S., Cimatti, A., Griggio, A., Irfan, A., Tonetta, S.: Implicit semi-algebraic abstraction for polynomial dynamical systems. In: Silva, A., Leino, K.R.M. (eds.) CAV 2021. LNCS, vol. 12759, pp. 529–551. Springer, Cham (2021). https://doi.org/10.1007/978-3-030-81685-8_25
23. Nethercote, N., Stuckey, P.J., Becket, R., Brand, S., Duck, G.J., Tack, G.: MiniZinc: towards a standard CP modelling language. In: Bessière, C. (ed.) CP 2007. LNCS, vol. 4741, pp. 529–543. Springer, Heidelberg (2007). https://doi.org/10.1007/978-3-540-74970-7_38
24. Niehren, J., Versari, C., John, M., Coutte, F., Jacques, P.: Predicting changes of reaction networks with partial kinetic information. Biosystems **149**, 113–124 (2016)
25. Paulevé, L., Sené, S.: Non-deterministic updates of Boolean networks. In: AUTOMATA: 27th International Workshop on Cellular Automata and Discrete Complex Systems, p. 2021. Marseille, France (2021)
26. Paulevé, L., Kolçà, J., Chatain, T., Haar, S.: Reconciling qualitative, abstract, and scalable modeling of biological networks. Nature Commun. **11**, 1–7 (2020)
27. Sankaranarayanan, S., Tiwari, A.: Relational abstractions for continuous and hybrid systems. In: Gopalakrishnan, G., Qadeer, S. (eds.) CAV 2011. LNCS, vol. 6806, pp. 686–702. Springer, Heidelberg (2011). https://doi.org/10.1007/978-3-642-22110-1_56
28. Thomas, R.: Boolean formalization of genetic control circuits. J. Theor. Biol. **42**(3), 563–585 (1973)

Abstraction-Based Segmental Simulation of Chemical Reaction Networks

Martin Helfrich[1]($\boxtimes$) (iD), Milan Češka[2]($\boxtimes$) (iD), Jan Křetínský[1] (iD),
and Štefan Martiček[2] (iD)

[1] Technical University of Munich, Munich, Germany
helfrich@in.tum.de
[2] Brno University of Technology, Brno, Czech Republic
ceskam@fit.vutbr.cz

Abstract. Simulating chemical reaction networks is often computationally demanding, in particular due to stiffness. We propose a novel simulation scheme where long runs are not simulated as a whole but assembled
from shorter precomputed segments of simulation runs. On the one hand,
this speeds up the simulation process to obtain multiple runs since we can
reuse the segments. On the other hand, questions on diversity and gen-
uineness of our runs arise. However, we ensure that we generate runs close
to their true distribution by generating an appropriate abstraction of the
original system and utilizing it in the simulation process. Interestingly,
as a by-product, we also obtain a yet more efficient simulation scheme,
yielding runs over the system's abstraction. These provide a very faithful
approximation of concrete runs on the desired level of granularity, at a
low cost. Our experiments demonstrate the speedups in the simulations
while preserving key dynamical as well as quantitative properties.

Keywords: Chemical reaction networks · Population models ·
Stochastic simulation algorithm · Model abstraction

1 Introduction

Chemical Reaction Networks (CRNs) are a versatile language widely used
for modeling and analysis of biochemical systems [10] as well as for high-level pro-
graming of molecular devices [6,34]. The time-evolution of CRNs is governed by
the Chemical Master Equation that leads to a (potentially infinite) discrete-space,
continuous-time Markov chain (CTMC) with "population" structure, describing
how the probability of the molecular counts of each chemical species evolve in time.
Many important biochemical systems feature complex dynamics, that are hard to
analyze due to *state-space explosion, stochasticity, stiffness, and multimodality* of
the population distributions [17,36]. This fundamentally limits the class of systems the existing techniques can handle effectively. There are several classes of

This work has been supported by the Czech Science Foundation grant GJ20-02328Y,
the German Research Foundation (DFG) projects 378803395 (ConVeY) and 427755713
(GOPro) as well as the ERC Advanced Grant 787367 (PaVeS).

I. Petre and A. Păun (Eds.): CMSB 2022, LNBI 13447, pp. 41–60, 2022.
https://doi.org/10.1007/978-3-031-15034-0_3

approaches that try to circumvent these issues, in particular, (i) *stochastic simulation* avoids the explicit construction of the state space by sampling trajectories in the CRN, and (ii) *abstraction* builds a smaller/simpler model preserving the key dynamical properties and allowing for an efficient numerical analysis of the original CRN. Over the last two decades, there has been very active research on improving the performance and precision of these approaches, see the related work below. Yet, running thousands of simulations to approximate the stochastic behavior often takes many hours; and abstractions course enough to be analyzed easily often fail to capture the complex dynamics, e.g. oscillations in the notorious tiny (two-species) predator-prey system.

Our Contribution. In this paper, in several simple steps, we uniquely *combine* the simulation methods with the abstraction methods for CTMC states, further narrowing the performance gap. As the first step, we suggest leveraging memoization, a general optimization technique that pre-computes and stores partial results to speedup the consequent computation. In particular, when simulating from a current state we reuse previously generated pieces of runs, called *segments*, that start in "similar enough" states. Thus, rather than spending time on simulating a whole new run, we quickly stitch together the segments. To ensure a high variety of runs and generally a good correspondence to the original probability space of runs, we not only have to generate a sufficiently large number of segments; but it is crucial to also consider their length and the similarity of their starting states. We show how the latter two questions can be easily answered using the standard *interval abstraction* on the populations, yielding faithful yet fast simulations of the CTMC for the CRN.

In a second step, we also produce simulation runs over the (e.g. interval) abstraction of the CTMC, not only fast but also with low memory requirements, allowing for efficient analysis on the desired level of detail. To this end, we drop all the concrete information of each segment, keeping only its abstraction plus its *concrete end state*. This surprising choice of information allows us to *define transitions on the abstraction using the simulation* in a rather non-standard way. The resulting semantics and dynamics of the abstraction are non-Markovian, but capture the dynamics of the analyzed system very precisely. From the methodological perspective, the most interesting point is that simulation and abstraction can *help each other* although typically seen as disparate tools.

Related Work. To speed up the standard Stochastic Simulation Algorithm (SSA) [15], several *approximate multi-scale* simulation techniques have been proposed. They include advanced τ-leaping methods [5,27], that use a Poisson approximation to adaptively take time steps leaping over many reactions assuming the system does not change significantly within these steps. Alternatively, various partitioning schemes for fast and slow reactions have been considered [4] allowing one to approximate the fast reactions by a quasi-steady-state assumption [17,32]. The idea of separating the slow and fast sub-networks has been further elaborated in hybrid simulations treating some appropriate species as continuous variables and the others as discrete ones [33]. As before, appropriate partitioning of the species is essential for the performance and accuracy, and thus several (adaptive) strategies have been proposed [14,24]. Recently a *deep learning* paradigm has been introduced to further shift the scalability of the CRN analysis. In [3], the authors learn from a set of stochastic simulations of the CRN a generator, in the form of a Generative Adversarial Network, that efficiently produces trajectories having a similar distribution as the trajectories in the original CRN. In [18] the authors go even further and learn from the simulations an estimator of the given statistic over the original CRN. The principal limitation of these approaches is the overhead related to the learning phase that typically requires a nontrivial number of the simulations of the original CRN.

To build a plausible and computationally tractable abstraction of CRNs, various *state-space reduction techniques* have been proposed that either truncate states of the underlying CTMC with insignificant probability [22,30,31] or leverage structural properties of the CTMC to aggregate/lump selected sets of states [1,2]. The *interval abstraction* of the species population is a widely used approach to mitigate the state-space explosion problem [13,29,37]. We define a segment of simulation runs as a sequence of transitions that can be seen as a single transition of the abstracted system. These "abstract" transitions in the interval abstractions are studied in [9] as "accelerated" transitions. Alternatively, several hybrid models have been considered levering a similar idea as the hybrid simulations. In [23], a pure deterministic semantic for large population species is used. The moment-based description for medium/high-copy number species was used in [19]. The LNA approximation and an adaptive partitioning of the species according to leap conditions (that is more general than partitioning based on population thresholds) were proposed in [7].

Advantages of Our Approach. We show that the proposed *segmental simulation scheme* preserves the key dynamical properties and its qualitative accuracy (with respect to the SSA baseline) is comparable with advanced simulation as well as deep-learning approaches. The scheme, however, provides a significant computational gain over these approaches. Consider a detailed analysis (including 100000 simulation runs) of the famous Toggle switch model reported in [24]. Using the τ-leaping implementation in StochPY [28] is not feasible and the state-of-the-art adaptive hybrid simulation method requires a day to perform such an analysis. However, our approach needs less than two hours. Moreover, our lazy strategy does not require a computationally demanding pre-computation and typically significant benefits from reusing segments after a small number of simulations. This is the key advantage compared to the learning approaches [3,18], where a large number of simulations of the original CRN are required, as well as to approaches based on (approximate) bisimulation/lumping [11,26], requiring a complex analysis of the original model. The approaches based on hybrid formal analysis of the underlying CTMC [7,19,23] have to perform a computationally demanding analysis of conditioned stochastic processes. For example, in [7] the authors report that an analysis of the Viral infection model took more than 1 h. The segmental simulation using 10000 runs provides the same quantitative information in 3 min.

In our previous work [9], we proposed a semi-quantitative abstraction and analysis of CRNs focusing on explainability of the results and low computational complexity, however, providing only limited quantitative accuracy. The proposed simulation scheme provides significantly better accuracy as it keeps track of the current concrete state and thus avoids "jumps" to the abstract state's representative. This kind of rounding is a major source of error as exemplified in the predator-prey model, where the semi-quantitative abstraction of [9] failed to accurately preserve the oscillation and our new approach captures it faithfully.

2   Preliminaries

Chemical Reaction Networks

CRN Syntax. A *chemical reaction network (CRN)* $\mathcal{N} = (\Lambda, \mathcal{R})$ is a pair of finite sets, where Λ is a set of *species*, $|\Lambda|$ denotes its size, and $\mathcal{R}$ is a set of reactions. Species in Λ interact according to the reactions in $\mathcal{R}$. A *reaction* $\tau \in \mathcal{R}$ is a triple $\tau = (r_\tau, p_\tau, k_\tau)$, where $r_\tau \in \mathbb{N}^{|\Lambda|}$ is the *reactant complex*, $p_\tau \in \mathbb{N}^{|\Lambda|}$ is the *product complex* and $k_\tau \in \mathbb{R}_{>0}$ is the coefficient associated with the rate of the reaction. r_τ and p_τ represent the stoichiometry of reactants and products. Given a reaction $\tau_1 = ([1, 1, 0], [0, 0, 2], k_1)$, we often refer to it as $\tau_1 : \lambda_1 + \lambda_2 \xrightarrow{k_1} 2\lambda_3$.

CRN Semantics. Under the usual assumption of mass action kinetics[1], the *stochastic* semantics of a CRN $\mathcal{N}$ is generally given in terms of a discrete-state, continuous-time stochastic process $\mathbf{X}(\mathbf{t}) = (X_1(t), X_2(t), \dots, X_{|\Lambda|}(t), t \geq 0)$ [12]. The *state change* associated with the reaction τ is defined by $\upsilon_\tau = p_\tau - r_\tau$, i.e. the state $\mathbf{X}$ is changed to $\mathbf{X}' = \mathbf{X} + \upsilon_\tau$. For example, for τ_1 as above, we have $\upsilon_{\tau_1} = [-1, -1, 2]$. A reaction can only happen in a state $\mathbf{X}$ if all reactants are present in sufficient numbers. Then we say that the reaction is enabled in $\mathbf{X}$. The behavior of the stochastic system $\mathbf{X}(\mathbf{t})$ can be described by the (possibly infinite) continuous-time Markov chain (CTMC). The transition rate corresponding to a reaction τ is given by a *propensity function* that in general depends on the stoichiometry of reactants, their populations and the coefficient k_τ.

Related Concepts

Population Level Abstraction. The CTMC is the accurate representation of CRN $\mathcal{N}$, but—even when finite—it is not scalable in practice because of the state space explosion problem [20,25]. Various (adaptive) population abstractions [1,13,29,37] have been proposed to reduce the state-space and preserve the dynamics of the original CRN. Intuitively, *abstract states* are given by intervals on sizes of populations (with an additional specific that the abstraction captures enabledness of reactions). In other words, the population abstraction divides the *concrete states* of the CTMC into hyperrectangles called abstract states. We chose one concrete state within each abstract state as its *representative*. Although our approach is applicable to very general types of abstractions, for simplicity and specificity we consider in this paper only the *exponential partitioning* for some

---

[1] We can handle alternative kinetics including Michaelis-Menten and Hill kinetics.

parameter $1 < c \leq 2$ given as $\{[0,0]\} \cup \{[\lfloor c^{n-1} \rceil, \lfloor c^n \rceil - 1] : n \in \mathbb{N}\}$ for all dimensions. For example with $c = 2$ the intervals are $[0, 0], [1, 1], [2, 3], [4, 7], [8, 15], \ldots$ i.e. they grow exponentially in c. While the structure of the abstract states is rather standard, the transitions between the abstract states are defined in different ways.

Stochastic Simulation. An alternative computational approach to the analysis of CRNs is to generate trajectories using stochastic algorithms for simulation. Gillespie's stochastic simulation algorithm (known as SSA) [15] is a widely used exact version of such algorithms, which produces statistically correct trajectories, i.e., sampled according to the stochastic process described by the Chemical Master Equation. To produce such a trajectory, SSA repeatedly applies one reaction at a time while keeping track of the elapsed time. This can take a long time if the number of reactions per trajectory is large. This is typically the case if (1) there are large numbers of molecules, (2) the system is stiff (with high differences in rates) or (3) we want to simulate the system for a time that is long compared to the rates of a single reaction. One of the approaches that mitigate the efficiency problem is τ-leaping [16]. The main idea is that for a given time interval (of length τ), where the reaction propensities do not change significantly, it is sufficient to sample (using Poisson distributions) only the number of occurrences for each reaction and not their concrete sequence. Having the numbers allows one to compute and apply the joint effect of the reactions at once. As detailed in the next section, instead of this time locality, we leverage a space locality.

3   The Plan: A Technical Overview

Since we shall work with different types of simulation runs, gradually building on top of each other, and both with the concrete system and its abstraction, we take the time here to overview the train of thoughts, the involved objects, and the four main conceptual steps:

- Section 4.1 introduces **segmental simulation** as a means to obtain simulation runs of the concrete system faster at the cost of (i) a significant memory overhead and (ii) skewing the probability space of the concrete runs, but only negligibly w.r.t. a user-given abstraction of the state space.
- Section 4.2 introduces **densely concrete simulation**, which eliminates the memory overhead but produces concrete simulation runs where only some of the concrete states on the run are known, however, frequently enough to get the full picture (again w.r.t. the abstraction), see Fig. 2 (bottom).
- Section 5.1 shows how to utilize Sect. 4 to equip the state-space abstraction (quotient) with a powerful transition function and semantics in terms of a probability space over abstract runs, i.e. runs over the abstract state space. The **abstraction is executable** and generates **abstract simulation runs** with extremely low memory requirements, yet allowing for transient analyses that are very precise (again w.r.t. the abstraction), see Fig. 3 (left).
- Section 5.2 considers the **concretization of abstract simulation runs** back to the concrete space, see Fig. 3 (right).

4 Segmental Simulation via Abstract States

4.1 Computing and Assembling Segments via Abstract States

Precomputing Segments. Assume we precomputed for each concrete state s a list of k randomly chosen short trajectories, called *segments*, starting in s. Note that we may precompute multiple segments with the same endpoint, reflecting that this evolution of the state is more probable. We can now obtain a trajectory of the system by repeatedly sampling and applying a precomputed segment for the current state instead of a single reaction.

Using Abstraction. While simulating with already precomputed segments would be faster, it is obviously inefficient to precompute and store the segments for each state separately. However, note that the rates of reactions in CRNs are similar for states with similar amounts of each species, in particular for states within the same abstract state of the population-level abstraction. Consequently, we only precompute k segments for one concrete state per abstract state: the abstract state's representative (typically its center). For other states, the distribution of the segments would be similar and our approximation assumes them to be the very same. While the exponential population-level abstraction is a good starting point for many contexts, the user is free to provide any partitioning (quotient) of the state space that fits the situation and the desired granularity of the properties in question. E.g. one could increase the number of abstract states in regions of the state space we want to study.

An example for $k = 3$ precomputed segments is depicted in Fig. 1 (left). We choose to terminate each segment when it leaves the abstract state. Intuitively, at this point, at least one dimension changes significantly, possibly inducing significantly different rates.

Assembling the Segments. In a *segmental simulation*, instead of sampling a segment for the current concrete state, we sample a segment for the current representative. Because the sampled segment may start at a different state, we apply the relative effect of the segment to the current concrete state. Note that this is a conceptual difference compared with our previous work [9] where the segments are applied to abstract states. The importance of this difference is discussed in [21, Appendix C]. Figure 1 (right) illustrates the segmental simulation for the segments on the left. The system starts in an initial state, which belongs to the bottom left abstract state. Thus, segment c was randomly chosen among the segments a, b, c belonging to that abstract state. After applying the effect of segment c, the system is in the bottom right abstract state and thus we sample from d, e, f and so on. Note that applying a segment might not change the current abstract state and might also only do so temporarily (like the application of l and h, respectively, in the figure). Once the segment leaves the current state a different set of reactions might be enabled. Thus, to make sure that we never

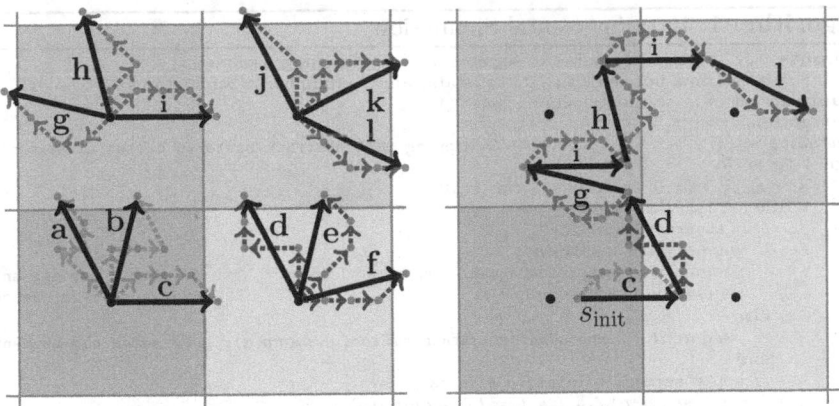

Fig. 1. (left) Four neighboring abstract states, drawn as squares. Each abstract state has $k = 3$ segments that start in their respective centers. Each segment is a sequence of reactions drawn as dotted arrows. The difference between the endpoint and the starting point is called a *summary* and is drawn in unbroken black. (right) A possible segmental simulation obtained by applying the segments c, d, g, i, h, i and l to the initial state s_{init}.

apply reactions that are not enabled, the population level abstraction has to satisfy some additional constraints.[2]

Lazy and Adaptive Computation of Segments. Instead of precomputing the segments for all abstract states, we generate them on the fly. When we need to sample the segments of an abstract state a and there are less than k segments, then we generate a new segment and store it. This new segment is the one we would have sampled from the k (not yet computed) segments for a. Thus, we enlarge the reservoir of segments lazily, only as we need it. Since many abstract states might be rarely reached we generate only few segments for them if any at all. In contrast, we only generate many segments for frequently visited states, which are thus reused many times, improving the efficiency without much overhead. Note that segments can be reused already for a single simulation if that simulation visits the same abstract state more than k times. However, the real benefit of our approach becomes apparent when we generate many simulations.

Algorithm. We summarize the approach in the pseudocode of Algorithm 1. It is already presented in a way that produces not one but m simulations and computes the segments lazily. We start with no precomputed segments. As we simulate, we always compute the current abstract state a (L. 6) and on L. 7 decide whether to simulate a new segment (L. 9) or uniformly choose from the previously computed ones (L. 11).

---

[2] We must choose a population abstraction such that applying any of the representative's possible segments to any corresponding concrete state may only change the enabledness of reactions with the last reaction. Similar constraints are needed if we want to avoid transitions to non-neighboring abstract states. For all presented models, the exponential population abstraction with $c \leq 2$ already has the desired properties.

Algorithm 1: Lazy Segmental Simulation

Inputs : $\mathcal{N}$ (CRN), k (number of segments), c (partitioning parameter),
      t_{end} (time horizon), s_{init} (start state) and m (number of simulations)

Output: list of m segmental simulations

1 $simulations := [\,]$;
2 $memory := \{\}$;   // mapping each abstract state to a list of segments
3 **for** 1 to m **do**
4     $s := s_{init}$; $t := 0$; $simulation := [(s, t)]$;
5   **while** $t < t_{end}$ **do**
6        $a :=$ abstractState$_c(s)$;
7   **if** $|memory(a)| < k$ **then**
8            $segment :=$ sampleNewSegment$(a.representative)$;   // sample new segment
9            $memory(a).$add$(segment)$;   // save it for reuse
10   **else**
11           $segment :=$ chooseUniformlyAtRandomFrom$(memory(a))$; // reuse old segment
12   **end**
        // apply segment's relative effects
13       $s := s + segment.\Delta_{state}$; $t := t + segment.\Delta_{time}$;
14       $simulation.$add$((s, t))$;
15   **end**
16     $simulations.$add$(simulation)$;
17 **end**
18 **return** $simulations$

4.2 Densely Concrete Simulations

Summaries. Storing and applying the whole segments can still be memory- and time-intensive. Therefore, we replace each segment with a single "transition", called a *summary*. It captures the overall effect on the state, namely the difference between the end state and the starting state, and the time the sequence of reactions took. The summaries of segments in Fig. 1 (left) are depicted as solid black arrows. Algorithm 1 applies the segment's summary in L. 13.

Assembling Summaries. Instead of segments, we can append their summaries to the simulation runs, see Fig. 1 (right). We call the result a *densely concrete simulation*. The effect of this modification can be seen in Fig. 2. The top part of the figure shows a typical oscillating run of the predator-prey model that was produced via SSA simulation. The middle part displays a segmental simulation based on our abstraction that uses segments. It exhibits the expected oscillations with varying magnitudes and is visually indistinguishable from an SSA simulation. On the bottom, we see the corresponding densely concrete simulation where the segments of the same segmental simulation have been replaced with their summaries. We still observe the same global behavior but lose the local detail. More precisely, we only see those concrete states of the middle simulation that are the *seams of the segments* (now displayed as dots). All the other concrete states are unknown. However, they are close to the dots since they can only be in the same or neighboring abstract states, meaning the known concrete states are arranged *densely* enough. Moreover, the distance between the dots corresponds to the lengths of the segments. Hence the dots are arranged sparsely only if the system entered a very stable abstract state. Altogether, all changes are reflected faithfully, relative to the level of detail of the abstraction.

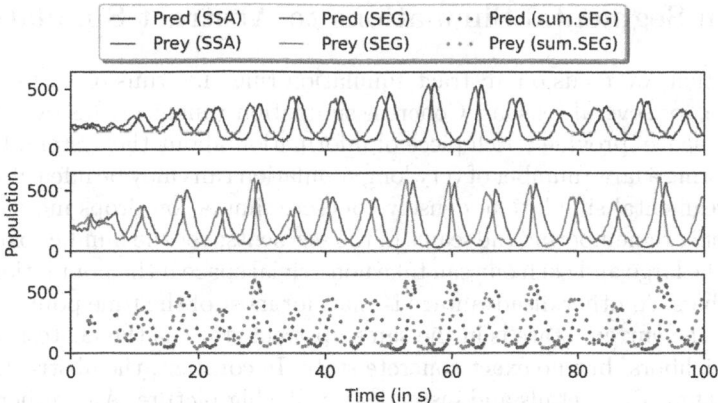

Fig. 2. Comparison of simulations for the predator-prey system: SSA simulation (top), segmental simulation (middle) and the same segmental simulation as densely concrete simulation where segments are replaced with their summaries (bottom).

4.3 Introduced Inaccuracy

We summarize the sources of errors our approach introduces.

(1) Number of Segments. Instead of sampling from all of the possible trajectories, we only sample from k segments and consequently lose some variance. Thus, if k is too small or if the trajectories are too long, our abstraction might miss important behavior of the original system. However, it is easy to see that this error vanishes for $k \to \infty$. It is thus crucial to choose an appropriate value for k and we discuss this choice in Sect. 6. However, note that sampling from segments instead of reactions cannot produce spurious behavior. In other words, all trajectories we obtain by sampling segments are possible trajectories of the original system. Further, if the segments we sample from are representative enough of the actual distribution of trajectories, we will exhibit the same global behavior.

(2) Size of the Abstract States. Recall that we do not sample the distribution of segments for the current state but instead sample the distribution for the representative. Because the propensities and thus the rates of the reactions are different in the current state and the representative state, this inherently introduces an error. However, this error is small if we assume that the distribution over segments does not significantly change within the abstract state. This assumption is reasonable since the propensity functions and thus the rates of the reactions are similar for similar populations and change only slowly; except when the number of molecules is close to zero but there the exponential abstraction provides very fine abstract states. Further, we can decrease the parameter c that determines the interval sizes of the exponential population-level abstraction. A discussion of the influence of parameter c on the accuracy of our method can be found in Sect. 6.

5   From Segmental Simulations to Abstract Simulations

In this section, we focus on abstract simulation runs, i.e. runs over the abstract state space, for several reasons. Concrete simulation runs, i.e. runs over the concrete state space, provide a rich piece of information about the system, however, already storing a large number of very long simulation runs may be infeasible. Compared to segmental simulation, densely concrete simulation drops most concrete states by only remembering the seams of the segments; yet the number of concrete states can be large and each one can take non-trivial space if the populations reach large numbers. Another disadvantage is that, for most of the time points, we only know that the current state is in the same abstract state as the nearest seams or in their neighbors, but no exact concrete state. In contrast, the abstraction may hide non-interesting details and instead show the big picture. Altogether, if only population levels are of interest, abstract simulations can be useful.

5.1   Segmental Abstraction of CTMC and Abstract Simulation

Population-level abstraction of the CTMC might be a lot more explainable than the complete CTMC. However, while the state space of the population abstraction of a CTMC is simply given by the population levels as the Cartesian product of the intervals over all the species, it is not clear what the nature of the transitions should be. There are two issues the previous literature faces. First, what should be the dynamics of an abstract state if each concrete state within behaves a bit differently? Second, what should be the dynamics of one abstract step between two abstract states when it corresponds to a varying number of concrete steps? Standard approaches either pick a representative state and copy its dynamics, e.g. the rate of the reaction, or take an over-approximation of all behaviors of all possible members of the class, e.g. take an interval of possible rates.

Here we reuse the segmental simulation and the concepts of Sect. 4 to formally define a transition function on the abstraction. This gives us the abstraction's semantics and makes it executable. Moreover, the resulting behavior is close to the original system (in contrast to, e.g. [9], we can even preserve the oscillations of the predator-prey models), but at the expense of making the abstract model non-Markovian. Intuitively, our *segmental abstraction* of the CTMC is given by the abstract state space and the segments, exactly as depicted in Fig. 1 (left). Similar to other non-Markovian systems, the further evolution of the system is not given only by its current state, but also by some information about the history of the run so far. For instance, in the case of semi-Markov processes, it is the time spent in the state so far; in the case of generalized semi-Markov processes, it is the times each event has been already scheduled for. In the case of the segmental abstraction, it is a vector forming a concrete state.

Formally, the *configuration* of the segmental abstraction is a triple (a, s, t) where a is an abstract state, s one of its concrete states, and t a time. The probability to move from (a, s, t) to (a', s', t') is then given by the probability to sample a segment with a summary $s' - s$ on states and taking $t' - t$ time. (Hence we can store the summaries only, as described in Sect. 4.2.) Given a concrete

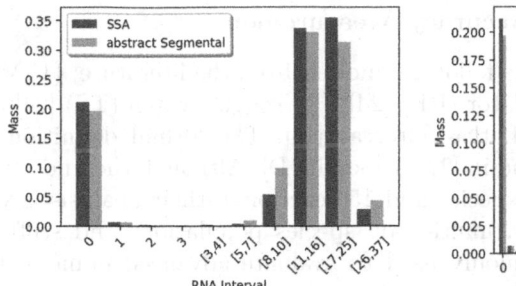 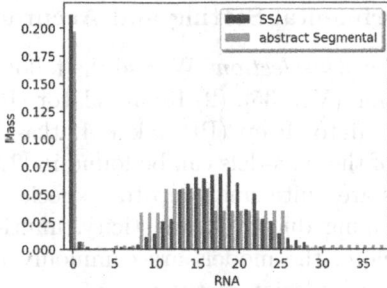

Fig. 3. RNA distribution in the viral infection model at $t = 200$ s predicted by SSA and abstract segmental simulation with $c = 1.5$ and $k = 100$: in the abstract domain (left) and the concrete domain (right) where the segmental simulation's abstract values were concertized using a uniform distribution over the interval.

state s to start in, there is a unique probability space over the abstract runs initiated in $(a, s, 0)$ obtained by dropping the second component.

The probability space coincides with the probability space introduced by the segmental simulation (in the variant with summaries of precomputed segments) when the concrete runs are projected by the population-level abstraction to the abstract runs. However, (i) the space needed to store the abstract simulations is smaller and (ii) transient analysis is well defined for every time point, while its results are still very faithful. Indeed, Fig. 3 (left) shows an example of the transient analysis (at a given time point t) obtained by (i) the states reached by real simulation runs and clustered according to the population-level abstraction, and (ii) abstract states reached by abstract segmental simulation runs. Given the granularity of the abstraction, the results are very close.

5.2   From Abstract Simulations Back to Concrete Predictions

Further, one can map the abstract states to sets of concrete states. Consequently, the results of the abstract transient analysis can be mapped to a distribution over concrete states, whenever we assume a distribution over the concrete states corresponding to one abstract state. For instance, taking uniform distribution as a baseline, we obtain a concrete transient analysis from the abstract one, see Fig. 3 (right), which already shows a close resemblance.

6   Experimental Evaluation

We evaluate the densely concrete version of segmental simulation and consider the following three research questions:

Q1 What is the accuracy of segmental simulation?
Q2 What are the trade-offs between accuracy and performance?
Q3 Are the achieved trade-offs competitive with alternative approaches?

Experimental Setting and Accuracy Measurement

Benchmark selection. We use the following models from the literature: (1) Viral Infection (VI) [35], (2) Repressilator (RP) [24], (3) Toggle Switch (TS) [24] and (4) Predator-Prey (PP, a.k.a. Lotka-Volterra) [15]. The formal definition for each of these models can be found in [21, Appendix D]. Although the underlying CRNs are quite small (up to 6 species and 15 reactions), their analysis is very challenging due to stochasticity, multi-scale species populations and stiffness. Therefore, the models are commonly used to evaluate advanced numerical as well as simulation methods.

Implementation and HW Configuration. Our approach is implemented as modification of SeQuaiA, a Java-based tool for semi-quantitative analysis of CRNs [8]. All experiments run on a 1.80 GHz Lenovo ThinkPad T580 with 8 GB of RAM. To report speedups, we use our own competitive implementation of the SSA method as a baseline. Depending on the model, it achieves between 1×10^6 and 7×10^6 reactions per second. We refer to the SeQuaiA repository[3] for all active development and provide an artefact[4] to reproduce the experimental results.

Assessing Accuracy. To measure the accuracy, we compare transient distributions for one species at a time. For this, we approximate the implied transient distribution of our approach and of SSA by running a large number of simulations. We used 1000 simulations for the models RP and TS as their simulations take longer, and 10000 for PP and VI. The resulting histograms for the studied species are then normalized to approximate the transient distribution of that species. To quantify the error, we compare the means of both distributions and report the earth-mover-distance (EMD) between them. Because EMD values are difficult to interpret without context, we additionally report the EMD between two different transient distributions that were computed with SSA. Intuitively, even if segmental simulation was as accurate as SSA, we would expect to see a EMD similar to the EMD of this "control SSA". Additionally, Fig. 12 in [21, Appendix B] compares the variance.

Q1 What is the Accuracy of Segmental Simulation?

In this section, we evaluate the accuracy of the segmental simulation scheme and the effects of parameters c (size of the abstraction) and k (the number of stored summaries) in a quantitative manner. For a more qualitative evaluation see Fig. 2 and [21, Appendix A] where you find exemplary simulations and trajectories for both segmental simulation and SSA. We consider three population abstractions given by $c \in \{2, 1.5, 1.3\}$ (recall that $c = 2$ is the most coarse abstraction as explained in Sect. 2) and three values of k, namely $k \in \{10, 100, 1000\}$.

We start with the VI model where one is typically interested in the distribution of the RNA population at a given time t. Figure 4 (left) shows the

---

[3] https://sequaia.model.in.tum.de (SeQuaiA).

[4] https://doi.org/10.5281/zenodo.6658924 (artifact).

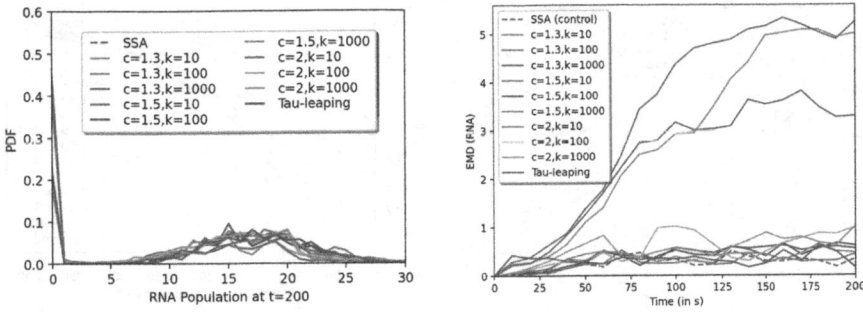

Fig. 4. Accuracy on the viral infection model using different abstractions.

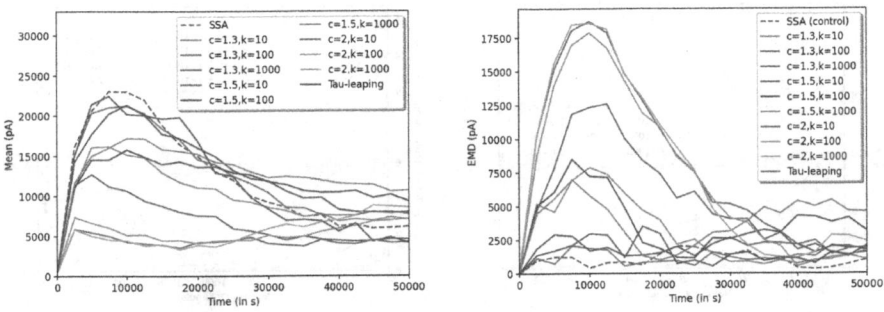

Fig. 5. Accuracy on the repressilator model using different abstractions.

distributions at $t = 200$ obtained by SSA simulation and by segmental simulations with different values of the parameters c and k. We observe that all distributions show the expected bi-modality [17]. If a less precise abstraction is used ($c = 2$ and/or $k = 10$), the probability that RNA dies out is significantly higher than the reference value. In Fig. 4 (right), we evaluate how the EMD (for the RNA) changes in time for particular settings. The results clearly confirm that $k = 10$ leads to significant inaccuracy. For all other settings, the EMD is very close to the SSA control demonstrating the very high accuracy of our approach. The only notable exceptions are the variants with $c = 2$, where the EMD fluctuates. We also observe that increasing k from 100 to 1000 does not bring any considerable improvement.

Different trends can be observed for the RP model. Figure 5 shows how the mean value (left) and the EMD (right) of the species pA change over time. We observe that the partitioning of the populations plays a more important role here, i.e., the coarse abstraction ($c = 2$) induces a notable inaccuracy. The fact that the accuracy is less sensitive with respect to the low values of k is a result of the very regular dynamics of the model where the populations of the proteins pA and pB oscillate and slowly decrease. Similar trends (not presented here) are observed also for the TS model.

Finally, we consider the PP model. Although very simple, it is notoriously difficult for abstraction-based approaches since they struggle to preserve the

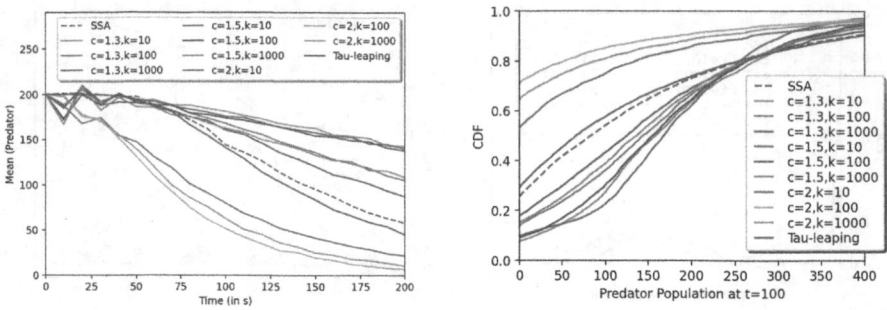

Fig. 6. Accuracy on the predator-prey model using different abstractions.

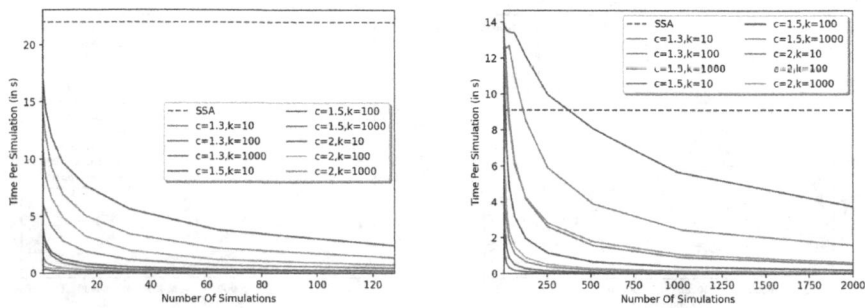

Fig. 7. Increasing speed of lazy segmental simulation for the toggle switch model (left) and repressilator model (right).

oscillation and the die-out time. Recall Fig. 2 of Sect. 4 where we clearly observe the expected oscillation with the correct frequency in a segmental simulation for $c = 2$ and $k = 100$. Figure 6 (left) shows how the mean population of the predators changes in time. We observe that the less precise abstractions do not accurately preserve the rate at which the mean population decreases. On the other hand, the most precise setting is close to the SSA reference and control curve. Figure 6 (right) shows the cumulative predator distributions at $t = 100$ demonstrating how the simulations using less precise abstractions deviate from the reference solution.

Q2 What are the Trade-Offs Between Accuracy and Performance?

Recall that our approach is based on re-using the segments generated in the previous simulation runs. Figure 7 shows how the average time per simulation decreases for a growing number of simulations. For some models (like TS) segmental simulation is always faster than SSA because it can reuse segments already during the first simulation. For other models (like RP), segmental simulation becomes faster after a number of simulations that depends on the precision of the abstraction. Table 1 shows the speedup factor we observe when running 10,000 segmental simulations instead of SSA simulations. Table 5 in [21, Appendix B] reports the average

Table 1. Average run-time of one SSA simulation and the speedup factor of segmental simulation when computing 10,000 simulations with different abstraction parameters.

Model	SSA	SEG $k = 10$			SEG $k = 100$			SEG $k = 1000$		
		$c = 2$	$c = 1.5$	$c = 1.3$	$c = 2$	$c = 1.5$	$c = 1.3$	$c = 2$	$c = 1.5$	$c = 1.3$
PP	0.014 s	70x	70x	70x	70x	70x	23x	28x	23x	12x
VI	0.88 s	730x	380x	180x	100x	48x	17x	8.6x	4.8x	2.9x
TS	22 s	360x	360x	340x	390x	350x	280x	250x	190x	110x
RP	9.1 s	760x	540x	320x	300x	140x	62x	54x	21x	7.4x

Table 2. Number of visited abstract states after 10,000 segmental simulations for different abstraction parameters.

Model	SEG $k = 10$			SEG $k = 100$			SEG $k = 1000$		
	$c = 2$	$c = 1.5$	$c = 1.3$	$c = 2$	$c = 1.5$	$c = 1.3$	$c = 2$	$c = 1.5$	$c = 1.3$
PP	163	398	832	170	391	888	170	397	885
VI	1,022	3,669	1,0337	1,269	3,797	11,524	1,353	4,018	11,218
TS	5,072	13,167	38,827	9,248	25,733	65,259	10,388	29,424	74,950
RP	12,921	42,241	124,280	18,014	56,312	155,394	21,460	64,385	146,126

number of reactions per SSA simulation and the average number of applied summaries per segmental simulation. Comparing these numbers highlights the source of the speedup: instead of sampling and applying many reactions, segmental simulation does the same for fewer summaries. This also gives an estimate for the speedup factor that can be reached once only precomputed summaries are used. Note that the speedup factors we report are smaller because we include the time needed for computing the summaries.

The PP model includes only two species and it is not stiff. Therefore, already the SSA simulation is quite fast. Our approach still achieves a stable speedup around 70x that drops to a factor of 12x for the most precise abstraction ($c = 1.3$ and $k = 1000$) providing accuracy close to the control SSA.

For the VI model, we observe a slowdown of the segmental simulation when improving the abstraction, namely, for $k - 1000$. Recall, we reported very good accuracy already for $c = 2$ and $k = 100$ which gives us a speedup factor of 100x. For an accuracy that is close to the control SSA, we archive a speedup factor of 50x.

The TS model exhibits regular oscillation, where a typical run repeatedly visits the same abstract states. This is very beneficial for our approach as we can very efficiently reuse segments. We observe a significant slowdown for $c = 1.3$ and $k = 1000$, since reusing segments is much less effective. The RP model has similar characteristics, but we observe an even more significant slowdown when the abstraction is refined. As we discussed in the previous section, very good accuracy for these models is achieved already for $c = 1.5$ and $k = 100$ which give us speedup factors 350x and 140x for the TS and RP model, respectively.

In general, segmental simulation is never slower than SSA by more than the constant factor that is the result of saving and loading segments and it eventually becomes faster if we compute enough simulations. We pay for the inevitable

Table 3. Size of segmental abstraction after 10,000 simulations for different parameters.

Model	SEG $k = 10$			SEG $k = 100$			SEG $k = 1000$		
	$c = 2$	$c = 1.5$	$c = 1.3$	$c = 2$	$c = 1.5$	$c = 1.3$	$c = 2$	$c = 1.5$	$c = 1.3$
PP	25 KB	61 KB	130 KB	250 KB	570 KB	1.3 MB	2.2 MB	4.8 MB	11 MB
VI	210 KB	730 KB	2.0 MB	1.8 MB	4.8 MB	13 MB	11 MB	25 MB	53 MB
TS	1.2 MB	3.0 MB	8.7 MB	15 MB	37 MB	85 MB	100 MB	250 MB	550 MB
RP	3.8 MB	12 MB	34 MB	43 MB	120 MB	300 MB	310 MB	760 MB	1.0 GB

speedup with increased memory consumption. Table 2 shows how the number of visited abstract states after 10,000 segmental simulations. The number of saved summaries [21, Appendix B, Table 6] is approximately the number of visited abstract states times k. The memory consumed by one summary is an integer vector and a floating-point number that describe the effect on state and time, respectively. The size of the abstraction is shown in Table 3.

We observe a trade-off between accuracy and performance: smaller abstract states results in segments with fewer steps and thus slower simulations and for more precise abstractions we need to save more summaries. Further, we can only reuse simulations if they visit the same abstract states. This implies that the presented approach does not scale favorably with the dimension of the studied system. To handle many species, one can use the available memory only for the most important abstract states, e.g. the most visited ones, and simulate normally in all other regions. But there is an inherent trade-off between memory consumption and simulation speed as we can only reuse segments if we save them.

Q3: Comparison with Alternative Approaches

Comparison with τ-Leaping. We first compare the performance and accuracy of our approach with the τ-leaping method implemented in StochPy [28], a widely used stochastic modeling and simulation package. τ-leaping achieves very good accuracy on the considered models. Quantitatively, it is very close to the SSA control runs and typically provides slightly better results than our best setting (compare Figs. 4, 5 and 6). On the other hand, we typically observe only a moderate speedup (around one order of magnitude) with respect to the SSA. Note that a direct comparison with our run-times is unfair as it is known that StochPy uses an inefficient python-based random number generator that can significantly slow down the simulation. For example, a single SSA simulation of the RP model takes in StochPy 1000 s and τ-leaping achieves a 16-times speedup while our SSA baseline takes around 9 s and the segmental simulation achieves the speedup of a factor over 140 (to our baseline) with only a small drop in the accuracy. On the other hand, for the PP model, τ-leaping provides only a negligible speedup (below factor 2). Recall we observed a speedup factor between 12x and 70x depending on the required precision.

Comparison with Advanced Simulation Methods. A fair comparison with slow-scale stochastic simulations [4, 17, 32] is problematic since, to our best knowledge,

Table 4. Runtime comparison with [24] for 100,000 simulations.

Model	results presented in [24]			Our results		
	SSA	Adaptive hybrid	Speedup	SSA	SEG($c = 1.5$, $k = 100$)	Speedup
RP	232 hours	3 hours	77x	252 hours	1.8 hours	140x
TS	47 days	1.1 days	43x	25 days	0.07 days	350x

there is no available implementation. Therefore, we focus on the comparison with the results presented in [24] representing the state-of-the-art hybrid simulation method. In Table 4, we compare the run-times of the adaptive hybrid simulation (100,000 runs) on the repressilator and toggle switch models reported in [24] (Fig. 1) with our approach. We report the runtimes only for the setting with $c = 1.5$ and $k = 100$ which already leads to very good accuracy for these models. The table also shows runtimes for the baseline SSA and the achieved speedup factor to make the comparison between the different hardware configurations fair. We observe that a significant computational gain (over two orders of magnitude) is achieved by our approach.

Comparison with the Deep Learning Approaches. Finally, we compare with the approach of [3] where a neural network is trained to provide a fast and accurate generator of simulations in the original CRN. To this end, we use a simpler variant of the toggle switch model considered in [3]. If we run more than 1000 simulations, a single simulation takes on average 0.0004 s which is comparable with the neural-based generator (the authors report 0.0008 s per simulation). Regarding the accuracy, we achieve comparable values of the EMD (note the EMD is scaled in [3]). The key benefit of our approach, however, lies in the fact that it does not require the computationally very demanding training phase.

7 Conclusion and Future Directions

We have proposed a novel simulation scheme enabling us to efficiently generate a large number of simulation runs for complex CRNs. It is based on reusing segments of the runs computed over abstract states but applied to concrete states. Already our initial experiments demonstrate that the simulation scheme preserves key dynamical and quantitative properties while providing a significant computational gain over the existing approaches. On the conceptual level, we define an executable abstraction of the CTMC, preserving the dynamics very faithfully. In particular, we have the machinery to generate abstract simulation runs, which take less space than the concrete ones, yet provide high precision on the level of detail given by the population levels defined by the user.

In future work, we want to investigate the error with the goal of giving formal error bounds. Further, we propose an adaptive version of the abstraction where the population abstraction and the number of precomputed segments are refined or learned. Alternatively, instead of memorizing a discrete distribution over precomputed segments, we can generalize to unobserved behavior by learning some continuous distribution.

Segmental simulation via abstract states can be understood as a general framework for accelerating the simulation of population models. As such, it can be combined with any method that predicts the evolution of such models. In particular, it can naturally leverage an adaptive multi-scale approach where different simulation techniques are used in different regions of the state-space.

References

1. Abate, A., Andriushchenko, R., Češka, M., Kwiatkowska, M.: Adaptive formal approximations of Markov chains. Perform. Eval. **148**, 102207 (2021)
2. Backenköhler, M., Bortolussi, L., Großmann, G., Wolf, V.: Abstraction-guided truncations for stationary distributions of Markov population models. In: Abate, A., Marin, A. (eds.) QEST 2021. LNCS, vol. 12846, pp. 351–371. Springer, Cham (2021). https://doi.org/10.1007/978-3-030-85172-9_19
3. Cairoli, F., Carbone, G., Bortolussi, L.: Abstraction of Markov population dynamics via generative adversarial nets. In: Cinquemani, E., Paulevé, L. (eds.) CMSB 2021. LNCS, vol. 12881, pp. 19–35. Springer, Cham (2021). https://doi.org/10.1007/978-3-030-85633-5_2
4. Cao, Y., Gillespie, D.T., Petzold, L.R.: The slow-scale stochastic simulation algorithm. J. Chem. Phys. **122**(1), 014116 (2005)
5. Cao, Y., Gillespie, D.T., Petzold, L.R.: Efficient step size selection for the tau-leaping simulation method. J. Chem. Phys. **124**(4), 044109 (2006)
6. Cardelli, L.: Two-domain DNA strand displacement. Math. Struct. Comput. Sci. **23**(02), 247–271 (2013)
7. Cardelli, L., Kwiatkowska, M., Laurenti, L.: A stochastic hybrid approximation for chemical kinetics based on the linear noise approximation. In: Bartocci, E., Lio, P., Paoletti, N. (eds.) CMSB 2016. LNCS, vol. 9859, pp. 147–167. Springer, Cham (2016). https://doi.org/10.1007/978-3-319-45177-0_10
8. Češka, M., Chau, C., Křetínský, J.: SeQuaiA: a scalable tool for semi-quantitative analysis of chemical reaction networks. In: Lahiri, S.K., Wang, C. (eds.) CAV 2020. LNCS, vol. 12224, pp. 653–666. Springer, Cham (2020). https://doi.org/10.1007/978-3-030-53288-8_32
9. Češka, M., Křetínský, J.: Semi-quantitative abstraction and analysis of chemical reaction networks. In: Dillig, I., Tasiran, S. (eds.) CAV 2019. LNCS, vol. 11561, pp. 475–496. Springer, Cham (2019). https://doi.org/10.1007/978-3-030-25540-4_28
10. Chellaboina, V., Bhat, S.P., Haddad, W.M., Bernstein, D.S.: Modeling and analysis of mass-action kinetics. IEEE Control Syst. Mag. **29**(4), 60–78 (2009)
11. Desharnais, J., Laviolette, F., Tracol, M.: Approximate analysis of probabilistic processes: logic, simulation and games. In: 2008 Fifth International Conference on Quantitative Evaluation of Systems, pp. 264–273. IEEE (2008)
12. Ethier, S.N., Kurtz, T.G.: Markov Processes: Characterization and Convergence, vol. 282. Wiley, Hoboken (2009)
13. Ferm, L., Lötstedt, P.: Adaptive solution of the master equation in low dimensions. Appl. Numer. Math. **59**(1), 187–204 (2009)

14. Ganguly, A., Altintan, D., Koeppl, H.: Jump-diffusion approximation of stochastic reaction dynamics: error bounds and algorithms. Multisc. Model. Simul. **13**(4), 1390–1419 (2015)
15. Gillespie, D.T.: Exact stochastic simulation of coupled chemical reactions. J. Phys. Chem. **81**(25), 2340–2361 (1977)
16. Gillespie, D.T.: Approximate accelerated stochastic simulation of chemically reacting systems. J. Chem. Phys. **115**(4), 1716–1733 (2001)
17. Goutsias, J.: Quasiequilibrium approximation of fast reaction kinetics in stochastic biochemical systems. J. Chem. Phys. **122**(18), 184102 (2005)
18. Gupta, A., Schwab, C., Khammash, M.: DeepCME: a deep learning framework for computing solution statistics of the chemical master equation. PLoS Comput. Biol. **17**(12), e1009623 (2021)
19. Hasenauer, J., Wolf, V., Kazeroonian, A., Theis, F.: Method of conditional moments (MCM) for the chemical master equation. J. Math. Biol. **69**, 1–49 (2013)
20. Heath, J., Kwiatkowska, M., Norman, G., Parker, D., Tymchyshyn, O.: Probabilistic model checking of complex biological pathways. Theoret. Comput. Sci. **391**(3), 239–257 (2008)
21. Helfrich, M., Češka, M., Křetínský, J., Martiček, Š.: Abstraction-based segmental simulation of chemical reaction networks. arXiv (2022). https://doi.org/10.48550/arXiv.2206.06677
22. Henzinger, T.A., Mateescu, M., Wolf, V.: Sliding window abstraction for infinite Markov chains. In: Bouajjani, A., Maler, O. (eds.) CAV 2009. LNCS, vol. 5643, pp. 337–352. Springer, Heidelberg (2009). https://doi.org/10.1007/978-3-642-02658-4_27
23. Henzinger, T.A., Mikeev, L., Mateescu, M., Wolf, V.: Hybrid numerical solution of the chemical master equation. In: CMSB 2010, pp. 55–65. ACM (2010)
24. Hepp, B., Gupta, A., Khammash, M.: Adaptive hybrid simulations for multiscale stochastic reaction networks. J. Chem. Phys. **142**(3), 034118 (2015)
25. Kwiatkowska, M., Thachuk, C.: Probabilistic model checking for biology. Softw. Syst. Saf. **36**, 165 (2014)
26. Larsen, K.G., Skou, A.: Bisimulation through probabilistic testing. Inf. Comput. **94**(1), 1–28 (1991)
27. Lester, C., Yates, C.A., Giles, M.B., Baker, R.E.: An adaptive multi-level simulation algorithm for stochastic biological systems. J. Chem. Phys. **142**(2), 01B612_1 (2015)
28. Maarleveld, T.R., Olivier, B.G., Bruggeman, F.J.: StochPy: a comprehensive, user-friendly tool for simulating stochastic biological processes. PLoS One **8**(11), e79345 (2013)
29. Madsen, C., Myers, C., Roehner, N., Winstead, C., Zhang, Z.: Utilizing stochastic model checking to analyze genetic circuits. In: Computational Intelligence in Bioinformatics and Computational Biology (CIBCB), pp. 379–386. IEEE (2012)
30. Mateescu, M., Wolf, V., Didier, F., Henzinger, T.A.: Fast adaptive uniformization of the chemical master equation. IET Syst. Biol. **4**(6), 441–452 (2010)
31. Munsky, B., Khammash, M.: The finite state projection algorithm for the solution of the chemical master equation. J. Chem. Phys. **124**, 044104 (2006)
32. Rao, C.V., Arkin, A.P.: Stochastic chemical kinetics and the quasi-steady-state assumption: application to the Gillespie algorithm. J. Chem. Phys. **118**(11), 4999–5010 (2003)
33. Salis, H., Kaznessis, Y.: Accurate hybrid stochastic simulation of a system of coupled chemical or biochemical reactions. J. Chem. Phys. **122**(5), 054103 (2005)

34. Soloveichik, D., Seelig, G., Winfree, E.: DNA as a universal substrate for chemical kinetics. Proc. Natl. Acad. Sci. U.S.A. **107**(12), 5393–5398 (2010)
35. Srivastava, R., You, L., Summers, J., Yin, J.: Stochastic vs. deterministic modeling of intracellular viral kinetics. J. Theor. Biol. **218**(3), 309–321 (2002)
36. Van Kampen, N.G.: Stochastic Processes in Physics and Chemistry, vol. 1. Elsevier, Amsterdam (1992)
37. Zhang, J., Watson, L.T., Cao, Y.: Adaptive aggregation method for the chemical master equation. Int. J. Comput. Biol. Drug Des. **2**(2), 134–148 (2009)

Qualitative Dynamics of Chemical Reaction Networks: An Investigation Using Partial Tropical Equilibrations

Aurélien Desoeuvres[1], Peter Szmolyan[2], and Ovidiu Radulescu[1]([⊠])

[1] LPHI UMR CNRS 5235, University of Montpellier, Montpellier, France
ovidiu.radulescu@umontpellier.fr
[2] Institute of Analysis and Scientific Computing, Technische Universität Wien, Vienna, Austria

Abstract. We discuss a method to describe the qualitative dynamics of chemical reaction networks in terms of symbolic dynamics. The method, that can be applied to mass-action reaction networks with separated timescales, uses solutions of the *partial tropical equilibration problem* as proxies for symbolic states. The partial tropical equilibration solutions are found algorithmically. These solutions also provide the scaling needed for slow-fast decomposition and model reduction. Any trace of the model can thus be represented as a sequence of local approximations of the full model. We illustrate the method using as case study a biochemical model of the cell cycle.

1 Introduction

Chemical reaction networks (CRN) are models of normal cell physiology and of disease and have multiple applications in biology and medicine. Rather generally, CRNs can be described as systems of polynomial differential equations that result from the mass action kinetics. In applications one would like to characterize these models in terms of attractors, their bifurcations, attraction basins, and of the sequence of states to and on these attractors. These questions belong to the qualitative theory of dynamical systems and are notoriously difficult.

In this paper we introduce a method to describe the qualitative dynamics of mass-action law CRNs, in situations when the dynamics involves processes on several well separated timescales. We have suggested that in these situations, the phase space of the CRN is patched with slow manifolds that are stable with respect to the fast dynamics [4,13,16,17]. The stable parts of the slow manifolds correspond to metastable states, defined as states whose lifetimes are relatively much longer than other states. Thus, the system stays repeatedly for relatively long time in some metastable state before switching to some other metastable state at points where slow manifolds lose their stability. In [13,17] we proposed to use tropical equilibrations as proxies for metastable states and finite-state machines as discrete abstractions for the ODE dynamics of the CRN.

I. Petre and A. Păun (Eds.): CMSB 2022, LNBI 13447, pp. 61–85, 2022.
https://doi.org/10.1007/978-3-031-15034-0_4

The concept of *tropical equilibration* comes from algebraic geometry and represents a constraint on the leading term of real Puiseux series solutions of systems of polynomial equations whose coefficients are powers or Puiseux series of some scaling parameter ϵ [14]. The concept is naturally related to the problem of finding scalings of differential equations needed in the *theory of singular perturbations* for systems with multiple timescales [7]. Singular perturbations allow to compute reduced models, valid approximately in the limit $\epsilon \to 0$, by using slow-fast [3] or multiscale decompositions [1,7]. We call the combination of geometry and singular perturbations, *tropical scaling*. Compared to other scaling methods based on nondimensionalization [5,18], our tropical scaling method is more comprehensive and provides more possible scalings and reduced models. The computation of tropical equilibrations was automatized using constraint programming [19], Newton polytopes [15] and SMT [8].

By revisiting the tropical scaling methodology we realized that the concept of *partial tropical equilibrations* is better suited to slow-fast decompositions than total tropical equilibrations [2,17,19]. The total tropical equilibration condition means that on slow manifolds each polynomial ODE has two dominant monomial terms of opposite signs that can equilibrate each other; the flow generated by the remaining, un-equilibrated monomial terms is slow. However, in slow-fast decompositions only fast variables need to be equilibrated; the ODE dynamics of slow variables may have only one dominant, but slow, monomial term. Although first mentioned in [19] and defined in [17], the partial tropical equilibrations have not yet been systematically computed and used for model reduction.

We provide an automatic method to compute partial tropical equilibrations, derived from the similar method for total tropical equilibrations based on SMT solvers, SMTcut [8], https://gitlab.com/cxxl/smtcut/-/tree/master/smtcut. Our code is available at https://github.com/Glawal/smtcutpartial. The partial tropical equilibrations are used for model order reduction using the method exposed in [7]. For given parameter values, a CRN can have several branches of partial tropical equilibrations, that are geometrically represented as polyhedra in the space of species concentrations (in logarithmic scale). Each branch corresponds to a reduced model whose validity is local, but spans several orders of magnitudes in species concentrations and in time. The intersection relations of these polyhedra define a connectivity graph providing possible transitions between reduced models.

As a case study for computing partial tropical equilibrations and model reductions we discuss a six variables biochemical network describing the cyclic phosphorylation of different substrates in the frog embryo cell cycle, proposed by J.J.Tyson [20]. We verify numerically that the partial equilibration branches of this model are reasonably well related to slow manifolds and that the allowed transitions from one slow manifold to another are all edges of the connectivity graph. Each branch corresponds to a reduced model that can be computed using the tropical scaling approach. Any trace of the model can be represented symbolically as a sequence of branches or reduced models. Also, any trace can be approximated locally by solutions of the corresponding reduced model. The cell cycle model has approximate conservation laws, i.e. conservation laws of the fast subsystem that we have to eliminate before applying the reduction methods from [7]. The local validity of the reduced models was tested with methods from [7]. We expect that the global validity of the patched together local approximations as approximations of solutions of the full model can also be shown rigorously; this is the subject of ongoing work.

2   Definitions and Methods

2.1   Tropical Geometry Concepts

We briefly recall here the tropical geometry concepts that we need in our tropical scaling method. We follow notations from [9].

We consider differential equations whose r.h.s. are multivariate polynomials $f = \sum c_u x^u$, where u are multi-indices and c_u coefficients. Here, c_u are considered to be rational powers (or more generally, Puiseux series, i.e. power series with negative and positive rational exponents multiple of $1/d$, where d is a positive integer) of a positive scaling parameter ϵ. We define the valuation of c_u as the limit

$$val(c_u) = \lim_{\epsilon \to 0} \log(c_u)/\log(\epsilon). \tag{1}$$

Another way to introduce valuations is directly via Puiseux series. If c_u is a Puiseux series of ϵ, then $c_u \sim \epsilon^{val(c_u)}$ at the lowest order. As $\epsilon^{val(c_u)}$ is the dominant term of c_u, the valuation of c_u is also the order of magnitude of c_u. For $\epsilon = 1/10$, valuations are decimal orders. $val(c_u)$ can be found from the numerical values of the coefficients (see Sects. 2.2, 3). The valuations of x are unknown, so they should result from a calculation. The rest of this subsection is about the constraints on $val(x)$, when x satisfies polynomial equations.

The *tropicalization* $trop(f)$ of $f = \sum c_u x^u$ is the piecewise-linear function

$$trop(f)(w) = \min_{u}(val(c_u) + u \cdot w). \tag{2}$$

The variety $V(f)$ is the set of all x solutions of $f(x) = 0$. The *tropical hypersurface* $trop(V(f))$ is the set of w where the minimum in $trop(f)$ is attained at least twice. A theorem of Kapranov relates the tropical hypersurface to the set of all possible valuations of x on $V(f)$, namely $trop(V(f))$ is the closure of $val(x)$, where $x \in V(f)$ [9]. In short, if we know the orders of c_u, the orders of x satisfying $f(x) = 0$ are given by $trop(V(f))$.

Kapranov's theorem refers to solutions of $f = 0$ in the complex field. If we are interested in the valuations of real positive solutions of $f = 0$, where c_u are all real, then one has to consider *tropical equilibrations*. A tropical equilibration is a w where the minimum in $trop(f)$ is attained at least twice, for at least one positive and at least one negative monomial [10,11,14]. The tropical equilibrations are thus possible valuations of the real positive solutions of $f(x) = 0$.

Let $f = (f_1, \ldots, f_n)$ be a polynomial vector field. Then we define

$$trop(f)(w_1, \ldots, w_n) = (trop(f_1)(w_1), \ldots, trop(f_n)(w_n)).$$

For $x = (x_1, \ldots, x_n)$, the valuations $val(x) = (val(x_1), \ldots, val(x_n))$ of the solutions of $f(x) = 0$ are in the intersection of the tropical hypersurfaces of the component polynomial f_i. This intersection is called *tropical prevariety*. By "abus de langage" we call tropical equilibration also an element of the tropical prevariety that is a tropical equilibration for each component.

2.2 Partial Tropical Equilibrations and Slow-Fast Decompositions

For simplicity of the presentation, we start by considering only two timescales, slow-fast systems. For slow-fast systems it is usually assumed that the governing equations have the form

$$\frac{dx}{dt} = f(x,y), \ \frac{dy}{dt} = \epsilon g(x,y), \tag{3}$$

i.e. the variables are a priori split into the fast variable x and the slow variable y; here $0 < \epsilon \ll 1$ is a suitable parameter measuring time-scale separation. The CRNs we have in mind are typically not given in this form which is a major obstacle in using slow-fast decompositions in their analysis. We will use partial tropical equilibrations to overcome these difficulties.

We start with an arbitrary splitting of the variables into two groups denoted by x and y

$$\frac{dx}{dt} = f(x,y;k), \ \frac{dy}{dt} = g(x,y;k), \tag{4}$$

where f and g are polynomial vector fields whose coefficients include the kinetic parameters k. Now we search for scalings and conditions leading to time scale separation similar to Eq. 3. To this purpose the model under study is considered to belong to a family of models indexed by ϵ. More precisely, the kinetic parameters k are considered to be powers of ϵ, $k(\epsilon) \sim \epsilon^\gamma$ (this implies that the coefficients of f, g are Puiseux series of ϵ). This is always possible by writing $k(\epsilon) = \bar{k}\epsilon^\gamma$, where $\gamma = \frac{1}{d}round(d\log_{\epsilon*}(k_*))$, where $round$ is the rounding to the nearest integer, ϵ_* is a fixed value of ϵ, d is a positive integer, and k_* are numerical parameters. The studied model is just a member of the family, obtained when $\epsilon = \epsilon_*$ and having kinetic parameters $k_* = k(\epsilon_*)$. Then, we consider the solutions of models from this family in the limit $\epsilon \to 0$. If the value ϵ_* placing the studied model in the family is small enough, one may expect that the limit solution is a good approximation for the model's solution.

The tropicalization is useful in the scaling process, because it allows to compute the lowest order terms of the Puiseux series expansions of the polynomials f and g. We have

$$\frac{d\bar{x}}{dt} = \epsilon^{\mu_x}\bar{f}(\bar{x},\bar{y};\bar{k}) + o(\epsilon^{\mu_x}), \ \frac{d\bar{y}}{dt} = \epsilon^{\mu_y}\bar{g}(\bar{x},\bar{y};\bar{k}) + o(\epsilon^{\mu_y}), \tag{5}$$

where $\mu_x = trop(f)(val(x),val(y)) - val(x)$, $\mu_y = trop(g)(val(x),val(y)) - val(y)$, and $\bar{x},\bar{y},\bar{f},\bar{g},\bar{k}$ have valuation zero. μ_x and μ_y are the timescale orders (in fact orders of reciprocal timescales) for the variations of x and y, respectively; smaller timescale orders mean faster variables. $\bar{f}$ and $\bar{g}$ are called tropically truncated versions of f and g.

Let us denote by $\mu_x \lhd \mu_y$ the set of inequalities $\{(\mu_x)_i < (\mu_y)_j \ \forall i, j\}$, meaning that variables x are faster than variables y.

If $\mu_x \lhd \mu_y$ classical results [3] show that the solutions of (5) converge to the solutions of the following reduced system

$$0 = \bar{f}(\bar{x}, \bar{y}; \bar{k}),$$
$$\frac{d\bar{y}}{dt} = \epsilon^{d_y} \bar{g}(\bar{x}, \bar{y}; \bar{k}), \tag{6}$$

provided that the solutions of $\bar{f}(\bar{x}, \bar{y}) = 0$ are hyperbolic attracting equilibria of the fast dynamics $\frac{d\bar{x}}{dt} = \bar{f}(\bar{x}, \bar{y}; \bar{k})$.

The first equation of (6) defines the quasi-steady state variety and imposes constraints on x, y. Using the tropical approach we transform these constraints into constraints on the order of magnitudes $val(x), val(y)$.

As a matter of fact, the valuations of x and y are constrained by

$$\begin{cases} (val(x), val(y)) \text{ is a tropical equilibration of } f, \\ trop(f)(val(x), val(y)) - val(x) \lhd trop(g)(val(x), val(y)) - val(y). \end{cases} \tag{7}$$

The first of the equations (7) follows from Kapranov's theorem because x satisfies $f(x, y) = 0$. The second equation is simply a condition on the timescales.

We call any solution of (7) *partial tropical equilibration*. Geometrically, (7) defines polyhedral complexes in the space of valuations.

If the system (4) has conservation laws, i.e. linear or polynomial functions $c(x, y)$ such that $\frac{\partial c}{\partial x} f + \frac{\partial c}{\partial y} g = 0$ identically, one needs to consider the quasi-state state equation $f(x, y) = 0$ together with the conservation equation $c(x, y) - c_0 = 0$ where c_0 is constant. In this case the problem (7) becomes

$$\begin{cases} (val(x), val(y)) \text{ is a tropical equilibration of } f \text{ and of } c(x, y) - c_0, \\ trop(f)(val(x), val(y)) - val(x) \lhd trop(g)(val(x), val(y)) - val(y). \end{cases} \tag{8}$$

We call *tropical scaling* of a polynomial ODE system, a fixed choice of the valuations of the polynomial coefficients and variables satisfying the partial tropical equilibration constraints. It is very important to keep in mind, that different scalings will be valid in different regions of the phase space and will lead to different reductions.

Like total tropical equilibrations, partial tropical equilibrations (7) can be grouped into branches [15,17]. In a branch, the tropically truncated functions $\bar{f}$ and $\bar{g}$ are fixed. In other words, the dominant terms, corresponding to the *min* value in $trop(f)$ and $trop(g)$ are the same for all solutions in a branch. Geometrically, a branch is a polyhedral face of the polyhedral complex.

2.3 Multiple Timescale Decompositions

In general, the tropical scaling method associates a timescale order to each species [7]. The CRN can have thus more than two timescales. In this case, a multiple timescale decomposition proposed by [1] and algorithmically formalized by [7] extends the ideas of Fenichel [3] valid for two timescales, to the case when the system has more than two timescales. For the sake of completeness, we briefly recall these results here.

Let us consider that the species concentrations are regrouped as $z = (z_1, \ldots, z_m)$ where z_1 regroups n_1 fastest species, z_2 regroups n_2 second fastest species, and so on to z_m that regroups n_m slowest species. One has $n_1 + \ldots + n_m = n$, where n is the total number of species in the CRN.

In this case the CRN dynamics is defined by the following rescaled system of ODEs (see [7]):

$$\dot{z}_1 = \epsilon^{b_1} \bar{f}_1(\bar{k}, z) + o(\epsilon^{b_1}),$$
$$\vdots$$
$$\dot{z}_m = \epsilon^{b_m} \bar{f}_m(\bar{k}, z) + o(\epsilon^{b_m}), \tag{9}$$

where b_l is the l-th timescale order, $1 \le l \le m$, and $b_1 = 0 < b_2 < \ldots < b_m$.

By redefining the time to $\tau = t\epsilon^{b_m}$, we get

$$\delta_1 \delta_2 \ldots \delta_{m-1} z_1' = \bar{f}_1(\bar{k}, z) + o(1),$$
$$\vdots$$
$$\delta_{m-1} z_{m-1}' = \bar{f}_{m-1}(\bar{k}, z) + o(1),$$
$$z_m' = \bar{f}_m(\bar{k}, z) + o(1). \tag{10}$$

This multiscale problem has $m - 1$ small parameters $\delta_k = \epsilon^{b_{k+1}-b_k}$, for $1 \le k \le m - 1$. Because $b_{k+1} > b_k$, $1 \le k \le m - 1$ the limit $\epsilon \to 0$ implies $\delta_k \to 0$, $1 \le k \le m - 1$.

The *reduced model at the slowest timescale* is obtained by setting $\delta_1 = \delta_2 = \ldots = \delta_{m-1} = 0$ in (10) and is defined by the following equations (see also [1,6,12]):

$$0 = \bar{f}_1(\bar{k}, z),$$
$$\vdots$$
$$0 = \bar{f}_{m-1}(\bar{k}, z),$$
$$z_m' = \bar{f}_m(\bar{k}, z). \tag{11}$$

Condition 1

i) *Consider that the system* $\bar{f}_1(\bar{k}, z) = 0, \ldots, \bar{f}_{l-1}(\bar{k}, z) = 0$ *can be solved for* $(z_1, \ldots, z_{l-1})$ *in the following, hierarchical way. First, define* $f_1^*(\bar{k}, z) = \bar{f}_1(\bar{k}, z)$. *Consider that there is a differentiable function* $\tilde{f}_1(\bar{k}, z_2, \ldots, z_m)$ *such that*

$$f_1^*(\bar{k}, \tilde{f}_1(\bar{k}, z_2, \ldots, z_m), z_2, \ldots, z_m) = 0.$$

Next, define $f_2^*(\bar{k}, z_2, \ldots, z_m) = \bar{f}_2(\bar{k}, \tilde{f}_1(\bar{k}, z_2, \ldots, z_m), z_2, \ldots, z_m)$ *and consider that there is a differentiable function* $\tilde{f}_2(\bar{k}, z_3, \ldots, z_m)$ *such that*

$$f_2^*(\bar{k}, \tilde{f}_2(\bar{k}, z_3, \ldots, z_m), \ldots, z_m) = 0.$$

Assuming that the procedure can go on, define $f_{m-1}^*(\bar{k}, z_{m-1}, \ldots, z_m)$ that is obtained from $\bar{f}_{m-1}(\bar{k}, z)$ by recursively replacing $z_1, z_2, \ldots, z_{m-2}$ by $\tilde{f}_2(\bar{k}, z_3, \ldots, z_m), \ldots, \tilde{f}_{m-2}(\bar{k}, z_{m-1}, z_m)$, respectively. Consider that there is a differentiable function $\tilde{f}_{m-1}(\bar{k}, z_m)$ such that

$$f_{m-1}^*(\bar{k}, \tilde{f}_{m-1}(\bar{k}, z_m), z_m) = 0.$$

The existence of these functions can be tested using the implicit function theorem by checking that the Jacobian matrices $D_{z_k} f_k^*$, $1 \le k \le m-1$ are non-singular.
ii) Consider that the real parts of all the eigenvalues of the Jacobian matrices $D_{z_k} f_k^*$, $1 \le k \le m-1$ are non-zero.

If the Condition 1 is satisfied then the solutions of the system (10) converge to the solutions of the system (11) when $\epsilon \to 0$ (Corrolary of Theorem A in [1]). In [7] the part ii) of the condition is tested algorithmically using the Hurwitz criterion.

2.4   Coarse Graining

The model has continuous parameters and variables which from the mathematical point of view can vary in $[0, \infty)$. Therefore, the valuations and in consequence the scales are in principle continuous. However, to obtain a finite number of useful approximating systems one has to use a suitable selection of discrete (often integer) scales, that cover the relevant domains in parameter- and phase space.

In order to do so, we use *logarithmic paper coarse graining*. The logarithmic paper simply means a representation of the numerical data (species concentrations, parameters) as logarithms with base ϵ_*, using the mapping $x \to \log_{\epsilon_*}(x)$ coordinate-wise [21]. Then, the coarse graining consists in defining a grid on the logarithmic paper and mapping all the points belonging to a grid cell onto the cell that contains them.

For instance, an example of logarithmic paper coarse graining is performed by the application

$$x \to \frac{1}{d} round(d \log_{\epsilon_*}(x)), \tag{12}$$

where x is any positive quantity, e.g. a kinetic parameter, concentration, monomial or polynomial of concentrations. The image of x via the mapping (12) represents the order of magnitude of x, which is an integer for $d = 1$ and a rational number from $\mathbb{Z}/d$ when $d > 1$.

Using (12) two values x_1 and x_2 have different images on logarithmic paper if

$$|\log(x_1) - \log(x_2)| > \frac{|\log \epsilon_*|}{d}. \tag{13}$$

Equation (13) specifies the grid cell-size on logarithmic paper. For a given ϵ_*, the largest cell-size is obtained for integer values $d = 1$. The cell-size increases when ϵ_* decreases. The limit $d \to \infty$ corresponds to the continuum.

In practice, we want to choose an intermediate cell-size. This should not be too small, to avoid continuous scaling, and not too large, to avoid loss of the structure. Although two parameters are in play, one can not change them independently. As a matter of fact, the value of ϵ_* is dictated by singular perturbations; we want this to be small enough. Therefore, the cell-size adjustment is performed by changing d after the choice of ϵ_*. The choice criteria are the error of the approximation and the robustness in terms of number of tropical branches, that can be tested numerically.

Given ϵ_*, d the optimal d corresponds to the largest cell-size that distinguishes the most robust structural features (geometry of branches of tropical equilibration solutions, differences between orders of magnitude of parameters, concentrations, monomials, see also Sect. 3.3 and Fig. 4).

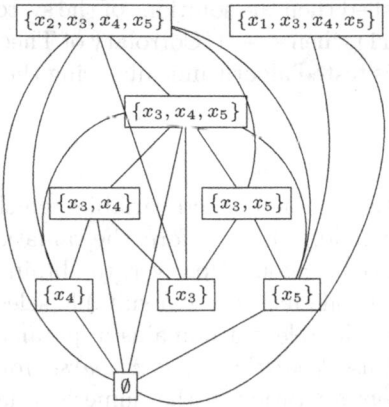

Fig. 1. Quotiented connectivity graph for $\epsilon = 1/11, d = 1$. Each node is the set of solution branches of a partial tropical equilibration problem (forming a polyhedral complex), denoted by the set of slow species. Each edge means that the intersection of two polyhedra (one from each complex) has the same dimension as one of the polyhedra (this does not necessarily mean that one polyhedron is included in the other).

3   Case Study: A Cell Cycle Model

The model presented is a modification of the original Tyson cell cycle model [20]. First, by considering constant concentrations as parameters, the model has been converted to an ODE system with six variables. In order to simplify the analysis we have also removed a variable x_6 that does not interact with the rest of the model. This gives us the following model:

$$\begin{aligned}
\dot{x}_1 &= k_1 x_3 - k_2 x_1 + k_3 x_2, \\
\dot{x}_2 &= k_2 x_1 - k_3 x_2 - k_4 x_2 x_5, \\
\dot{x}_3 &= k_{10} x_4 - k_1 x_3 + k_9 x_3^2 x_4, \\
\dot{x}_4 &= k_4 x_2 x_5 - k_9 x_3^2 x_4 - k_{10} x_4, \\
\dot{x}_5 &= k_6 - k_4 x_2 x_5,
\end{aligned} \tag{14}$$

with the parameter values (from [20]) $k_1 = 1$, $k_2 = 1000000$, $k_3 = 1000$, $k_4 = 200$, $k_6 = 3/200$, $k_8 = 3/5$, $k_9 = 180$, $k_{10} = 9/500$, $k_{14} = 1$, where k_{14} is the total concentration associated to the conservation law $k_{14} = x_1 + x_2 + x_3 + x_4$.

3.1 Tropical Scaling of the Cell Cycle Model

The reaction rate constants and the concentrations in this model have very different orders of magnitude which suggests dynamics on many well separated time scales. To identify these we rescale the model as a first step of using the procedure described abstractly in Sect. 2:

- Consider $0 < \epsilon_* < 1$.
- Write $k_i = \bar{k}_i \epsilon_*^{\gamma_i}$ and $x_i = \bar{x}_i \epsilon_*^{a_i}$; thus $\gamma_i = val(k_i)$ and $a_i = val(x_i)$.
- The exponents γ_i are computed from the numerical values of the parameters. We use $\gamma_i = \mathrm{round}(d \log(|k_i|)/\log(\epsilon_*))/d$ to obtain rational (integer if $d = 1$) exponents. These exponents become valuations when we view the cell cycle model as being part of a family of models with the same structure and with parameters $k_i(\epsilon) = \bar{k}_i \epsilon^{\gamma_i}$ in the limit $\epsilon \to 0$.
- Using this scaling we compute the rescaled system

$$
\begin{aligned}
\epsilon^{a_1} \dot{\bar{x}}_1 &= \bar{k}_1 \bar{x}_3 \epsilon^{\gamma_1 + a_3} - \bar{k}_2 \bar{x}_1 \epsilon^{\gamma_2 + a_1} + \bar{k}_3 \bar{x}_2 \epsilon^{\gamma_3 + a_2}, \\
\epsilon^{a_2} \dot{\bar{x}}_2 &= \bar{k}_2 \bar{x}_1 \epsilon^{\gamma_2 + a_1} - \bar{k}_3 \bar{x}_2 \epsilon^{\gamma_3 + a_2} - \bar{k}_4 \bar{x}_2 \bar{x}_5 \epsilon^{\gamma_4 + a_2 + a_5}, \\
\epsilon^{a_3} \dot{\bar{x}}_3 &= \bar{k}_{10} \bar{x}_4 \epsilon^{\gamma_{10} + a_4} - \bar{k}_1 \bar{x}_3 \epsilon^{\gamma_1 + a_3} + \bar{k}_9 \bar{x}_3^2 \bar{x}_4 \epsilon^{\gamma_9 + 2a_3 + a_4}, \\
\epsilon^{a_4} \dot{\bar{x}}_4 &= \bar{k}_4 \bar{x}_2 \bar{x}_5 \epsilon^{\gamma_4 + a_2 + a_5} - \bar{k}_9 \bar{x}_3^2 \bar{x}_4 \epsilon^{\gamma_9 + 2a_3 + a_4} - \bar{k}_{10} \bar{x}_4 \epsilon^{\gamma_{10} + a_4}, \\
\epsilon^{a_5} \dot{\bar{x}}_5 &= \bar{k}_6 \epsilon^{\gamma_6} - \bar{k}_4 \bar{x}_2 \bar{x}_5 \epsilon^{\gamma_4 + a_2 + a_5},
\end{aligned}
\tag{15}
$$

and the timescale orders of each variable:

$$
\begin{aligned}
\mu_1 &= \min(\gamma_1 + a_3,\ \gamma_2 + a_1,\ \gamma_3 + a_2) - a_1, \\
\mu_2 &= \min(\gamma_2 + a_1,\ \gamma_3 + a_2,\ \gamma_4 + a_2 + a_5) - a_2, \\
\mu_3 &= \min(\gamma_{10} + a_4,\ \gamma_1 + a_3,\ \gamma_9 + 2a_3 + a_4) - a_3, \\
\mu_4 &= \min(\gamma_4 + a_2 + a_5,\ \gamma_9 + 2a_3 + a_4,\ \gamma_{10} + a_4) - a_4, \\
\mu_5 &= \min(\gamma_6,\ \gamma_4 + a_2 + a_5) - a_5.
\end{aligned}
\tag{16}
$$

- In this setting the slow-fast decomposition follows from the timescale orders. Instead of renaming slow variables as y like in Sect. 2.2, we define instead a subset $S \subset \{1, \ldots, n\}$ containing indices of the slow components. Thus, all variables x_j with indices $j \in S$ are slow and the remaining variables x_i are faster iff $\mu_i < \mu_j$ for all $i \notin S$, $j \in S$.
- Self-consistently, the valuations a_i are solutions of the partial tropical equilibration problem for a fast/slow splitting with slow variables S (see the following Subsect. 3.2 and Appendix 1).

3.2   Calculation of the Partial Tropical Equilibrations

The solutions of the partial tropical equilibration problem for S form a polyhedral complex, each face encoding one combinatorial possibility in the equations (i.e. a choice of minima for each equation). The partial equilibration problem for S can be decomposed into two kinds of constraints: i) equilibration of fast species and conservation laws, and ii) timescale orders constraints resulting from the slow/fast decomposition.

For example, suppose that we are interested in the partial tropical equilibration when $S = \{x_3\}$, then the problem is given by:

$$
\begin{aligned}
\gamma_1 + a_3 &= \min(\gamma_2 + a_1, \ \gamma_3 + a_2), \\
\gamma_2 + a_1 &= \min(\gamma_3 + a_2, \ \gamma_4 + a_2 + a_5), \\
\gamma_4 + a_2 + a_5 &= \min(\gamma_9 + 2a_3 + a_4, \ \gamma_{10} + a_4), \\
\gamma_6 &= \gamma_4 + a_2 + a_5, \\
\gamma_{14} &= \min(a_1, \ a_2, \ a_3, \ a_4), \\
\mu_3 > \mu_1, \ \mu_3 &> \mu_2, \ \mu_3 > \mu_4, \ \mu_3 > \mu_5.
\end{aligned}
\tag{17}
$$

The first four equations in (17) come for the tropicalization of the polynomial vector field for the fast species $\{x_1, x_2, x_4, x_5\}$. The fifth equation results from the tropicalization of the linear conservation law $k_{14} = x_1 + x_2 + x_3 + x_4$. The remaining equations simply mean that the species x_3 is slower than all the other.

The real solutions of (17), $\boldsymbol{a} \in \mathbb{R}^5$, form a polyhedral complex. Coarse graining means that we look for solutions in $(\mathbb{Z}/d)^5$, in which case we obtain a discrete set of points in the polyhedral complex. To solve this problem we treat each constraint separately. For each constraint, choosing a minima defines a polytope P. We define a *bag* B_i as the union of each polytope P_{ij} resulting from a choice j in the constraint i. The solution of the problem is the intersection of these bags.

For example, the first equation in (17) leads to two possibilities that each gives a polytope. When $\min(\gamma_2 + a_1, \ \gamma_3 + a_2) = \gamma_2 + a_1$, the polytope is given by

$$
P_{1,1} := \{ \boldsymbol{a} \in \mathbb{R}^5, \ \gamma_1 + a_3 = \gamma_2 + a_1, \gamma_2 + a_1 \leq \gamma_3 + a_2 \},
\tag{18}
$$

whereas when $\min(\gamma_2 + a_1, \ \gamma_3 + a_2) = \gamma_3 + a_2$, the polytope is given by

$$
P_{1,2} := \{ \boldsymbol{a} \in \mathbb{R}^5, \ \gamma_1 + a_3 = \gamma_3 + a_2, \gamma_3 + a_2 \leq \gamma_2 + a_1 \}.
\tag{19}
$$

So, for this equation, we get two polytopes. The bag B_1 is defined by (18) $\cup$ (19).

The last equation in (17) comes from the timescale constraints and has the form:

$$
\min(\gamma_{10} + a_4, \ \gamma_1 + a_3, \ \gamma_9 + 2a_3 + a_4) - a_3 > \min(\gamma_6, \ \gamma_4 + a_2 + a_5) - a_5,
\tag{20}
$$

which gives, if $\min(\gamma_{10} + a_4, \ \gamma_1 + a_3, \ \gamma_9 + 2a_3 + a_4) = \gamma_1 + a_3$ and $\min(\gamma_6, \ \gamma_4 + a_2 + a_5) = \gamma_4 + a_2 + a_5$ the polytope given by equations:

$$
\gamma_1 > \gamma_4 + a_2, \ \gamma_1 + a_3 \leq \gamma_{10} + a_4, \ \gamma_1 + a_3 \leq \gamma_9 + 2a_3 + a_4, \ \gamma_4 + a_2 + a_5 \leq \gamma_6. \tag{21}
$$

The partial tropical equilibration problem is then to compute the intersection of these bags (see Appendix 1). Generally, the problem reads:

$$A_S = \bigcap_{i=1}^{n_e} B_i = \bigcap_{i=1}^{n_e} \bigcup_{j=1}^{c_i} P_{ij},$$

where n_e is the number of equations in the partial tropical equilibration problem, c_i is the number of choices for the equation i. As intersections and unions are in finite numbers we can reverse them and get:

$$A_S = \bigcap_{i=1}^{n_e} \bigcup_{j=1}^{c} P_{ij} = \bigcup_{j=1}^{c} \bigcap_{i=1}^{n_e} P_{ij},$$

where c is the total number of choices $c = \sum c_i$, and a *branch*, which is a face of the polyhedral complex, is given by the intersection of each polytopes for a given choice: $\bigcap_{i=1}^{n_e} P_{ij}$.

For our example, given $S = \{x_3\}$, most of the choices lead to an empty set. There are thus only two branches (for $\epsilon_* = 1/11, d = 1$): $\{x_3\}_{00}$ and $\{x_3\}_{01}$.

If we consider the five variables cell cycle model, there are $2^5 = 32$ possible partial equilibration problems. One of them will not be considered as it consists of no constraints, this is when S is the set of all species.

We have tested the remaining 31 possibilities: the total tropical equilibrations when $S = \emptyset$, and 30 partial equilibrations. Only 9 are non-empty: 8 partial tropical equilibrations and the total one.

Denoting a partial tropical equilibration problem by the associated set of slow species S, the list of solutions of all partial equilibration problems is given by $\Big\{\emptyset, \{x_3\}, \{x_4\}, \{x_5\}, \{x_3, x_4\}, \{x_3, x_5\}, \{x_3, x_4, x_5\}, \{x_1, x_3, x_4, x_5\}, \{x_2, x_3, x_4, x_5\}\Big\}$.

The solutions of each partial tropical equilibration problem are grouped in a number of branches. Each branch is denoted by an index number, for instance $\{x_3\} = \Big\{\{x_3\}_{00}, \{x_3\}_{01}\Big\}$.

Although each partial tropical equilibration is a polyhedral complex, their union may not be a complex (the intersection of two polyhedra of different partial tropical equilibrations can be just part of a face). In [17] we have introduced the connectivity graph describing adjacency of branches as faces of the polyhedral complex of total tropical equilibrations: two branches are connected if they share a face. We introduce here a different connectivity graph for partial tropical equilibrations. The vertices of this graph are partial tropical equilibration branches. Two partial tropical equilibration branches are connected if their intersection has maximum dimension (the dimension of the intersection is equal to the dimension of one of the attached polyhedra).

The Fig. 1 shows the connectivity graph for all the partial tropical equilibrations of the cell cycle model, quotiented over partial equilibration problems (all branches of one partial equilibration problem are gathered in one node).

Remark: this graph has been made for $\epsilon_* = 1/11$ and $d = 1$. We found that the quotiented graph is robust and does not change for different (ϵ_*, d) despite the fact that some polytope branches may be different for two different (ϵ_*, d). Indeed, for $\epsilon_* = 1/11$, $d = 1$, the total tropical equilibrations form a segment plus a half-line whereas for $d = 1000$ the solution is a point.

3.3 Symbolic Dynamics by Tropicalization

We expect that the traces of the system (14) are most of the time in proximity of partial tropical equilibration solutions. Therefore we can use the tropical equilibrations for symbolic coding of these traces.

In order to test this property, we have simulated for 200 min (using the solver ode23s of Matlab R2021b) 375 numerical traces of (14) starting from the different sets of initial conditions respecting $x_1 + x_2 + x_3 + x_4 = 1$. For each point $x = (x_1, x_2, x_3, x_4, x_5) \in \mathbb{R}^5$ of the numerical trace, we have computed a valuation $a = (a_1, a_2, a_3, a_4, a_5) \in \mathbb{Q}^5$ using $a_i = round(d \log(|x_i^{(l)}|)/\log(\varepsilon))/d$. This allows to compute the time-scale order of each species for this point, but also to check if a species is equilibrated or not. With these informations, we can first determine in which partial tropical equilibration the point is (if it lives in a equilibration), and then we can obtain the truncated system, and so, the branch.

Making a projection of these traces on the space $(\log_{\epsilon_*}(x_3), \log_{\epsilon_*}(x_4), \log_{\epsilon_*}(x_5))$ we obtain Fig. 2. On this figure, we can see that trajectories are first converging to low dimension manifolds (dimension two or one in projection) that lead to a limit cycle. In Fig. 3 we have symbolically coded the points of a

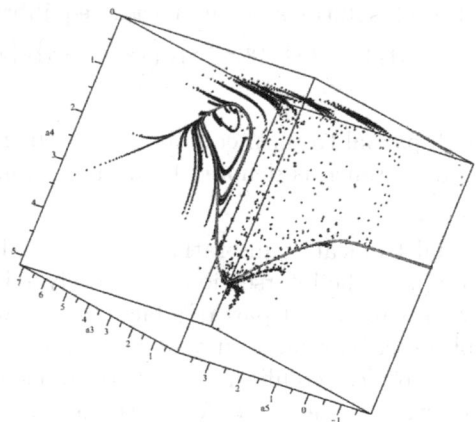

Fig. 2. 375 traces on the space $(\log_{\epsilon_*}(x_3), \log_{\epsilon_*}(x_4), \log_{\epsilon_*}(x_5))$, $\epsilon_* = 1/11$. In red the trace starting from $(1, 0, 0, 0, 100)$. (Color figure online)

particularly long trajectory (marked in red), using the method described above. This figure shows that the trace follows constrained transitions guided by some partial tropical equilibrations.

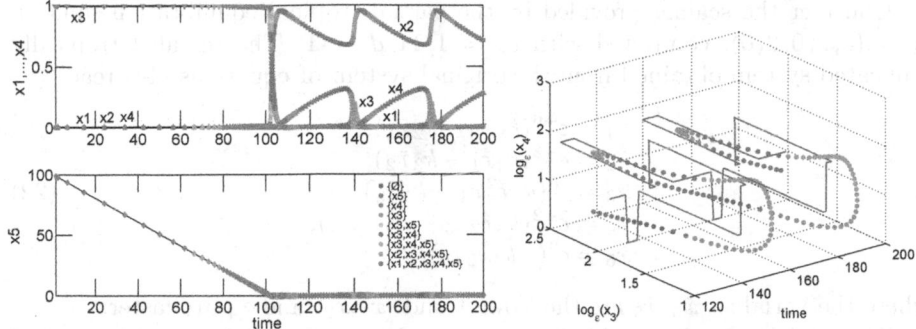

Fig. 3. Symbolic coding of the trace starting at $(1, 0, 0, 0, 100)$ (a few very rapid transients states at the beginning of the trace are not represented). On the left side, the marker colors indicate the tropical equilibration branch close to which the trajectory point lies. On the right side, the tropicalization (red line) is represented together with the continuous trace on the limit cycle part (for $\epsilon^* = 1/11, d = 1$). (Color figure online)

In order to obtain more insight, in Fig. 4 we projected the traces on the plane $(\log_{\epsilon_*}(x_3), \log_{\epsilon_*}(x_4))$. We used this "logarithmic paper" representation to show also the tropical equilibration solutions and their polyhedral branches. A tropical equilibration $\boldsymbol{a} = (a_1, a_2, a_3, a_4, a_5) \in (\mathbb{Z}/d)^5$ is represented as a point of coordinates (a_1, a_2) in this representation. Partial tropical equilibration branches are line segments or colored polygonal domains containing tropical equilibration points. As shown in Fig. 4b), a large value of d ensures a small cell-size on the logarithmic paper. This ensures a precise representation of the limit cycle, but with a scaling that varies almost continuously and with new polygonal domains. As we are interested only in the robust features of the tropical solutions, we favor the value $d = 1$ corresponding to Fig. 4a).

3.4 Model Reduction Using Partial Tropical equilibrations

Each partial tropical equilibration solution provides a scaling of the variables. This scaling is used for identification of slow and fast variables and for automatic model reduction with algorithms introduced in [2,7]. The algorithms from [7] work when the quasi-steady state equations of fast variables satisfy hyperbolicity conditions. Very often, the hyperbolicity conditions fail because of the existence of exact or approximate conservation laws, which are conservation laws of the full and tropically truncated systems, respectively. We showed in [2] that systems with full or approximate conservation laws can be transformed into systems without conservation laws by a change of variables. This extends the applicability of reduction algorithms from [7] to the case when there are conservation laws. We illustrate these techniques on an example and refer the reader to [2,7] for the complete algorithmic solutions.

Consider the scaling provided by the partial tropical equilibration solution $a = (5, 2, 0, 2, 0)$, computed with $\epsilon_* = 1/11, d = 1$. The rescaled tropically truncated system obtained from the original system of equations (14) reads

$$\begin{aligned}
\dot{\bar{x}}_1 &= \epsilon^{-6}(\bar{k}_3\bar{x}_2 - \bar{k}_2\bar{x}_1), \\
\dot{\bar{x}}_2 &= \epsilon^{-3}(\bar{k}_2\bar{x}_1 - \bar{k}_3\bar{x}_2), \\
\dot{\bar{x}}_3 &= \epsilon^0(\bar{k}_9\bar{x}_3^2\bar{x}_4 - \bar{k}_1\bar{x}_3), \\
\dot{\bar{x}}_4 &= \epsilon^{-2}(\bar{k}_4\bar{x}_2\bar{x}_5 - \bar{k}_9\bar{x}_3^2\bar{x}_4), \\
\dot{\bar{x}}_5 &= \epsilon^0(-\bar{k}_4\bar{x}_2\bar{x}_5),
\end{aligned} \tag{22}$$

where the variables $\bar{x}_3, \bar{x}_5$ are the slowest and $\bar{x}_1, \bar{x}_2$, and $\bar{x}_4$ are faster.

The model reduction at the slowest timescale from Sect. 2.3 can not be applied here because the truncated nonrescaled system describing the fast dynamics

$$\begin{aligned}
\dot{x}_1 &= k_3x_2 - k_2x_1, \\
\dot{x}_2 &= k_2x_1 - k_3x_2, \\
\dot{x}_4 &= k_4x_2x_5 - k_9x_3^2x_4,
\end{aligned} \tag{23}$$

has a conservation law $x_1 + x_2$ which is an approximate conservation law of the full system (14). In this case, the part i) of the Condition 1 does not hold (one can easily test that the Jacobian matrix $D_{(x_1,x_2)}(k_3x_2 - k_2x_1, k_2x_1 - k_3x_2)$ is singular).

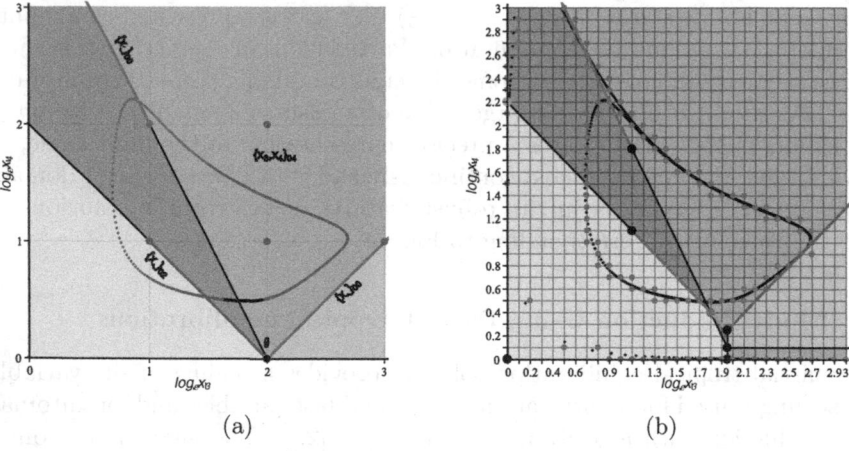

(a)   (b)

Fig. 4. a) Tropicalizing each point of the trace for $\epsilon_* = 1/11$, $d = 1$, we obtain six points with integer components closest to the limit cycle, each of them corresponding to a partial tropical equilibration: $(0, 0)$ black, $(2, 1)$, $(2, 2)$ green, $(1, 1)$, $(3, 1)$ blue, and $(1, 2)$ red. We use the same color for the points of the trace to indicate which of the 6 points is the closest one. Black lines are intersection of two polytopes. b) For $(\epsilon_* = 1/11, d = 10)$, we observe that we have more points for the limit cycle, and more polyhedra in the tropical structure. This is due to the coarse graining: the grid cell is smaller for larger d. (Color figure online)

The system (14) has also the exact conservation law $x_1 + x_2 + x_3 + x_4$.

However, we can eliminate all the conservation laws (approximate and exact) by using the following change of variable:

$$x_3 \leftarrow k_{14} - x_1 - x_2 - x_4,$$
$$x_2 \leftarrow x_1 + x_2. \tag{24}$$

After this change of variables we obtain a transformed system of equations that has no conservation laws:

$$\dot{x}_1 = k_1 k_{14} - (k_2 + k_3)x_1 - k_1 x_4 + (k_3 - k_1)x_2,$$
$$\dot{x}_2 = k_1 k_{14} - k_1 x_2 - k_1 x_4 + k_4 x_1 x_5 - k_4 x_2 x5,$$
$$\dot{x}_4 = -(k_9 k_{14}^2 + k_{10})x_4 + 2k_9 k_{14} x_2 x_4 + 2k_9 k_{14} x_4^2 \tag{25}$$
$$\quad -k_9 x_2^2 x_4 - 2k_9 x_2 x_4^2 + k_4 x_5 x_2 - k_9 x_4^3 - k_4 x_1 x_5,$$
$$\dot{x}_5 = k_6 + k_4 x_1 x_5 - k_4 x_2 x_5.$$

The rescaled tropically truncated system obtained from the transformed system of equations (25) reads

$$\dot{\bar{x}}_1 = \epsilon^{-6}(\bar{k}_3 \bar{x}_2 - \bar{k}_2 \bar{x}_1),$$
$$\dot{\bar{x}}_2 = \epsilon^{-2}(\bar{k}_1 \bar{k}_{14} - \bar{k}_4 \bar{x}_2 \bar{x}_5),$$
$$\dot{\bar{x}}_4 = \epsilon^{-2}(\bar{k}_4 \bar{x}_2 \bar{x}_5 - \bar{k}_9 \bar{k}_{14}^2 \bar{x}_4), \tag{26}$$
$$\dot{\bar{x}}_5 = \epsilon^{0}(-\bar{k}_4 \bar{x}_2 \bar{x}_5).$$

The fast variables x_1, x_2, x_4 can be eliminated successively (first x_1, then x_2 and x_4).

The elimination of x_1 is possible as $D_{x_1}(\bar{k}_3 \bar{x}_2 - \bar{k}_2 \bar{x}_1) = -\bar{k}_2 < 0$. The elimination of x_2 and x_4 is possible when $x_5 \neq 0$, because $D_{(x_2, x_4)}(\bar{k}_1 \bar{k}_{14} -$
$\bar{k}_4 \bar{x}_2 \bar{x}_5, \bar{k}_4 \bar{x}_2 \bar{x}_5 - \bar{k}_9 \bar{k}_{14}^2 \bar{x}_4) = \begin{bmatrix} -\bar{k}_4 \bar{x}_5 & 0 \\ \bar{k}_4 \bar{x}_5 & -\bar{k}_9 \bar{k}_{14}^2 \end{bmatrix}$ has eigenvalues $-\bar{k}_4 \bar{x}_5$ and $-\bar{k}_9 \bar{k}_{14}^2$.

The reduced model at the slowest timescale reads

$$\dot{x}_5 = -k_1 k_{14},$$
$$x_1 = (k_1 k_3 k_{14})/(k_2 k_4 x_5),$$
$$x_2 = (k_1 k_{14})/(k_4 x_5), \tag{27}$$
$$x_4 = k_1/(k_9 k_{14}),$$

in nonrescaled variables.

The reduced model is one dimensional and describes the decrease at constant rate of x_5 as can be observed in the first part of the trace shown in the Fig. 3. However, as shown above, this reduced model loses validity when x_5 approaches 0.

We have computed (using $\epsilon_* = 1/11, d = 1$) all the reduced models at the slowest timescale for the sequence of partial tropical equilibration solutions obtained from the trace starting at $(1, 0, 0, 0, 100)$ (see Fig. 5 for a schematic representation of this sequence). The reduced models are given in the Table 1.

We found that scalings from the same branch lead to the same reduced model. This important property shows the robustness of the reduction because a branch can span several orders of magnitude of the concentrations.

Furthermore, reduced models are nested in the sense that reduced models for scalings on a face of the polyhedral branch are supermodels (contain all the monomial terms) of reduced models originating from scalings at the interior of polyhedral branches.

Finally, all the reduced models for solutions on the limit cycle are submodels of the reduced model obtained from the total tropical equilibration that reads:

$$
\begin{aligned}
\dot{x}_2 &= k_1 x_3 - k_6, \\
\dot{x}_3 &= k_{10} x_4 - k_1 x_3 + k_9 x_3^2 x_4, \\
\dot{x}_4 &= k_6 - k_{10} x_4 - k_9 x_3^2 x_4, \\
x_1 &= (k_3 x_2)/k_2, \\
x_5 &= k_6/(k_4 x_2).
\end{aligned}
\tag{28}
$$

Indeed, as can be seen in Fig. 4a), the total tropical equilibration is at the intersection of all polyhedral domains corresponding to partial tropical equilibrations in the limit cycle. This reduced model is in fact two dimensional (x_3 and x_4 are decoupled from x_2) and can be used to replace in simulation and further analysis the original five dimensional cell cycle model.

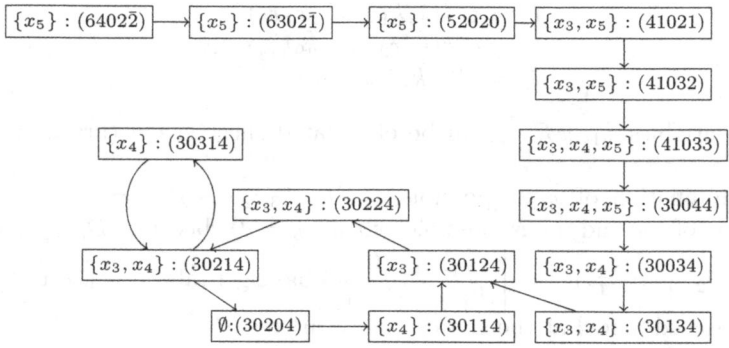

Fig. 5. Sequence of partial tropical equilibration solutions on the trace starting with $(1,0,0,0,100)$. The branch symbol is followed by the valuation a. Here the symbol $\bar{j}$ means $-j$, e.g. the valuation $(6402\bar{2})$ denotes $(6, 4, 0, 2, -2)$.

4   Conclusion and Future Work

The tropicalization method decomposes the phase space into polyhedra within which the dynamics of the system is simpler and can be represented by simpler models, with less dynamical variables and parameters. These domains can span several orders of magnitude and therefore the resulting reduced models are robust.

A first possible application of this approach is to find the most robust reduced model, that applies to the largest domain of interest. For the case study discussed in this paper, we found a two variables reduced model covering the entire limit cycle and suitable for describing system's oscillatory dynamics on this attractor.

The reduced models described in this paper are valid locally. The construction of a global approximation matching these local approximations cannot be fully justified by the underlying algebraic process alone and needs to be investigated analytically by methods from dynamical systems theory and perturbation theory. The justification of the method for small to medium size systems, e.g. the cell cycle model used in the case study, in terms of a hierarchy of slow manifolds connected by fast jumps or other transitions at states where normal hyperbolicity is lost, is the subject of ongoing work of the authors. We expect that this will substantialy increase the understanding of and the confidence in the method for larger systems.

The method also provides possible ways to approach attractors, in terms of sequences of symbolic states and local reductions. These abstractions can be used to model adaptation behavior of biological systems, described as switching from one attractor to another, passing through transient states of different lifetimes. Our approach provides the timescales of each state and the sequence of variables that are active and relax in each transient. The predicted timescales and relaxing variables are important biologically. In the case of the cell cycle, they could be used to discuss interesting dynamical regimes. In medical applications, network perturbations are used in targeted therapies. Dynamical information is rarely taken into account when designing such therapies. However, as shown in this paper, orders of magnitude of variables and parameters and the associated multiple timescales strongly influence and structure the possible dynamics and response of the system. Thus, together with network topology, orders of magnitude and timescale information must be considered together for predicting the effect of network perturbation.

Acknowledgements. The work of A.D. and O.R. was funded by the ANR-17-CE40-0036 project SYMBIONT and by the Campus France/BMBWF Program Amadeus 2020. The work of P.S. was partially funded by OEAD as WTZ-project FR 04/2020.

Appendix 1. Calculation of Partial Tropical Equilibrations: Algorithms and Implementation

The input of our algorithms is a set of polynomial ODE system describing the CRN kinetics:

$$\dot{x}_i = f_i(\boldsymbol{x}), \text{ where}$$

$$f_i(\boldsymbol{x}) = \sum_{j=1}^{r} C_{ij} k_j \boldsymbol{x}^{\boldsymbol{\alpha}_j} \in \mathbb{R}[x_1, \ldots, x_n], 1 \le i \le n, \tag{29}$$

C_{ij} are integers representing stoichiometric coefficients and k_j are real, positive kinetic parameters.

Let $0 < \epsilon_* < 1$ and define $\gamma_j := \text{round}(d\frac{\log(|k_j|)}{\log(\epsilon_*)})/d$.

Following the general approach of the Sects. 2.1, 2.2, we tropicalize the polynomials f_i:

$$Trop(f_i) = \min_{j, C_{ij} \neq 0} (\gamma_j + \langle \boldsymbol{a}, \boldsymbol{\alpha}_j \rangle).$$

Let us consider that a $S \subset \{1, ..., n\}$ subset represents the slow species of the (29). Then, according to the definition introduced in the Sect. 2.2, the partial tropical equilibration problem for S consists in finding a vector $\boldsymbol{a}$ such that:

$$\min_{j,C_{ij}>0}(\gamma_j + \langle \boldsymbol{a}, \boldsymbol{\alpha}_j \rangle) = \min_{j,C_{ij}<0}(\gamma_j + \langle \boldsymbol{a}, \boldsymbol{\alpha}_j \rangle), \ i \notin S \tag{30}$$

$$\min_{j,C'_{lj}\neq 0}(a_j) = \gamma'_l, \ 1 \leq l \leq n_c \tag{31}$$

$$\min_{j,C_{ij}\neq 0}(\gamma_j + \langle \boldsymbol{a}, \boldsymbol{\alpha}_j \rangle) - a_i < \min_{j,C_{ij}\neq 0}(\gamma_j + \langle \boldsymbol{a}, \boldsymbol{\alpha}_j \rangle) - a_{i'}, \ i \notin S, \ i' \in S. \tag{32}$$

The solution of the partial tropical equilibration for S is a polyhedral complex, each maximal polyhedron (called branches and denoted S_{index}) being the solution of one set of choices as described below.

Let

$$\rho_i = argmin_{j,C_{ij}>0}(\gamma_j + \langle \boldsymbol{a}, \boldsymbol{\alpha}_j \rangle)$$
$$\eta_i = argmin_{j,C_{ij}<0}(\gamma_j + \langle \boldsymbol{a}, \boldsymbol{\alpha}_j \rangle)$$
$$\zeta_i = argmin_{j,C_{ij}\neq 0}(\gamma_j + \langle \boldsymbol{a}, \boldsymbol{\alpha}_j \rangle - a_i) \tag{33}$$
$$\omega_{i'} = argmin_{j,C_{i'j}\neq 0}(\gamma_{i'j} + \langle \boldsymbol{a}, \boldsymbol{\alpha}_j \rangle - a_{i'})$$
$$\sigma_l = argmin_{j,C'_{lj}\neq 0}(a_j)$$

For each $i \notin S$, by fixing ρ_i and η_i, we get a polyhedron consisting in the equation and inequations

$$\gamma_{\rho_i} + \langle \boldsymbol{a}, \boldsymbol{\alpha}_{\rho_i} \rangle = \gamma_{\eta_i} + \langle \boldsymbol{a}, \boldsymbol{\alpha}_{\eta_i} \rangle$$
$$\gamma_{\rho_i} + \langle \boldsymbol{a}, \boldsymbol{\alpha}_{\rho_i} \rangle \leq \gamma_j + \langle \boldsymbol{a}, \boldsymbol{\alpha}_j \rangle, \quad C_{ij} \neq 0 \tag{34}$$

We have also a polyhedron for each $1 \leq l \leq n_c$ by fixing σ_l, given by the equation and inequations

$$a_{\sigma_l} = \gamma'_l$$
$$a_{\sigma_l} \leq a_j, \quad C_{lj} \neq 0. \tag{35}$$

For each $i \notin S, i' \in S$, by fixing ζ_i and $\omega_{i'}$, we get a polyhedron consisting in the equation and inequations

$$\gamma_{\zeta_i} + \langle \boldsymbol{a}, \boldsymbol{\alpha}_{\zeta_i} \rangle - a_i \leq \gamma_{\omega_{i'}} + \langle \boldsymbol{a}, \boldsymbol{\alpha}_{\eta_{i'}} \rangle - a_{i'}$$
$$\gamma_{\zeta_i} + \langle \boldsymbol{a}, \boldsymbol{\alpha}_{\zeta_i} \rangle \leq \gamma_j + \langle \boldsymbol{a}, \boldsymbol{\alpha}_j \rangle, \quad C_{ij} \neq 0 \tag{36}$$
$$\gamma_{\omega_{i'}} + \langle \boldsymbol{a}, \boldsymbol{\alpha}_{\omega_{i'}} \rangle \leq \gamma_j + \langle \boldsymbol{a}, \boldsymbol{\alpha}_j \rangle, \quad C_{i'j} \neq 0$$

For each i in (30), by rolling the choices (ρ_i, η_i), we get a union of polyhedra, called bag. For each l in (31), by rolling the choices (σ_l), we get a union of polyhedra, also called bag. For each i, i' in (32), by rolling the choices $(\zeta_i, \omega_{i'})$, we get a union of polyhedra, also called bag.

If we note B_i these bags, the problem reads:

$$A = \bigcap_{i=1}^{n_e} B_i = \bigcap_{i=1}^{n_e} \bigcup_{j=1}^{c_i} P_{ij}$$

where n_e is the number of equations in the partial tropical equilibration problem, and c_i is the number of choices for the equation i. As intersections and unions are in finite numbers we can reverse them and get:

$$A = \bigcap_{i=1}^{n_e} \bigcup_{j=1}^{c} P_{ij} = \bigcup_{j=1}^{c} \bigcap_{i=1}^{n_e} P_{ij}$$

where c is the total number of choices ($c = \sum c_i$). A branch is then given by $\bigcap_{i=1}^{n_e} P_{ij}$ for one choice j.

To solve the tropical equilibration problem, we follow the same process as described in [8], except that we have a different set of bags. We give here how we implement it, the code can be found at https://github.com/Glawal/smtcutpartial.

We have tested it on the model (14) under python 3.7, with the solver mathsat, each solution was given under $0.5\,\mathrm{s}$, for a total time under $2\,\mathrm{s}$. The complexity of finding each tropical equilibration is exponential in the number of variable, due to the combinatorial choice of S. And the complexity of solving one equilibration problem is theoretically exponential in the number of bags, but the smt method used, with a preprossessing, allow to reduce this complexity by removing some choices.

As each polyhedron is given by a choice of $(\rho_i, \eta_i, \zeta_i, \omega_{i'}, \sigma_l)$, we can associate to the polyhedron a tropically truncated system E' defined by:

$$
\begin{aligned}
\bar{f}_i(\boldsymbol{x}) &= C_{i\rho_i} k_{\rho_i} \boldsymbol{x}^{\alpha_{\rho_i}} + C_{i\eta_i} k_{\eta_i} \boldsymbol{x}^{\alpha_{\eta_i}}, \quad i \notin S \\
\bar{f}_{i'}(\boldsymbol{x}) &= C_{i'\omega_{i'}} k_{\omega_{i'}} \boldsymbol{x}^{\alpha_{\omega_{i'}}}, \quad i' \in S \\
k'_l &= C'_{l\sigma_l} x_{\sigma_l}, \quad 1 \le l \le n_c,
\end{aligned}
\tag{37}
$$

The tropically truncated system describe the dominant dynamics of each species.

Suppose now that you have a point $\boldsymbol{x}$ coming from a simulation. We can associate to this point a polyhedron of one of the tropical equilibration, if possible, using the following procedure. We compute the species concentration orders $a_i = \mathrm{round}(d\frac{\log(x_i)}{\log(\epsilon_*)})/d$ and check if these orders are solution of one of the tropical equilibration problem (30), (31), (32). To check if $\boldsymbol{a}$ is a solution, we need to compute the timescale of each species (that allow a list of S choice), and to compute orders of each monomials (given by $\gamma_j + \langle \boldsymbol{a}, \boldsymbol{\alpha}_j \rangle$), which give us the list of equilibrated species. Then we choose S as being minimal (each equilibrated species that are faster that non equilibrated species are considered outside S), this gives us the final choice $(S, \rho_i, \eta_i, \zeta_i, \omega_{i'}, \sigma_l)$, denoted S_{index}, the index being the index of P in rr in Algorithm 7.

Algorithm smtcutpartial*((29) , S)*
 $bb \leftarrow$ makePolyhedraForPTE$((29), S)$
 $rr \leftarrow$ computePolyhedronDnf(bb)
 return rr

Algorithm 1: The algorithm used to find the partial tropical equilibration for S.

Algorithm makePolyhedraForPTE*((29), S)*
 $bb \leftarrow \emptyset$
 foreach $i \in \{1, ..., n\}\backslash S$ **do**
  $bb \leftarrow bb \cup$ equilibrate$(\dot{x}_i)$
  foreach $j \in S$ **do**
   $bb \leftarrow bb \cup$ slowFastPol$(\dot{x}_i, \dot{x}_j)$
 return bb

Algorithm 2: This algorithm makes the list of bags, each bag representing a list of polyhedra, such that each polyhedron is linked to an equilibration for a fast species or a conservation law, or a slow fast decomposition.

Algorithm equilibrate*($\dot{x}_i$)*
 $pp, np, b \leftarrow \emptyset$
 foreach $1 \leq j \leq r_i$ **do**
  $t = trop(k_j \boldsymbol{x}^{\alpha_j})$
  if $S_{ij} < 0$ **then**
   $np \leftarrow np \cup t$
  else
   $pp \leftarrow pp \cup t$
 foreach $(a, c) \in pp \times np$ **do**
  $p =$ makePolyhedron(a, c, pp, np)
  $b \leftarrow b \cup p$
 return b

Algorithm 3: This algorithm computes the bag b linked to an equation of the system, that represent each possible equilibration. It splits negative and positive monomials, computes their tropicalization and makes the bag.

Algorithm makePolyhedron(a, c, pp, np)
  $eq \leftarrow a = c$
  $ieq \leftarrow \emptyset$
  foreach $i \in pp \cup np$ **do**
    $\llcorner$ $ieq \leftarrow ieq \cup a \leq i$
  $p \leftarrow$ make(eq, ieq)
  return p

Algorithm 4: p is a polyhedron defined by Eq. (34) or (35).

Algorithm slowFastPol$(\dot{x}_i, \dot{x}_j)$
  $sp, fp, b \leftarrow \emptyset$
  foreach $1 \leq m \leq r_i$ **do**
    $t = trop(k_m \boldsymbol{x}^{\alpha_m - a_i})$
    $fp \leftarrow fp \cup t$
  foreach $1 \leq m \leq r_j$ **do**
    $t = trop(k_m \boldsymbol{x}^{\alpha_m - a_j})$
    $sp \leftarrow sp \cup t$
  foreach $(a, c) \in sp \times fp$ **do**
    $p = $ makePolyhedronSF(a, c, sp, fp)
    $b \leftarrow b \cup p$
  return b

Algorithm 5: This algorithm computes the bag b linked to a slow fast decomposition between two species x_i (fast) and x_j (slow). As the order the species has an impact on the slow fast decomposition, we multiply each monomial in $\dot{x}_q$ by $\frac{1}{x_q}$. Then we split each monomials, compute their tropicalization and make the bag.

Algorithm makePolyhedronSF(a, c, sp, fp)
  $eq \leftarrow \emptyset$
  $ieq \leftarrow a \geq c$
  foreach $i \in sp$ **do**
    $\llcorner$ $ieq \leftarrow ieq \cup a \leq i$
  foreach $i \in fp$ **do**
    $\llcorner$ $ieq \leftarrow ieq \cup c \leq i$
  $p \leftarrow$ make(eq, ieq)
  return p

Algorithm 6: p is a polyhedron defined by Eq. (36).

```
Algorithm computePolyhedronDnf (bb)
    solver ← getSolveur(incremental = true)
    f ← convertToSMTFormula(bb)
    rr ← ∅, bool ← true
    while bool do
        solver.addAssertion(f)
        (x, bool) = solver.solve(f) /*x is a point that satisfy the constraints,
        bool is false if no x*/
        if Not(bool) then
            └ Break
        else
            R = ∅
            foreach b ∈ bb do
                foreach P ∈ b do
                    if x ∈ P then
                        │ R ← R ∪ P.constraints()
                        └ Break
            f ← Not(R)
            └ rr ← rr ∪ R
    └ return rr
```

Algorithm 7: This algorithm computes the intersection of a set of bags *bb*. Each polyhedron is transformed to a logical constraint. x represents a point that satisfy the set of constraints, then, if a such x is found, it is contained by a polyhedron P common to each equilibration. We remove this polyhedron from search by adding a constraint and continue the search until there is no feasible point. At the end we get a list of polyhedra rr, that is the tropical equilibration.

Appendix 2. Sequence of Reduced Models

Table 1. The sequence of reduced models corresponding to different partial tropical equilibrations along the trace starting at $(1,0,0,0,100)$, computed with $\epsilon_* = 1/11, d = 1$. A few rapid transient states at the beginning of the trace where not analysed and we start with the partial tropical equilibration solution $(6,4,0,2,-2)$. We indicate the change of variables needed for the elimination of some exact and approximate conservation laws, the truncated rescaled system indicating the local timescales of the variables and the local reduced system in nonscaled variables and parameters. The change of variables is included in the definition of the reduced system.

Tropical solution	Truncated rescaled system	Change of variables	Reduced system
$\{x_5\}_{02}$ $(6,4,0,2,-2)$	$\dot{\bar{x}}_1 = \epsilon^{-6}(\bar{k}_1\bar{k}_{14} - \bar{k}_2\bar{x}_1)$ $\dot{\bar{x}}_2 = \epsilon^{-4}(\bar{k}_2\bar{x}_1 - \bar{k}_4\bar{x}_2\bar{x}_5)$ $\dot{\bar{x}}_4 = \epsilon^{-2}(\bar{k}_4\bar{x}_2\bar{x}_5 - \bar{k}_9\bar{k}_{14}^2\bar{x}_4)$ $\dot{\bar{x}}_5 = \epsilon^2(-\bar{k}_4\bar{x}_2\bar{x}_5)$	$x_3 \leftarrow k_{14} - x_1 - x_2 - x_4$	$\dot{x}_5 = -k_1k_{14}$ $x_1 = (k_1k_{14})/k_2$ $x_2 = (k_1k_{14})/(k_4x_5)$ $x_4 = k_1/(k_9k_{14})$
$\{x_5\}_{02}$ $(6,3,0,2,-1)$	$\dot{\bar{x}}_1 = \epsilon^{-6}(\bar{k}_1\bar{k}_{14} + \bar{k}_3\bar{x}_2 - \bar{k}_2\bar{x}_1)$ $\dot{\bar{x}}_2 = \epsilon^{-3}(\bar{k}_2\bar{x}_1 - \bar{k}_3\bar{x}_2 - \bar{k}_4\bar{x}_2\bar{x}_5)$ $\dot{\bar{x}}_4 = \epsilon^{-2}(\bar{k}_4\bar{x}_2\bar{x}_5 - \bar{k}_9\bar{k}_{14}^2\bar{x}_4)$ $\dot{\bar{x}}_5 = \epsilon^1(-\bar{k}_4\bar{x}_2\bar{x}_5)$	$x_3 \leftarrow k_{14} - x_1 - x_2 - x_4$	$\dot{x}_5 = -k_1k_{14}$ $x_1 = (k_1k_{14}(k_3 + k_4x_5))$ $/(k_2k_4x_5)$ $x_2 = (k_1k_{14})/(k_4x_5)$ $x_4 = k_1/(k_9k_{14})$
$\{x_5\}_{03}$ $(5,2,0,2,0)$	$\dot{\bar{x}}_1 = \epsilon^{-6}(\bar{k}_3\bar{x}_2 - \bar{k}_2\bar{x}_1)$ $\dot{\bar{x}}_2 = \epsilon^{-2}(\bar{k}_1\bar{k}_{14} - \bar{k}_4\bar{x}_2\bar{x}_5)$ $\dot{\bar{x}}_4 = \epsilon^{-2}(\bar{k}_4\bar{x}_2\bar{x}_5 - \bar{k}_9\bar{k}_{14}^2\bar{x}_4)$ $\dot{\bar{x}}_5 = \epsilon^0(-\bar{k}_4\bar{x}_2\bar{x}_5)$	$x_3 \leftarrow k_{14} - x_1 - x_2 - x_4$ $x_2 \leftarrow x_1 + x_2$	$\dot{x}_5 = -k_1k_{14}$ $x_1 = (k_1k_3k_{14})/(k_2k_4x_5)$ $x_2 = (k_1k_{14})/(k_4x_5)$ $x_4 = k_1/(k_9k_{14})$
$\{x_3,x_5\}_{01}$ $(4,1,0,2,1)$	$\dot{\bar{x}}_1 = \epsilon^{-6}(\bar{k}_3\bar{x}_2 - \bar{k}_2\bar{x}_1)$ $\dot{\bar{x}}_2 = \epsilon^{-1}(\bar{k}_1\bar{k}_{14} - \bar{k}_4\bar{x}_2\bar{x}_5)$ $\dot{\bar{x}}_4 = \epsilon^{-2}(\bar{k}_4\bar{x}_2\bar{x}_5 - \bar{k}_9\bar{k}_{14}^2\bar{x}_4)$ $\dot{\bar{x}}_5 = \epsilon^{-1}(-\bar{k}_4\bar{x}_2\bar{x}_5)$	$x_3 \leftarrow k_{14} - x_1 - x_2 - x_4$ $x_2 \leftarrow x_1 + x_2$	$\dot{x}_2 = k_1k_{14} - k_4x_2x_5$ $\dot{x}_5 = -k_4x_2x_5$ $x_1 = (k_3x_2)/k_2$ $x_4 = (k_4x_2x_5)/(k_9k_{14}^2)$
$\{x_3,x_5\}_{01}$ $(4,1,0,3,2)$	$\dot{\bar{x}}_1 = \epsilon^{-6}(\bar{k}_3\bar{x}_2 - \bar{k}_2\bar{x}_1)$ $\dot{\bar{x}}_2 = \epsilon^{-1}(\bar{k}_1\bar{k}_{14})$ $\dot{\bar{x}}_4 = \epsilon^{-2}(\bar{k}_4\bar{x}_2\bar{x}_5 - \bar{k}_9\bar{k}_{14}^2\bar{x}_4)$ $\dot{\bar{x}}_5 = \epsilon^{-1}(-\bar{k}_4\bar{x}_2\bar{x}_5)$	$x_3 \leftarrow k_{14} - x_1 - x_2 - x_4$ $x_2 \leftarrow x_1 + x_2$	$\dot{x}_2 = k_1k_{14}$ $\dot{x}_5 = -k_4x_2x_5$ $x_1 = (k_3x_2)/k_2$ $x_4 = (k_4x_2x_5)/(k_9k_{14}^2)$
$\{x_3,x_4,x_5\}_{22}$ $(4,1,0,3,3)$	$\dot{\bar{x}}_1 = \epsilon^{-6}(\bar{k}_3\bar{x}_2 - \bar{k}_2\bar{x}_1)$ $\dot{\bar{x}}_2 = \epsilon^{-1}(\bar{k}_1\bar{k}_{14})$ $\dot{\bar{x}}_4 = \epsilon^{-2}(-\bar{k}_9\bar{k}_{14}^2\bar{x}_4)$ $\dot{\bar{x}}_5 = \epsilon^{-1}(\bar{k}_6 - \bar{k}_4\bar{x}_2\bar{x}_5)$	$x_3 \leftarrow k_{14} - x_1 - x_2 - x_4$ $x_2 \leftarrow x_1 + x_2$	$\dot{x}_2 = k_1k_{14}$ $\dot{x}_4 = -k_9k_{14}^2x_4$ $\dot{x}_5 = k_6 - k_4x_2x_5$ $x_1 = (k_3x_2)/k_2$
$\{x_3\}_{00}$ $(3,0,0,4,4)$	$\dot{\bar{x}}_1 = \epsilon^{-6}(\bar{k}_3\bar{x}_2 - \bar{k}_2\bar{x}_1)$ $\dot{\bar{x}}_2 = \epsilon^0(\bar{k}_1\bar{k}_{14} - \bar{k}_1\bar{x}_2)$ $\dot{\bar{x}}_4 = \epsilon^{-2}(\bar{k}_4\bar{x}_2\bar{x}_5 + 2\bar{k}_9\bar{k}_{14}\bar{x}_2\bar{x}_4 - \bar{k}_9\bar{k}_{14}^2\bar{x}_4 - \bar{k}_9\bar{x}_2^2\bar{x}_4)$ $\dot{\bar{x}}_5 = \epsilon^{-2}(\bar{k}_6 - \bar{k}_4\bar{x}_2\bar{x}_5)$	$x_3 \leftarrow k_{14} - x_1 - x_2 - x_4$ $x_2 \leftarrow x_1 + x_2$	$\dot{x}_2 = k_1k_{14} - k_1x_2$ $x_1 = (k_3x_2)/k_2$ $x_4 = k_6/(k_9k_{14}^2 + k_9x_2^2 - 2k_9k_{14}x_2)$ $x_5 = k_6/(k_4x_2)$
$\{x_3,x_4\}_{05}$ $(3,0,0,3,4)$	$\dot{\bar{x}}_1 = \epsilon^{-6}(\bar{k}_3\bar{x}_2 - \bar{k}_2\bar{x}_1)$ $\dot{\bar{x}}_2 = \epsilon^0(\bar{k}_1\bar{k}_{14} - \bar{k}_1\bar{x}_2)$ $\dot{\bar{x}}_4 = \epsilon^{-2}(2\bar{k}_9\bar{k}_{14}\bar{x}_2\bar{x}_4 - \bar{k}_9\bar{k}_{14}^2\bar{x}_4 - \bar{k}_9\bar{x}_2^2\bar{x}_4)$ $\dot{\bar{x}}_5 = \epsilon^{-2}(\bar{k}_6 - \bar{k}_4\bar{x}_2\bar{x}_5)$	$x_3 \leftarrow k_{14} - x_1 - x_2 - x_4$ $x_2 \leftarrow x_1 + x_2$	$\dot{x}_2 = k_1k_{14} - k_1x_2$ $\dot{x}_4 = 2k_9k_{14}x_2x_4 - k_9x_2^2x_4 - k_9k_{14}^2x_4$ $\dot{x}_5 = k_6 - k_4x_2x_5$ $x_1 = (k_3x_2)/k_2$

(*continued*)

Table 1. (*continued*)

Tropical solution	Truncated rescaled system	Change of variables	Reduced system
$\{x_3, x_4\}_{04}$ $(3,0,1,3,4)$	$\dot{\bar{x}}_1 = \epsilon^{-6}(\bar{k}_3\bar{x}_2 - \bar{k}_2\bar{x}_1)$ $\dot{\bar{x}}_2 = \epsilon^1(\bar{k}_1\bar{x}_3)$ $\dot{\bar{x}}_3 = \epsilon^0(-\bar{k}_1\bar{x}_3)$ $\dot{\bar{x}}_4 = \epsilon^{-1}(\bar{k}_4\bar{x}_2\bar{x}_5)$ $\dot{\bar{x}}_5 = \epsilon^{-2}(\bar{k}_6 - \bar{k}_4\bar{x}_2\bar{x}_5)$	$x_2 \leftarrow x_1 + x_2$	$\dot{x}_2 = k_1 x_3$ $\dot{x}_3 = -k_1 x_3$ $\dot{x}_4 = k_6$ $x_1 = (k_3 x_2)/k_2$ $x_5 = k_6/(k_4 x_2)$
$\{x_3\}_{00}$ $(3,0,1,2,4)$	$\dot{\bar{x}}_1 = \epsilon^{-6}(\bar{k}_3\bar{x}_2 - \bar{k}_2\bar{x}_1)$ $\dot{\bar{x}}_2 = \epsilon^1(\bar{k}_1\bar{x}_3)$ $\dot{\bar{x}}_3 = \epsilon^0(-\bar{k}_1\bar{x}_3)$ $\dot{\bar{x}}_4 = \epsilon^0(\bar{k}_4\bar{x}_2\bar{x}_5 - \bar{k}_9\bar{x}_3^2\bar{x}_4)$ $\dot{\bar{x}}_5 = \epsilon^{-2}(\bar{k}_6 - \bar{k}_4\bar{x}_2\bar{x}_5)$	$x_2 \leftarrow x_1 + x_2$	$\dot{x}_2 = k_1 x_3$ $\dot{x}_3 = -k_1 x_3$ $\dot{x}_4 = k_6 - k_9 x_3^2 x_4$ $x_1 = (k_3 x_2)/k_2$ $x_5 = k_6/(k_4 x_2)$
$\{x_3, x_4\}_{04}$ $(3,0,2,2,4)$	$\dot{\bar{x}}_1 = \epsilon^{-6}(\bar{k}_3\bar{x}_2 - \bar{k}_2\bar{x}_1)$ $\dot{\bar{x}}_2 = \epsilon^2(\bar{k}_1\bar{x}_3 - \bar{k}_4\bar{x}_2\bar{x}_5)$ $\dot{\bar{x}}_3 = \epsilon^0(-\bar{k}_1\bar{x}_3)$ $\dot{\bar{x}}_4 = \epsilon^0(\bar{k}_4\bar{x}_2\bar{x}_5)$ $\dot{\bar{x}}_5 = \epsilon^{-2}(\bar{k}_6 - \bar{k}_4\bar{x}_2\bar{x}_5)$	$x_2 \leftarrow x_1 + x_2$	$\dot{x}_2 = k_1 x_3 - k_6$ $\dot{x}_3 = -k_1 x_3$ $\dot{x}_4 = k_6$ $x_1 = (k_3 x_2)/k_2$ $x_5 = k_6/(k_4 x_2)$
$\{x_3, x_4\}_{04}$ $(3,0,2,1,4)$	$\dot{\bar{x}}_1 = \epsilon^{-6}(\bar{k}_3\bar{x}_2 - \bar{k}_2\bar{x}_1)$ $\dot{\bar{x}}_2 = \epsilon^2(\bar{k}_1\bar{x}_3 - \bar{k}_4\bar{x}_2\bar{x}_5)$ $\dot{\bar{x}}_3 = \epsilon^0(-\bar{k}_1\bar{x}_3)$ $\dot{\bar{x}}_4 = \epsilon^1(\bar{k}_4\bar{x}_2\bar{x}_5)$ $\dot{\bar{x}}_5 = \epsilon^{-2}(\bar{k}_6 - \bar{k}_4\bar{x}_2\bar{x}_5)$	$x_2 \leftarrow x_1 + x_2$	$\dot{x}_2 = k_1 x_3 - k_6$ $\dot{w}_3 = -k_1 x_3$ $\dot{x}_4 = k_6$ $x_1 = (k_3 x_2)/k_2$ $x_5 = k_6/(k_4 x_2)$
$\{x_4\}_{00}$ $(3,0,3,1,4)$	$\dot{\bar{x}}_1 = \epsilon^{-6}(\bar{k}_3\bar{x}_2 - \bar{k}_2\bar{x}_1)$ $\dot{\bar{x}}_2 = \epsilon^2(-\bar{k}_4\bar{x}_2\bar{x}_5)$ $\dot{\bar{x}}_3 = \epsilon^0(\bar{k}_{10}\bar{x}_4 - \bar{k}_1\bar{x}_3)$ $\dot{\bar{x}}_4 = \epsilon^1(\bar{k}_4\bar{x}_2\bar{x}_5)$ $\dot{\bar{x}}_5 = \epsilon^{-2}(\bar{k}_6 - \bar{k}_4\bar{x}_2\bar{x}_5)$	$x_2 \leftarrow x_1 + x_2$	$\dot{x}_2 = -k_6$ $\dot{x}_3 = k_{10} x_4 - k_1 x_3$ $\dot{x}_4 = k_6$ $x_1 = (k_3 x_2)/k_2$ $x_5 = k_6/(k_4 x_2)$
$\{\emptyset\}_{00}$ $(3,0,2,0,4)$	$\dot{\bar{x}}_1 = \epsilon^{-6}(\bar{k}_3\bar{x}_2 - \bar{k}_2\bar{x}_1)$ $\dot{\bar{x}}_2 = \epsilon^2(\bar{k}_1\bar{x}_3 - \bar{k}_4\bar{x}_2\bar{x}_5)$ $\dot{\bar{x}}_3 = \epsilon^0(\bar{k}_{10}\bar{x}_4 + \bar{k}_9\bar{x}_3^2\bar{x}_4 - \bar{k}_1\bar{x}_3)$ $\dot{\bar{x}}_4 = \epsilon^2(\bar{k}_4\bar{x}_2\bar{x}_5 - \bar{k}_{10}\bar{x}_4 - \bar{k}_9\bar{x}_3^2\bar{x}_4)$ $\dot{\bar{x}}_5 = \epsilon^{-2}(\bar{k}_6 - \bar{k}_4\bar{x}_2\bar{x}_5)$	$x_2 \leftarrow x_1 + x_2$	$\dot{x}_2 = k_1 x_3 - k_6$ $\dot{x}_3 = k_{10} x_4 - k_1 x_3 +$ $+ k_9 x_3^2 x_4$ $\dot{x}_4 = k_6 - k_{10} x_4 -$ $k_9 x_3^2 x_4$ $x_1 = (k_3 x_2)/k_2$ $x_5 = k_6/(k_4 x_2)$
$\{x_4\}_{02}$ $(3,0,1,1,4)$	$\dot{\bar{x}}_1 = \epsilon^{-6}(\bar{k}_3\bar{x}_2 - \bar{k}_2\bar{x}_1)$ $\dot{\bar{x}}_2 = \epsilon^1(\bar{k}_1\bar{x}_3)$ $\dot{\bar{x}}_3 = \epsilon^0(\bar{k}_9\bar{x}_3^2\bar{x}_4 - \bar{k}_1\bar{x}_3)$ $\dot{\bar{x}}_4 = \epsilon^0(-\bar{k}_9\bar{x}_3^2\bar{x}_4)$ $\dot{\bar{x}}_5 = \epsilon^{-2}(\bar{k}_6 - \bar{k}_4\bar{x}_2\bar{x}_5)$	$x_2 \leftarrow x_1 + x_2$	$\dot{x}_2 = k_1 x_3$ $\dot{x}_3 = k_9 x_3^2 x_4 - k_1 x_3$ $\dot{x}_4 = -k_9 x_3^2 x_4$ $x_1 = (k_3 x_2)/k_2$ $x_5 = k_6/(k_4 x_2)$

References

1. Cardin, P.T., Teixeira, M.A.: Fenichel theory for multiple time scale singular perturbation problems. SIAM J. Appl. Dyn. Syst. **16**(3), 1425–1452 (2017). https://doi.org/10.1137/16M1067202
2. Desoeuvres, A.: Tropical geometry and interval arithmetic methods for the analysis of biochemical networks: homeostasis research and model reduction in the presence of conservation laws. Ph.D. thesis, I2S, University of Montpellier (2021)
3. Fenichel, N.: Geometric singular perturbation theory for ordinary differential equations. J. Differ. Equ. **31**(1), 53–98 (1979)
4. Gorban, A.N., Radulescu, O.: Dynamic and static limitation in multiscale reaction networks, revisited. Adv. Chem. Eng. **34**(2008), 103–173 (2007). https://doi.org/10.1016/S0065-2377(08)00003-3

5. Heineken, F.G., Tsuchiya, H.M., Aris, R.: On the mathematical status of the pseudo-steady state hypothesis of biochemical kinetics. Math. Biosci. **1**(1), 95–113 (1967)
6. Hoppensteadt, F.: On systems of ordinary differential equations with several parameters multiplying the derivatives. J. Differ. Equ. **5**(1), 106–116 (1969)
7. Kruff, N., Lüders, C., Radulescu, O., Sturm, T., Walcher, S.: Algorithmic reduction of biological networks with multiple time scales. Math. Comput. Sci. **15**(3), 499–534 (2021). https://doi.org/10.1007/s11786-021-00515-2
8. Lüders, C.: Computing tropical prevarieties with satisfiability modulo theories (SMT) solvers. arXiv preprint arXiv:2004.07058 (2020)
9. Maclagan, D., Sturmfels, B.: Introduction to Tropical Geometry. Graduate studies in mathematics. vol. 161, American Mathematical Society, Providence (2009)
10. Noel, V., Grigoriev, D., Vakulenko, S., Radulescu, O.: Tropical geometries and dynamics of biochemical networks application to hybrid cell cycle models. Electron. Notes Theor. Comput. Sci. **284**, 75–91 (2012). https://doi.org/10.1016/j.entcs.2012.05.016
11. Noel, V., Grigoriev, D., Vakulenko, S., Radulescu, O.: Tropicalization and tropical equilibration of chemical reactions. Trop. Idempotent Math. Appl. **616**, 261–277 (2014). https://doi.org/10.1090/conm/616/12316
12. O'Malley, R.: On initial value problems for nonlinear systems of differential equations with two small parameters. Arch. Ration. Mech. Anal. **40**(3), 209–222 (1971)
13. Radulescu, O., Swarup Samal, S., Naldi, A., Grigoriev, D., Weber, A.: Symbolic dynamics of biochemical pathways as finite states machines. In: Roux, O., Bourdon, J. (eds.) CMSB 2015. LNCS, vol. 9308, pp. 104–120. Springer, Cham (2015). https://doi.org/10.1007/978-3-319-23401-4_10
14. Radulescu, O., Vakulenko, S., Grigoriev, D.: Model reduction of biochemical reactions networks by tropical analysis methods. Math. Modell. Nat. Phenom. **10**(3), 124–138 (2015). https://doi.org/10.1051/mmnp/201510310
15. Samal, S.S., Grigoriev, D., Fröhlich, H., Weber, A., Radulescu, O.: A geometric method for model reduction of biochemical networks with polynomial rate functions. Bull. Math. Biol. **77**(12), 2180–2211 (2015). https://doi.org/10.1007/s11538-015-0118-0
16. Samal, S.S., Krishnan, J., Esfahani, A.H., Lüders, C., Weber, A., Radulescu, O.: Metastable regimes and tipping points of biochemical networks with potential applications in precision medicine. In: Liò, P., Zuliani, P. (eds.) Automated Reasoning for Systems Biology and Medicine. Computational Biology, vol. 30, pp. 269–295. Springer, Cham (2019). https://doi.org/10.1007/978-3-030-17297-8_10
17. Samal, S.S., Naldi, A., Grigoriev, D., Weber, A., Théret, N., Radulescu, O.: Geometric analysis of pathways dynamics: application to versatility of TGF-β receptors. Biosystems **149**, 3–14 (2016). https://doi.org/10.1016/j.biosystems.2016.07.004
18. Segel, L.A., Slemrod, M.: The quasi-steady-state assumption: a case study in perturbation. SIAM Rev. **31**(3), 446–477 (1989). https://doi.org/10.1137/1031091
19. Soliman, S., Fages, F., Radulescu, O.: A constraint solving approach to model reduction by tropical equilibration. Algorithms Mol. Biol. **9**(1), 1–11 (2014)
20. Tyson, J.J.: Modeling the cell division cycle: cdc2 and cyclin interactions. Proc. Natl. Acad. Sci. **88**(16), 7328–7332 (1991)
21. Viro, O.: Dequantization of real algebraic geometry on logarithmic paper. In: Casacuberta, C., Miró-Roig, R.M., Verdera, J., Xambó-Descamps, S. (eds.) European Congress of Mathematics, Progress in Mathematics, vol. 201, pp. 135–146. Springer, Heidelberg (2001). https://doi.org/10.1007/978-3-0348-8268-2_8

Boolean Networks

Prioritization of Candidate Genes Through Boolean Networks

Clémence Réda[1]([✉])(iD) and Andrée Delahaye-Duriez[1,2,3](iD)

[1] Univ. Paris Cité, Neurodiderot, Inserm, 75019 Paris, France
{clemence.reda,andree.delahaye}@inserm.fr
[2] Univ. Sorbonne Paris Nord, UFR de santé, médecine et biologie humaine,
93000 Bobigny, France
[3] Unité fonctionnelle de médecine génomique et génétique clinique,
Hôpital Jean Verdier, AP-HP, 93140 Bondy, France

Abstract. The *in silico* detection of master regulator genes is a popular attempt at speeding up drug development. These genes might be directly related to the onset of the disease, or may act on one pathway which counteracts the associated symptoms. Then, one could perhaps screen drugs to select chemical compounds targeting these genes. In prior works, the detection of these candidates was performed through the identification of the regulatory interactions between genes of interest for the disease. Indeed, system biology approaches have proven a useful tool to integrate transcriptomic data and predict transcriptional profiles under gene perturbations. However, for rare or tropical neglected diseases, building such a regulatory model can become a tedious and time-consuming task. In this work, we show how to build, in a reproducible and transparent fashion, a gene regulatory network using publicly available data. Then, we describe a method to identify master regulatory genes, which have an impact on the dynamics of the gene regulation in a specific disease-related transcriptional context. We showed that our novel method for the identification of master regulatory genes was consistent with network controllability measures, while targeting genes that were significantly enriched for epilepsy-related terms. Our pipeline allows for systematic and transparent Boolean network synthesis, and identification of master re-gulators, which might help tackle the issue of rare or tropical neglected diseases.

Keywords: master regulator prioritization · drug-resistant epilepsy · boolean network · influence maximization · machine learning application

1 Introduction

We propose a novel generic method for the detection of master regulator genes, which can be applied to any disease, and relies on a dynamical interplay between a gene regulatory network and gene expression data. We focus here, as a proof-of-concept, on an application to epilepsy.

© The Author(s), under exclusive license to Springer Nature Switzerland AG 2022
I. Petre and A. Păun (Eds.): CMSB 2022, LNBI 13447, pp. 89–121, 2022.
https://doi.org/10.1007/978-3-031-15034-0_5

Epilepsy actually encompasses various neurological diseases and syndromes, which can originate from brain injury or genetic background, that have in common a propensity to trigger chronic epileptic crises. Epileptic crises are characterized by a transitory abnormal neuron electric discharge, which might lead to unconsciousness, seizures, and/or body stiffness. Epilepsy is one of the most common neurological diseases worldwide, with around 50 million people living with this disease [73]. Moreover, more than 25% of epileptic patients are afflicted with drug-resistant epilepsy [28] –also called refractory epilepsy– that is, symptoms in those patients could not be managed by at least two different antiepileptic therapies. This shows the limits of conventional antiepileptic medication, which are often molecules with antiseizure effects, and emphasizes the need to look for novel therapeutic candidates. Epilepsy-related genes are shown to be usually mainly expressed in a specific brain region, called hippocampus [46], which is also affected by morphological changes linked to neuronal discharges in some epileptic patients [51]. The exact relationship between lesions in the hippocampus and epilepsy-associated symptoms is still unclear, but might be related to the fact that hippocampus is one of the most excitable parts of the brain [38]. Several animal models of epilepsy exist, including a mouse model where injection of pilocarpine induce symptoms similar to temporal lobe epilepsy [62], or another involving sodium channels, which are used to convey electric potentials (Dravet syndrome model, by knocking out gene *Scn1a* [35]).

In prior works, the identification of master regulators in gene networks has been a powerful method to detect novel genes of interest for a given disease. Master regulator genes are DNA sequences which might have a large, global influence on the expression of a group of genes in a specific pathway. For instance, *SESTRIN3* [33] and *CSF1R* [61] were prioritised as candidate antiepileptic drug targets using different systems-biology approaches dedicated to identifying master regulators of epilepsy-associated networks of gene expression. Such genes might be forcibly expressed or knocked out –*i.e.*, no more expressed– by molecules, which might be interesting antiepileptic drug candidates. Other approaches exploit the location of a given gene inside a gene regulatory network –the more central it is, the most regulatory it should be– or the concept of "network controllability" [43]. Yet, most of the cited approaches for the detection of interesting regulatory genes only leverage topological knowledge about the network, without considering the actual dynamics of the regulatory system. A notable exception is the work by [75], which aimed at finding master regulator genes related to rheumatoid arthritis based on expression data. Their approach combines a transcription factor (TF) co-regulatory network and gene expression in fibroblast-like synoviocytes in patients afflicted with rheumatoid arthritis. TF influence in these samples was assessed using the tool CoRegNet [50], which computes a score of influence for a given TF on a set of transcriptional profiles. This score is defined for any TF t, that activates a set $\mathcal{A}_t$ of genes and inhibits a set $\mathcal{I}_t$ of genes, and for a given matrix of transcriptional profiles M^1 as follows

---

[1] In this notation, rows are genes, and columns are samples.

$$\text{Influence}(t) := \frac{\left(\frac{1}{|\mathcal{A}_t|}\sum_{a\in\mathcal{A}_t} M[a,:]\right) - \left(\frac{1}{|\mathcal{I}_t|}\sum_{i\in\mathcal{I}_t} M[i,:]\right)}{\sqrt{(s_{\mathcal{A}_t})^2/|\mathcal{A}_t| + (s_{\mathcal{I}_t})^2/|\mathcal{I}_t|}} \tag{1}$$

where $s_{\mathcal{A}_t}$ (resp., $s_{\mathcal{I}_t}$) is the standard deviation of expression levels of all genes in $\mathcal{A}_t$ (resp., $\mathcal{I}_t$) across all profiles in M. However, such a computation does not take into account downstream transcriptional cascades [9], that is, regulatory effects which trickle down the network, beyond the genes directly regulated by the TF. However, taking into account these regulatory cascades might allow to control for off-target genes, which are genes subject to non specific and involuntary changes, for which perturbation might lead to serious side effects [32]. In order to model these regulatory cascades, we are interested in Boolean networks, which model discrete gene regulatory interactions, for their increased interpretability. Indeed, in this type of network, the expression level of a gene is reduced to binary values (genes are either expressed, or not expressed), and is the product of a unique logical function, which takes inputs from the expression states of *direct* regulators of this gene [36,67]. Yet building a Boolean network for a large number of genes is a painful task without automation, as further discussed in Appendix A.

In order to tackle these issues, we developed a fully automated pipeline to infer a Boolean network which models the regulatory interactions in a well-chosen cell line. This pipeline connects existing databases to inference methods from the literature, and is an important contribution towards speeding up and easily adapting network identification to other diseases. This procedure is based on gene perturbation experiments, and on the integration of supplementary biological information to further constraint our inference procedure. Transcriptional profiles are extracted from the LINCS L1000 database, which collects a large number of profiles for several cell types and genetic perturbations [65]. In our application, we focused on the gene module M30, which global expression was shown to be anticorrelated with various epileptic profiles and with the severity of epilepsy [19]. Using the Boolean network selected by this method, we ranked genes in M30 according to their ability to permanently modify the global expression of the network, and prioritized top genes. In favor of their important role in epilepsy-related biological processes, this set of candidate master regulators was significantly enriched in terms associated with epilepsy and neurodevelopmental issues with respect to the M30 module.

2 Methods

2.1 Reproducible Inference of a Cell-Line Specific Boolean Network

This part of our work aims at designing a method which, given a subset of genes of interest, is able to retrieve a gene regulatory network on these genes that allows the prediction of transcriptomic profiles under perturbation. We consider the formalism of Boolean networks, introduced in [36,67], which are popular models to describe gene regulations.

Boolean Networks. A Boolean network is first characterized by a graph –that is, the network– which connects genes by their regulatory interactions. Such connections are enriched with the direction of the interaction, which distinguishes between regulator and regulated genes, and with the sign of this interaction, that is, whether the regulator inhibits or activates the expression of its target. Second, the dynamics of the system are described by logical functions, called "gene regulatory functions", where variables correspond to the binary expression level (or state) of genes in the network. A single function is assigned to each gene. The expression of a given variable is set to 1 if the associated gene is expressed, otherwise 0. For a given gene g, its associated formulæ contains in its premise the variables corresponding to *direct* regulators of g –*i.e.*, direct predecessors of the node in the network– and in its conclusion the variable associated with the expression level of g. Then, given the expression states of the regulators at a given time step, one can obtain the expression state of the considered gene at the next time step, by evaluating the corresponding formulæ. The network state (or configuration) is the concatenation of all gene expression states. The order of evaluation of regulatory functions to go from one network state to another is called "update step". From a Boolean network, one can build a *state-transition diagram*, where an edge goes from a given network state A to another network state B if and only if one can reach state B from state A in a single update step. One can read from this diagram attractor states, that is, self-looping nodes, which are defined as steady stable network configurations. That is, the application of the update step to this configuration will lead to itself. Attractor states are interesting because they are commonly related to observable biological phenotypes [8,72]. This diagram also displays cycles of configurations, which correspond to unsteady stable configurations; the application of the update step makes the system oscillate between a set of configurations in a cyclic way, and can also have a biological interpretation [68]. A state-transition diagram is associated with a given model *and* a type of update. Several types have been suggested in the literature [13], the most well-known ones being the synchronous and the asynchronous updates. In the former, all regulatory functions are evaluated in a single step, whereas in the asynchronous update, only one regulatory function is evaluated at one update step. Recently, [54] have introduced new dynamics for Boolean networks, which was shown to be flexible enough to represent (a)synchronous dynamics as well as multi-level formalisms, that is, beyond boolean values for gene expression.

Building a Boolean Network from Scratch. A detailed state-of-the-art is available in Appendix A.1. Our work focused on combining several data sources and methods for the design of an end-to-end pipeline for Boolean network synthesis, as represented in Fig. 1. This network models the regulatory dynamics on a subset of genes, in the absence of external perturbation, in a well-chosen cell line; for instance, the regulations between M30 genes in brain cell lines for our application to epilepsy. Contrary to the contemporaneous work of [47] applied to cancer, here we do not have any access to a generic model which could model any type of epilepsy to start with. Moreover, relying too much on prior epilepsy-oriented knowledge might lead us to find already known gene candidates, whereas finding novel master regulators might help investigating refractory epilepsy. The big picture of this pipeline comprises the following three main steps in chrono-logical order, respectively denoted (A), (B) and (C) in Fig. 1:

(A) Data Collection. Step **(A)** encompasses the collection and filtering of information from public, large databases: measurements of transcriptomic data are retrieved from the LINCS L1000 database [65] using careful filtering and qual-ity control measures; known unsigned, undirected protein pairwise regulatory interactions involving genes in M30 are obtained from the STRING database [66].

(B) Data Processing. Then, step **(B)** comprises the processing of this infor-mation into appropriate inputs for the inference of Boolean networks. First, a set of binarized phenotypes is built, corresponding to profiles from single gene perturbations and their associated controls, from LINCS L1000. Then, a signed network of possible valid regulatory interactions is constructed from the protein-protein interactions in STRING, by filtering out and signing edges based on gene pairwise expression correlations computed on LINCS L1000 profiles.

(C) Network Inference. Finally, step **(C)** starts by the inference of a set of Boolean networks which satisfy all the experimental and topological constraints given by the phenotypes and the signed network. The experimental constraints comprise knockout or overexpression experiments, where the control phenotype is considered the initial condition, and the perturbed phenotype the final con-figuration reached after the gene perturbation.

A Boolean network solution should satisfy all of these time-series constraints by only considering a set of regulatory interactions present in the signed network. The final step of the procedure is the selection of an optimal Boolean network among these solutions, according to its topology. This final inferred network is selected through a desirability function maximization [5], which depends on several topological measures. Details about the implementation and tools are available in Appendix A. Note that our method can also guess an appropriate gene subset associated with a given disease, although this issue did not arise in our application to epilepsy. If no gene set is provided, the method automati-cally retrieves genes from DisGeNet [57] using the disease Concept ID (CID) [21]. Table 7 in Appendix reports the values used to filter out genes from the DisGeNet database. The single network obtained at the end of step **(C)** is a dynamical sys-tem which can predict the behavior of gene expression under one or several gene perturbations, by considering the stable states (attractors and cycles) reachable from a given initial state under these perturbations.

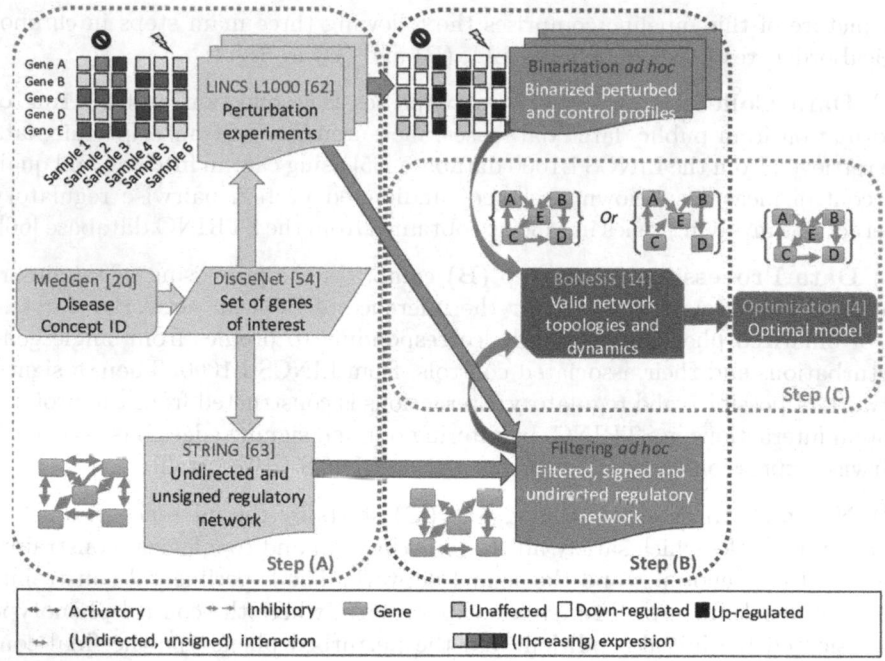

Fig. 1. Overview of the Boolean network identification pipeline. From left to right, step **(A)** (data collection), **step (B)** (processing of gene expression and interaction data), and step **(C)** (inference of an optimal valid network topology and dynamics). Databases and methods used in the pipeline (with their associated references) are shown at the top of blocks.

We now dwell on how to use this model to rank genes according to their regulatory influence on the remainder of the network.

2.2 Detection of Master Regulators in a Specific Disease-Context

As mentioned in Sect. 1, when looking for therapeutic candidates, one might be interested in master regulators, that is, genes at the top of the gene regulation hierarchy, which change in expression induces the largest downstream gene expression change; for instance, by encoding for a transcription factor which affects the transcription of other genes [45]. In practice, it is frequently quantified using the node (outgoing) degree and detection of hub nodes in the network. Many measures defining such a "centrality" property can be computed using Cytoscape [59], through modules NetworkAnalyzer [4] and CytoCtrlAnalyser [74]: for instance, control centrality [44], which has been recently used to identify regulations between *NFATC4* and Type 2 diabetes-associated genes [60]. However, these measures only use the topological information in the network, whereas our network inference pipeline allows –along with the identification of regulatory connections– the inference of gene regulatory functions, adding interesting dynamical information. This is why we designed a master regulator detection method which leverages this information to model regulatory cascades. In [27], a Machine Learning technique called "influence maximization" was exploited to identify key genes in a continuous model of the yeast regulatory network. In our work, we adapted influence maximization to Boolean networks. For long, online recommendation and advertising researchers have been interested in influence maximization [37], which aims at finding a node subset of fixed size which influences most the remainder of the network. In order to make this technique applicable to Boolean networks, we need to explicitly define the concept of influence on gene expression in these models. This influence is called "spread process", and is the quantity that propagates along the edges of the network from any node. We define influence in an iterative way; first we consider a single gene and one initial network state. Then, we proceed to define multi-gene influence starting from an initial state. We eventually describe influence for several genes across a set of initial network states.

Genewise Influence in a Boolean Network. The most intuitive definition of influence –denoted in the remainder of the paper "spread value"– $I(n)$ of a given node n on the other nodes in a Boolean network with initial state i, would be that any perturbation of this node would "greatly change" the attractor states reachable from state i, compared to attractor states reachable from that state in the absence of perturbation. We define this great change, yielding a positive spread value, by the fact that those two sets of attractors have an empty intersection. Let us denote $\mathcal{A}(i, P)$ the set of attractor states reachable from state i, under the set of perturbations P. Set P contains pairs of gene names and their associated perturbation (either 0 for knockout, or 1 for overexpression). Let us also denote $\mathcal{O}$ the set of output genes, that we define here as the set of genes with a positive ingoing-degree, and a given similarity measure $\mathcal{S}$ between network states. Then, we define the spread value for node n, initial state i, and Boolean network $\mathcal{B}$ as the *minimum* change incurred by the perturbation of n

$$\mathrm{SV}_{\mathcal{B}}(\{n\}, i) = 1 - \max\left\{\mathcal{S}\left(a^1_{|\mathcal{O}}, a^2_{|\mathcal{O}}\right) : a^1, a^2 \in \mathcal{A}(i, \emptyset) \times \in \mathcal{A}(i, \{(n, \neg i[n])\})\right\},$$

where $a^1_{|\mathcal{O}}$ and $a^2_{|\mathcal{O}}$ are the respective restrictions of network states a^1 and a^2 to the set of output genes $\mathcal{O}$. The perturbation denoted by $(n, \neg i[n])$ means that gene n is perturbed in the opposite direction to its expression state $i[n]$ in i: for instance, if n is expressed in state i, then we consider knockouts of gene n. The restriction to output genes in $\mathcal{O}$ is actually important in order to have consistent results when considering isolated nodes.

Note that, if n does not have a determined expression state in initial state i, we set the associated perturbation set to $\emptyset$. This implies that some genes with individual spread value equal to 0 can either have no true influence on the network, or have no determined expression state in initial state s, which means that they are not measured during the generation of transcriptomic data. In the latter case, this most likely means that the gene is not expressed in the considered cell line(s). Most importantly, $\mathrm{SV}_{\mathcal{B}}(\{n\}, i)$ is equal to 0 if and only if

$$\mathcal{A}(i, \emptyset) \cap \mathcal{A}(i, \{(n, \neg i[n])\}) \neq \emptyset,$$

that is, if there is any attractor in common. However, if the intersection is empty, $SV_\mathcal{B}(\{n\}, i)$ is not necessarily equal to 1, as reachable attractors might still be close to those obtained without any external perturbation.

Geneset Influence in a Boolean Network. When considering a set $\mathcal{N}$ of nodes instead of a single node n, the influence of $\mathcal{N}$ is the spread value computed over all attractors reachable under simultaneous perturbations of these nodes

$$SV_\mathcal{B}(\mathcal{N}, i) = 1 - \max \left\{ \mathcal{S}\left(a^1_{|\mathcal{O}}, a^2_{|\mathcal{O}}\right) : a^1, a^2 \in \mathcal{A}(i, \emptyset) \times \mathcal{A}(i, \{(n, \neg i[n]) : n \in \mathcal{N}\}) \right\}.$$

Aggregation of Values for Several Initial States. Finally, if we consider a whole set of initial states $\mathcal{I}$ and a gene set $\mathcal{N}$, the associated influence is defined as the geometric mean of spread values across initial states

$$SV_\mathcal{B}(\mathcal{N}, \mathcal{I}) = \left(\prod_{i \in \mathcal{I}} (SV_\mathcal{B}(\mathcal{N}, i) + 1) \right)^{1/|\mathcal{I}|} - 1,$$

where $|\mathcal{I}|$ is the number of initial states. Note that we need to correct for zeroes to order to avoid the collapse of this measure when one perturbation does not trigger a change in reachable attractors for one of the initial states, while keeping spread values between 0 and 1 for better interpretability.

Once the spread process is defined, we propose the following greedy influence maximization algorithm to prioritize master regulators.

Algorithm 1. Greedy Influence Maximization Algorithm for Boolean Networks

---

Input: $\mathcal{B}$ a Boolean network on node set V ; K the minimal number of simultaneous perturbations on the network ; $\mathcal{I}$ set of initial Boolean states
Initialize $\mathcal{N} = \emptyset$, $k = 0$
repeat
    $k \leftarrow k + 1$
    # Adding to set $\mathcal{N}$ nodes that maximize the spread value

$$\mathcal{N} \leftarrow \mathcal{N} \cup N_k, \text{ where } N_k \leftarrow \arg \max_{n \in V \setminus \mathcal{N}} SV_\mathcal{B}(\mathcal{N} \cup \{n\}, \mathcal{I})$$

until

$$k = K \text{ or } \max_{n \in V \setminus \mathcal{N}} SV_\mathcal{B}(\mathcal{N} \cup \{n\}, \mathcal{I}) \leq SV_\mathcal{B}(\mathcal{N}, \mathcal{I})$$

Output: $\mathcal{N}$

Influence Maximization Algorithm on Boolean Networks. We describe how to leverage spread values to identify master regulators in the network. Current literature on influence maximization [55], which is a NP-hard problem, relies on the fact that the spread function is submodular: roughly, as the considered subset increases, the difference in the value of this function due to adding another single element to the subset decreases. However, no such property can be assessed for the definition of spread defined in the previous paragraph. We then slightly adapted the greedy algorithm in [37] in Algorithm 1. This algorithm determines the set of nodes of minimal size K which are the most influent, where K is a predefined fixed value. It goes as follows: starting from an empty set of nodes $\mathcal{N}_0$, a fixed set of initial states $\mathcal{I}$, and a Boolean network $\mathcal{B}$, at each step $k \in \{1, 2, \ldots, K\}$, the algorithm selects the node $n \notin \mathcal{N}_k$ which maximizes spread value $SV_\mathcal{B}(\mathcal{N}_k \cup \{n\}, \mathcal{I})$ and computes the set $\mathcal{N}_{k+1} = \{n\} \cup \mathcal{N}_k$. The algorithm stops at $k = K$, or at the first step k when the spread value $SV_\mathcal{B}(\mathcal{N}_k, \mathcal{I})$ is no longer increasing, that is,

$$\max\{SV_\mathcal{B}(\mathcal{N}_k \cup \{n\}, \mathcal{I}) : n \notin \mathcal{N}_k\} \leq SV_\mathcal{B}(\mathcal{N}_k, \mathcal{I}).$$

This condition is necessary to compensate for the fact that the function might not be submodular. If, at a given step k, several nodes maximize the spread value, they are all added to set $\mathcal{N}_{k+1}$. The iteratively built set $\mathcal{N}_K$ is then the set of possible K-sized gene subsets to simultaneously perturb on the network, such that the set of attractors reachable from initial set $\mathcal{I}$ is greatly modified. In this work, $K = 1$, that is, we only looked at individual contributions of genes to the changes, and we ranked gene n among genes in the network according to its spread value $SV_\mathcal{B}(\{n\}, \mathcal{I})$.

Finally, we explicit the implementation in the application to epilepsy:

Set of Initial Network States ($\mathcal{I}$). We consider transcriptomic profiles from human hippocampi afflicted with temporal lobe epilepsy (TLE) in [46] for the initial states, such that genes are ranked according to their influence in an epileptic context. Temporal lobe epilepsy is one of the most common forms of partial epilepsy, where seizures affect one part of the brain, and is often associated with cases of refractory epilepsy that cannot be surgically treated [30]. Details about the implementation and initial states are available in Appendix E. 207 genes out of 232 genes both from the M30 module and present in the network are mapped to expression levels in these states, which means that for these genes, we are sure that any spread value equal to 0 for any of these genes truly means that the gene has no influence over the remainder of the network.

Similarity Between Attractor States ($\mathcal{S}$). The definition of the spread process relies on a similarity function $\mathcal{S}$ defined between two network states, that was left to be defined. In our implementation, we wanted to compute the differences in the presence of ones *and* zeroes, which prevented us from directly using Jaccard's score. Based on previous surveys of the state-of-the-art on binary distances [16], we implemented a "normalized" ℓ_1-norm distance. That is, if a^1 and a^2 are the two binary vectors to compare (of size d), then the resulting similarity $\mathcal{S}(a^1, a^2)$ between a^1 and a^2 is:

$$\mathcal{S}(a^1, a^2) = 1 - \frac{1}{d} \sum_{i=1}^{d} |a^1[i] - a^2[i]|.$$

This expression is exactly the ratio of row-wise equal coefficients in a^1 and a^2 to the total number of coefficients, and yields 1 when $a^1 = a^2$, and 0 for $a^2 = (a^1 + 1) \equiv [2]$ (modulo 2). It penalizes in a symmetric way differences in ones and zeroes.

Genericity of the Method. Note that this methodology, combining the synthesis of a Boolean network and influence maximization, can generically be applied to any disease. To adapt this pipeline to another disease, one needs to change the gene subset and the cell line(s) on which the network should be built, as well as the set of initial network states for the detection of master regulators. In the application to epilepsy, we considered the M30 gene module, and the two brain cell lines present in the LINCS L1000 database. Code for the synthesis of the Boolean network and the detection of master regulators is available at https://github.com/clreda/PrioritizationMasterRegulators.

3   Results

3.1   Networks Obtained from the Inference Procedure

We discuss the network solutions resulting from step **(C)**.

The Final Network Compiles Several Sources of Regulation. The final network obtained at the end of step (C) is shown in Fig. 2. In this figure, nodes are colored by their degree; the darker the color, the higher the degree. Edges in Fig. 2 are colored according to their source of evidence as reported by the STRING database [66]. One can notice that there are a lot of undirect gene-to-gene regulatory interactions in this network. This actually is not very surprising, since few gene pairwise interactions are experimentally tested compared to all possibly existing interactions. Moreover, our model does not aim at taking into account exclusively transcriptomic interactions, but possibly non-physical, post-transcriptomic effects. This network is not connected (36 weakly connected components), which might be explained by the weak regulatory role of genes from the smaller components in the considered perturbation experiments.

The Synthesis Outputs Similar Network Solutions. Now, we consider all 25^2 network solutions generated at step **(C)**. We assess how far they are from each other, in terms of node degree distribution, edge numbers, redundancy in interactions, unicity of regulatory functions for each node across those solutions, and values of general topological parameter (GTP). GTP is a value comprised between 0 and 1 that is used to select the final network among the 25 ones (as further described in Subsect. A.7 in Appendix) and characterizes the proximity of a network topology to a scale-free-like one. Table 1 shows distribution statistics about the values of GTP and the unicity of gene regulatory functions across solutions. Note that all solutions present similar topologies, with similar GTP

---

[2] That number was chosen for reasons related to computational cost and time. Note that in Appendix we discuss how adding 25 additional network solutions neither changes the final network, nor the conclusions made in this section.

scores quite close to 1, which matches what can be expected from biochemical interaction networks in non-fungi systems [11]. Moreover, except for less than 25% of the genes in the network, genes are assigned at most 3 different regulatory functions across all solutions, which shows that their function in the network is globally preserved. Figure 3 displays the boxplots of edge number and node degree distributions across solutions. These two plots show that, as mentioned before, the typical scale-free topology, with a few "hub nodes" with large degree and a large number of genes with few regulatory interactions, is present in all solutions. There are 74 interactions (that is, around 30 − 34% of edges) which are present in at least 75% of the solutions, among which 25 are present in all of them. They are shown in Table 8 located at the Appendix. These numbers are confirmed by plotting the network comprising of all genepairwise interactions which are present in at least one solution, shown in Fig. 4. All in all, the networks obtained just before the model selection step mostly seem similar, both functionally – at the level of regulatory functions– and topologically –considering the node degrees, the number of edges, and the GTP scores.

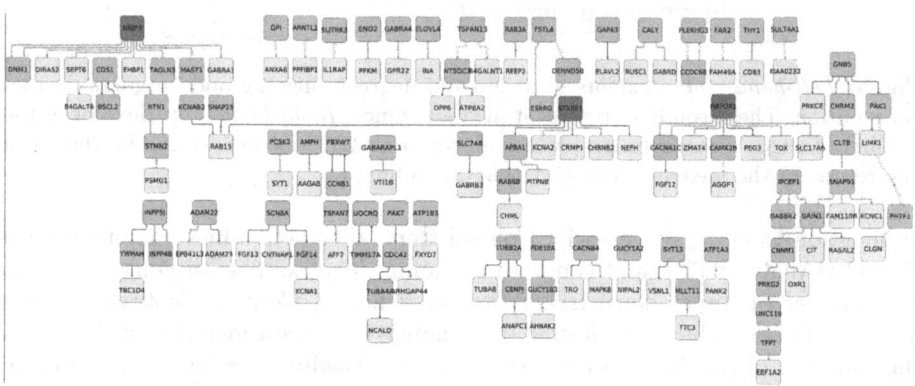

Fig. 2. Inferred network resulting from our pipeline applied to the M30 gene module. Tee-headed arrows represent inhibitory regulatory interactions, whereas triangle-headed arrows are activatory regulations. Edges are drawn according to their source of evidence (on the *undirected* interactions) as reported by the STRING database: contiguous arrows denote coexpression, solid line denote experimental proofs of interaction, and sinewave interactions are derived from text-mining procedures. Gene nodes are colored according to their out-degree: lightest color for genes with outdegree equal to 0, darkest for genes with an outdegree higher than 5. Isolated nodes are not shown. (Color figure online)

3.2   Recommended Master Regulator Candidates

We now study the master regulator candidates ranked by spread value.

Spread Values Are Correlated With Network Centrality and Gene Pathogenicity. We computed the correlation between spread values for our

Table 1. Distribution statistics on the number of *unique* regulatory functions (RFs) across solutions per gene, and on the value of the general topological parameter (GTP) used for network selection in step (C) of the inference procedure.

	Min.	25th quantile	Median	Mean	75th quantile	Max.
# RFs	1	1	2	2.202	3	11
GTP	0.796	0.798	0.800	0.800	0.800	0.802

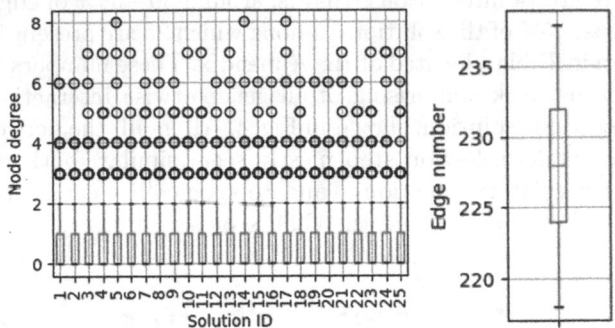

Fig. 3. *Left-hand plot*: Boxplots of node total degrees (ingoing and outgoing degree) per solution. The green lines represent median values. *Right-hand plot*: Boxplot of the number of edges across solutions (which all comprise 232 M30 genes). Again, the green line represent the median value. (Color figure online)

application to epilepsy, genewise Control Centrality [44] values computed with CytoCtrlAnalyser [74], and genewise outgoing degrees. The outgoing degree is the number of direct downstream targets, whereas Control Centrality is the number of nodes which are affected by a change in the considered node, based on the topology of the directed network. More specifically, in order to compute the Control Centrality for any gene g, at some time step t, (continuous) expression levels $x(t) \in \mathbb{R}^N$ are time-invariant and depend linearly on those at the previous time step $t-1$ $x(t-1) \in \mathbb{R}^N$, where N is the number of genes in the network

$$\frac{\partial x(t)}{\partial t} = Ax(t) + u_g(t), \tag{2}$$

where A is the adjacency matrix in $\mathbb{R}^{N \times N}$ associated with the network, and $u_g(t) \in \mathbb{R}$ is the external signal imposed on node g at time t (either overexpression if it is positive, or knockout otherwise). In such a system, computing the number of nodes which can be controlled by gene g boils down to getting the rank of the so-called controllability matrix related to A and g, that is a function of powers of matrix A. This rank can be computed by solving a combinatorial optimization problem described in Equation (3) in [44]. Moreover, since the true nonzero values in A as well as $u_g(\cdot)$ are often unknown, Control Centrality aims at quantifying *structural* controllability, independently from the values of

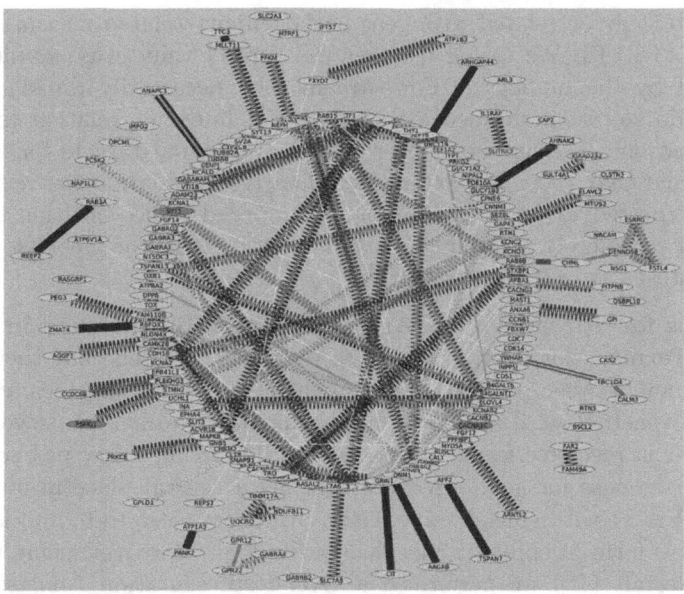

Fig. 4. Network comprising of all gene-to-gene interactions which are present in at least one solution. The darker and thicker an edge is, the more frequent it is across solutions. Sinewave edges are inhibitory interactions, whereas solid lines denote activatory interactions. Orange nodes correspond to the genes which are perturbed in the LINCS L1000 experimental profiles used for inference. (Color figure online)

nonzero coefficients in A and $u_g(\cdot)$. All in all, Control Centrality is a solid counterpart to our method. It does not take into account neither the set of regulatory functions nor the gene expression levels in patients, but models regulatory cascades through the differential equation in Eq. 2. We also compared spread values to scores associated with the pathogenicity of genes:

- probability of loss of function intolerance (pLI) [40], which quantifies the intolerance to the loss of function of a given gene in patient and control cohorts.
- enhancer-domain score (EDS) [71], which studies the conservation of the regulatory domain around genes.
- residual variation intolerance score (RVIS) [56], which is related to the presence of functional genetic variation in patient exomes, and is anticorrelated with gene pathogenicity.

Finally, we computed "TF" influence scores [50] as well, which expression is reported in Eq. 1. Figure 5 displays the correlation heatmap between these different measures. We observed that, contrary to (TF) influence values, spread values were consistent and strongly correlated with network controllability measures, that is Control Centrality and the outgoing degree. Moreover, spread values are more strongly correlated with gene pathogenicity-related measures pLI and (opposite of) RVIS. We tested whether the spread value was actually totally determined by the number of downstream (not necessarily direct) regulated genes. To do so, we performed a Spearman's ρ linear correlation test on the spread values and the number of downstream regulated targets for each gene. We confirmed that there is a strong, significative correlation between the two –which is expected, given the definition of the spread value– but that the spread value is not completely determined by this value; that is, the associated statistic is not equal to 1 ($\rho = 0.82$, $p = 3.10^{-57}$).

Top Genes for Spread Values Are Significatively Enriched in Disease-Related Terms. Moreover, from Fig. 6, it can be noticed that there is a lot of discrepancy between pLI scores and spread values on M30 genes (correlation $\rho \approx 0.11$). Nonetheless, it should be noted that [77] warns against genes which are involved in recessive forms of diseases, while having a low pLI score. That is actually the case for gene *GNB5*, which has a central place in our network (shown in Fig. 1) with spread value 0.024, pLI score close to 0, and is involved in a recessive form of epileptic encephalopathy [58]. Moreover, using the online tool WebGestalt [41], we performed a Over-Representation Analysis (ORA), in order to check if the shortlist of 14 genes with spread value greater than 0.01 was significantly enriched in epilepsy-associated terms, compared to the 232 genes present in the network. The disease terms were annotations from the DisGeNet database [57].[3] Indeed, this shortlist is (weakly) significantly enriched in genes related to the term "Epileptic encephalopathy" at level 5% (odds ratio $OR = 7.5$, Benjamini-Hochberg (BH) [7]-adjusted $p \approx 0.038$), and more strongly enriched with (neuro)developmental issues, for instance, "Loss of developmental

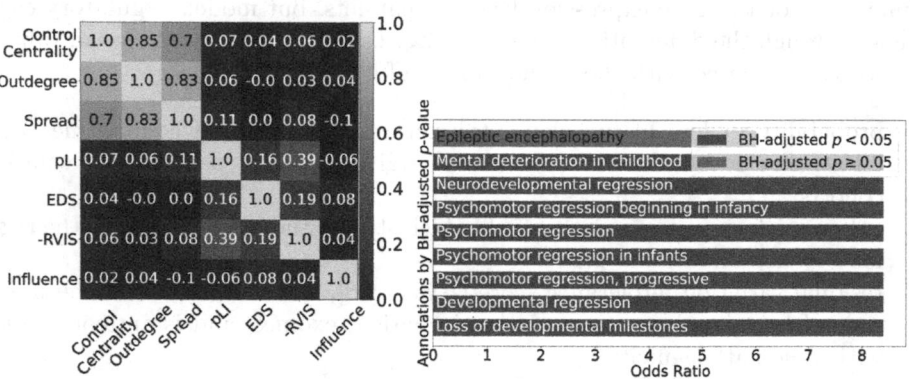

Fig. 5. *Left*: Spearman's ρ correlation heatmap between different gene measures either related to the influence of a node on a network, or to the genetic variations associated with pathogenicity. *Right*: Enrichment results from the ORA analysis. All reported adjusted p-values are lower than 20%.

---

[3] Remember that, in the application to epilepsy, we *did not* use genes from DisGeNet, but the preselected set of genes M30.

milestones" ($OR = 10.5$, BH-adjusted $p \approx 0.012$), as reported in Fig. 5. Similar results can be observed on another family of gene annotations, GLAD4U [34], as shown in Fig. 11 in Appendix. The considered shortlist of genes is shown on Fig. 6. These enrichment results go beyond the fact that M30 is globally enriched in epilepsy-related *de novo* mutations compared to the *whole* measured genome in brain cell lines, as shown in [19]: what is shown is that, among genes *in the M30 module*, ranking by spread values still prioritized interesting genes.

Most Top Genes Cause Human Epilepsies. Based on the results shown in Fig. 5 and Table 2, a shorter list of candidate genes is selected. It comprises genes with rather large spread value (greater than 0.01) and pLI score (greater than 0.9), *and* of genes with very large spread value (greater than 0.02). The last condition holds in order to avoid the previously mentioned shortcoming incurred by pLI scores. These candidate genes are *CACNA1A*, *RBFOX1*, *STXBP1*, *DNM1*, *NRIP3*, *SCN8A*, *CHRM2*, *GNB5*, *TUBB2A*, *PAK7*, and *GRIN1*, shown on Fig. 6. Most of these candidates –except for *NRIP3*, which is notably mainly expressed in the hippocampus– have a relationship to epilepsy-related symptoms in humans shown in prior works [2,3,12,18,39,49,52,58,63], as expected due to their membership to the M30 module. Some of these genes may have never been investigated in the research related to epilepsy, such as *NRIPS3*. For other genes, for instance, *STXBP1* and *GRIN1*, knockouts of orthologuous genes were associated with epileptic seizures in zebrafish [29].

Table 2. Distribution statistics (rounded up to the 5th decimal place) of the spread values obtained for M30 genes present in the inferred network.

Minimum	25th quantile	Median	Mean	75th quantile	Maximum
0.0	0.0	0.0	0.00254	0.0	0.0556

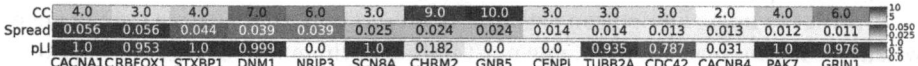

CC	4.0	3.0	4.0	7.0	6.0	3.0	9.0	10.0	3.0	3.0	3.0	2.0	4.0	6.0
Spread	0.056	0.056	0.044	0.039	0.039	0.025	0.024	0.024	0.014	0.014	0.013	0.013	0.012	0.011
pLI	1.0	0.953	1.0	0.999	0.0	1.0	0.182	0.0	0.0	0.935	0.787	0.031	1.0	0.976
	CACNA1C	RBFOX1	STXBP1	DNM1	NRIP3	SCN8A	CHRM2	GNB5	CENPJ	TUBB2A	CDC42	CACNB4	PAK7	GRIN1

Fig. 6. Genes ranked by decreasing spread value, restricted to spread value greater than 0.01 (*center bar*), with their associated Control Centrality (CC) (*top bar*), and pLI scores (*bottom bar*).

4 Discussion

We introduced in this work two main contributions to *in silico* disease research. First, we designed a method for the automated inference of a gene regulatory network from scratch, starting from a subset of genes. This method carefully combines information from several public databases and methods, and infers a dynamical model of gene regulation adapted to specific cell lines. This method yields quite robust network solutions, and is easily reproducible. Then, we showed how to exploit this dynamical system to detect master regulator genes. We applied our methodology to investigate epilepsy, and to find novel candidate genes to hopefully tackle drug-resistant epilepsy. A list of candidate genes was prioritized, which perturbations greatly impact the whole network in an epileptic transcriptomic context. This methodology allows reproducible and transparent research, while reducing the amount of data needed as input, which is one of the main caveats of researching on rare or tropical neglected diseases.

Acknowledgements. This work was supported by the Institut National pour la Santé et la Recherche Médicale (Inserm, France), the French Ministry of Higher Education and Research (#ENS.X19RDTME-SACLAY19-22) (C.R.), Université Paris Cité, Université Sorbonne Paris Nord, the French National Research Agency (#ANR-19-CE23-0026-04) (C.R.), (#ANR-18-CE17-0009-01) (A.D.-D., C.R.), (#ANR-18-CE37-0002-03) (A.D.-D.,C.R.). The supporting bodies played no role in any aspect of study design, analysis, interpretation or decision to publish this data. Authors thank the anonymous reviewers for their feedback.

A   Building the Boolean Network

This section of the Appendix further describes the procedure used to infer a Boolean network in our application to epilepsy.

A.1   State-of-the-Art in Boolean Network Inference

Many approaches to infer Boolean networks associated with a given biological pathway or disease mechanisms already exist. For instance, a Boolean network can be straightforwardly built from a known static gene regulatory network, or maps of molecular interactions, by automatically assigning gene regulatory functions based on the activators, resp. the inhibitors, of each gene, for instance using CaSQ [1].

In order to infer the Boolean network "from scratch", that is, from a set of genes when a static map is unknown, one must define both an interaction graph and an set of gene regulatory functions which match experimental observations under a selected update type, as described in the main text. One way to find the appropriate boolean network requires to know a set of possible genepairwise interactions (sometimes known as the "prior knowledge network" [53,69]) and a set of time-series (binarized) transcriptional profiles which (hopefully) exhibits enough of the true dynamics of the biological system. Then gene regulatory functions are tuned until all the experimental constraints are satisfied by the model.

However, there are two main hurdles in this inference problem. First, the binarization step to go from transcriptional profiles of continuous gene expression to binary profiles (with potentially genes which have no known binary state, if their expression is not high or low enough) incurs an unavoidable loss of information, and should be performed carefully. Binary profiles might also been built using prior knowledge on the expected behavior of the system, for instance, the phenotypes that should be observed under specific perturbations, but this requires supplementary access to custom wet-lab experiments or to the literature, at the risk of propagating scientific bias. Second, the problem of inferring a network, which satisfies both the topological (related to interactions) and the dynamical (related to observations) constraints, is usually underdetermined.

As the number of genes in the network grows, so does the amount of experimental data needed in order to pinpoint a single model which can satisfy all the constraints. This fact makes unbiased model building difficult when done manually. Some tools allow the automation of the tuning step, such as Re:In [22] or BoNeSiS [15]. These methods are based on enumerating putative network solutions, and testing whether the network satisfies all constraints using a solver of Boolean equations (SAT solver).

This state-of-the-art justifies our first contribution in this paper, that is, a pipeline which automatically connects known databases and methods to identify a disease-associated Boolean network, with novel and data-adaptive binarization method of gene expression and filtering of regulatory interactions.

A.2   Step (A): Building an Undirected Unsigned Graph

This step builds an undirected, unsigned network of putative gene-to-gene regulations. The gene module M30, as defined by [19], comprises 320 genes which global expression anticorrelates with epileptic phenotypes. We retrieved all 320 genes of the M30 module from the additional file 1 in [19]. Undirected and unsigned protein-pairwise interactions involving two proteins encoded by M30 genes were extracted from the STRING database [66] for the human (NCBI taxon ID 9606). In order to perform the inference, it is necessary for computational reasons to restrict the set of edges to consider; however, (weak) connectivity in the graph of interactions should also be preserved to fully exploit the dynamical constraints provided later on.

Considering the full network retrieved from the STRING database, we trimmed out isolated genes –*i.e.*, without any interactions with any gene, not even themselves. 318 genes out of 320 were retained after this procedure, with a total of 14, 662 edges. The STRING database also provides scores associated with each undirected edge named "combined scores", which are comprised between 0 and 1, 000, and aggregate various scores related to the type of evidence supporting these edges [70]. The higher this score is, the more strongly supported the associated edge is. Provided a user-provided threshold η on this combined score,

(a) First, we built a protein-protein interaction (PPI) network by preserving all edges with a combined score greater than η;

(b) Then, all edges in the initial network from STRING that do not appear in the PPI at the previous step, and which involve at least one isolated gene in the PPI, were sorted in the order of decreasing combined score, and added them sequentially (adding simultaneously edges with the same score) to the network, until the number of weakly connected components is 1. We tested (weak) connectivity by performing a Depth-First Search [17], which is a well-known procedure that explores all the nodes in a graph by favoring the exploration of child nodes instead of sibling nodes, until all nodes have been visited.

In order to select the threshold $\eta = 400$, we performed a grid search on $[\![100; 1,000]\!]$ with a step of 5, and selected the value which minimized the number of edges. The pseudocode for parameter selection is shown in Algorithm 2, whereas the plot of the number of edges depending on the value of η is displayed in Fig. 7. This step is automatically performed the first time that the inference procedure in the repository[4] is run, such that the user can use the threshold value η recommended by the grisearch. Choosing $\eta = 400$ allowed reducing the number of undirected edges from $14,662$ to $1,633$.

```
eta = float("inf")
value = float("inf")
for e in range(100,1000,5)
        # Application of
        # steps (a), (b)
        # with eta=e
        ppi = filtering(e)
        # Number of edges
        v = len(ppi)
        if (v < value)
                value = v
                eta = e
return eta
```

Algorithm 2: Pseudocode in Python for the Grid Search for Parameter η.

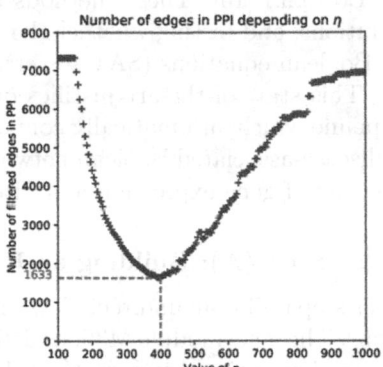

Fig. 7. Number of edges (after filtering) depending on the choice of η, and value when $\eta = 400$ (minimizer of the number of edges).

A.3 Step (A): Gene Perturbation Experiments

At this step, we restricted the set of genes –subsequently, of interactions– to genes present in the database of transcriptional profiles LINCS L1000 [65]. To do so, we converted all gene identifiers in M30 into EntrezGene IDs using BioDBnet [48]. Then, we filtered out genes for which the EntrezGene ID was not present in LINCS L1000. After this step, 236 genes, out of 318, were retained. We selected experiments present in LINCS L1000 such that at least one gene from M30 has been perturbed in a genetic experiment (knockdown or overexpression, along with control samples) on a brain cell line. Unfortunately, there are no hippocampal neuron human (HN-h) cell lines in LINCS L1000; These cells are able to differenciate into neurons and glial cells as shown in the rat [24], which is why we assumed that the neural progenitor cell (NPC) line present in LINCS L1000 might be appropriate. Furthermore, among the genetic perturbation experiments listed in the database, we selected those which satisfy all following conditions.

– which are associated to the highest metric distil_ss (as provided by the LINCS L1000) which is correlated with the number of significantly differentially expressed transcripts found in the differential analysis between the

---

[4] https://github.com/clreda/PrioritizationMasterRegulators.

matching genetically treated and the control groups. In practice, this measure is correlated with the reproducibility of a drug signature [42].
- where there is at least two replicates from the same plate for the perturbed and control (of type ctl_vector) conditions.
- which interference scale (computed as described in [14]) is positive. This ensures that the associated genetic perturbation experiment was successful. That means that a gene which has been perturbed by a knockdown (resp., an overexpression) has an expression lower (resp., greater) in treated profiles than in controls, compared to an appropriate housekeeping gene. The expression levels in these housekeeping genes should not dramatically change in both groups of profiles.
- where the associated experiment is either using shRNA (knockdown perturbation), cDNA, also known as knock-in (overexpression perturbation), or CRISPR (knockout perturbation).
- where the associated cell line is either SHSY5Y (neuroblastoma) or NPC (neural progenitor cells), which are the only brain cell lines in LINCS L1000.

The result of this step is a matrix of M30 genes by experimental profiles, which contains Level 3 LINCS L1000 data (normalized expression data for the whole genome) for each perturbation experiment. See Table 5 in Appendix for the list of experimental profiles retained in the application to epilepsy.

A.4   Step (B): Binarization of Experiments into Binary Profiles

Although there are known methods for the binarization of (single) RNA-seq data [6,25], probably due to the fact Level 3 LINCS L1000 data is a combination of measured and inferred expression data, for different platforms (RNA-sequencing data for the most recent version, microarray for the first generated profiles), there were issues with the model fitting; only a few genes were assigned a binary value 0 or 1 –the alternative being that they are not considered expressed "enough", according to the thresholds computed by these methods, to be assigned a state equal to 1, nor too weakly expressed to be assigned a state equal to 0). A data-driven method to tune the granularity of the binarization, adaptive to the selected perturbation expression data, was necessary in order to explicitly enforce a trade-off between a full reliance on undirected edges provided by the STRING database, and on experimental profiles from LINCS L1000.

Background Expression Data. However, this method relies on having enough data to compute reliable statistics of expression for each gene, which is why, for each cell line, we automatically retrieved from LINCS L1000 a "background" expression matrix, which we concatenated to the set of profiles before binarization. After binarization, we removed samples associated with the background dataset. In order to collect the background expression matrix, we selected all experiments in the considered cell line, with type pert_sh (knockdown experiments), and we filtered out experiments with less than two replicates, with metric distil_cc_q75 greater or equal to 0.2, and with metric pct_self_rank_q25 lower than 0.05. Metrics distil_cc_q75 and `pct_self_rank_q25` are two measures associated with experimental profiles which quantify the reproducibility based on the correlation between the same technical replicates (`distil_cc_q75`) and the diversity of profiles for a given experimental setting (`pct_self_rank_q25`). These rules correspond to the requirements for reproducible and distinct (so-called "gold") profiles according to LINCS L1000 documentation. Finally, we selected the same-plate replicates with the highest value of `distil_ss`.[5]

Binarization. Then, the following *ad hoc* binarization procedure was independently applied to experimental and background profiles from the same cell line. Gene expression data (as normalized RNA counts) was first quantile-normalized and clipped to the interval $[0, 1]$. Control samples (for the same cell line) were aggregated by considering the genewise median expression value. Given the threshold ζ, all genes with expression greater than $1 - \zeta$ were considered greatly expressed (with assigned state 1), whereas genes with expression lower than ζ were considered non-expressed (with assigned state 0). Genes which expression levels were in the interval $[\zeta, 1 - \zeta]$ had an undetermined expression state. Finally, background profiles are dropped. The higher ζ is, the more constrained the experiments are, as more genes have a determined expression state 0 or 1. Lower thresholds mean less constrained experiments, and a higher preference for the regulatory interactions filtered from the STRING database over expression data from LINCS L1000. Using a bisection method in interval $[0; 0.5]$ with precision 0.0005, we identified $\zeta = 0.265$ as the maximum threshold such that the inference of Boolean networks satisfying these experimental constraints admits at least one solution. We recommend using this bisection method to determine the threshold ζ when using the pipeline with another dataset.

One might wonder why Level 3 LINCS L1000 gene expression data is quantile-normalized *once again*, or why Level 4 data (which already are z-scores) had not been used instead. Actually, quantile normalization at Level 3 and differential expression (z-scoring) at Level 4 is performed *at well plate level*, that is, across sample profiles from the same plate [65]. In both cases, that means that the gene expression data is not normalized *across plates in the same cell line*, which is what we want to achieve. Based on the work in [76], we apply the normalization "by class", here, the cell line.

A.5   Step (B): Implementation of Topological Constraints

The inference of a Boolean network relies on a set of admissible interactions and a set of time-series expression constraints. Indeed, solution networks only comprise a subset of these admissible interactions, such that all constraints provided by the observations are satisfied.

In order to build the set of admissible interactions, we considered the PPI network extracted from the STRING database. Since these interactions are unsigned, we decided to reduce the number of possible interactions –and thus, the computational cost of the method– by using the gene perturbation expression matrix retrieved from LINCS L1000 (Table 5)

---

[5] These measures are further described at https://clue.io/connectopedia/glossary.

- First, a Pearson's r [10] gene correlation matrix was computed from these profiles, and raised to the power of $\beta = 1$ coefficient-wise, which allowed signing the interactions using pairwise correlation signs.
- Then, to preserve connectivity, we built the filtered signed undirected network similarly to what we previously did, using a threshold on the correlation values equal to $\tau = 0.4$.

β was chosen as it is known that raising an adjacency matrix A to the power of β yields coefficients $A[i, j]$ in position (i,j) equal to the number of paths (with eventually repeated edges) between node i and node j of length β. Selecting $\beta = 1$ means that the gene expression correlation value $A[i, j]$ along the path of length β between genes i and j (that is, for $\beta = 1$, the direct correlation value between i and j) was considered for filtering edges. Selecting a larger value of β puts more emphasis on indirect regulation.

τ was chosen as a compromise between richness of the network (number of edges) and computational cost, by a bisection search in interval $[0.01; 1]$ with step 0.005, which would be the way to go to apply our method to other datasets. That is, we set τ to the smallest value in $[0.01; 1]$ that allowed a successful run of the inference in step **(C)** without triggering a memory overload.

After this procedure, we removed isolated genes in the network (that is, with both ingoing-degree and outgoing-degree equal to 0. After this step, 232 genes were left in the network, with 637×2 putative genepairwise interactions (one for each direction between two genes). We stress on the fact that preserving connectivity will be crucial for properly exploiting the experimental data, which is why we trim out isolated genes.

A.6   Step (B): Implementation of Experimental Constraints

After building the set of admissible interactions, we turned to building dynamical constraints, that is, the binary expression states of genes in the network according to experimental profiles from LINCS L1000.

The experiments shown in Table 5 comprise control and perturbed profiles in single gene perturbation experiments –by knockdown through shRNA in our application to epilepsy. First, these profiles were binarized using the binarization procedure described above. Then, for each knockdown experiment, we considered as initial condition the profile obtained from control samples, and as final condition the one obtained from perturbed samples, which is set as a (steady) attractor state.

In order to implement the new dynamics in [54], we used the Python package BoNeSiS [15], which encodes the experimental constraints and infers by answer-set programming Boolean networks –both the set of regulatory interactions and regulatory functions– that satisfy the experimental constraints with a subset of admissible interactions. We use the procedure in BoNeSiS which randomizes the search for network solutions. Moreover, in order to avoid trivial solutions without interactions, we also implemented the constraint that the state where all genes were not expressed (i.e., with expression state 0) cannot lead to any of the reported final attractor states. This constraint can be challenged, as one might assume that a network could end up in the state where all genes are turned off in a transient way, if there are some genes which are only regulated by inhibitors. However, in practice, the inference procedure without this constaint yields singularly trivial and poorly connected solutions (*i.e.*, most genes ending up without any regulators). We conjecture that it is linked to the procedure of answer-set programming, as similar methods, for instance Re:In [23], give the option of adding supplementary constraints about the presence of an activator for some genes.

A.7   Step (C): Inference Solutions and Model Selection

Inference of Boolean Network Solutions. In BoNeSiS, we asked for the enumeration of at most 1 solution to the set of topological and experimental constraints defined above. In the implementation of the most permissive semantics in [54] by [15], the size of the Boolean function specification can be upper-bounded by a prespecified value. In our application, we have used the maximum total (*i.e.*, ingoing and outgoing) degree of the underlying network, in order to avoid spurious gene regulatory functions. Due to the intrinsic randomness stemming from the solver clingo [26], and the randomized search procedure used in BoNeSiS, we iterated this enumeration, such that we obtained 25 Boolean network solutions (among which 25 are unique in terms of regulatory functions).

Selection of an Optimal Model. In order to select a "representative" network consistent with what is known about the topology of biological networks, authors in [5] compiled a list of network measures to maximize in biological networks, and computed a single scalar criterion value comprised in the interval $[0, 1]$ to maximize through the Harrington desirability index [31]. This value was called "general topological parameter". In practice, using the notation from [5], we considered the following weights

$$a_{\mathrm{DS}} = 3, a_{\mathrm{CL}} = 3, a_{\mathrm{Centr}} = 3, \text{ and } a_{\mathrm{GT}} = 1,$$

where

- DS corresponds to the *network density*, that is, the ratio of the number of edges to the maximum number of possible connections between the nodes in the network (that is, if the network was fully connected); for a network of n nodes, this maximum number is equal to $(n-1)n/2$.
- CL corresponds to the *network clustering coefficient* which is the average of node-wise clustering coefficients. The clustering coefficient of a node is the ratio of the degree of the considered node and the maximum possible number of connections such that this node and its current neighbors form a clique (*i.e.*, form a fully connected graph).
- Centr corresponds to the *network centralization*, which is correlated with the similarity of the network to a graph with a star topology.
- GT corresponds to the *network heterogeneity*, which quantifies the nonuniformity of the node degrees across the network by computing the ratio between the standard deviation of the node degrees and the average degree across the network.

Note that all of these values are comprised between 0 and 1. The higher the weights, the more importance is given to having a large associated coefficient. These weights were selected among the set of values $[\![-3; 3]\!]$, which were interpreted as "strongly minimized (in scale-free networks)", "moderately minimized", "weakly minimized", "unaffected", "weakly maximized", "moderately maximized", "strongly maximized". All of these measures should be maximized for scale-free networks, however the network heterogeneity (GT) contradicts the objective of network density, which should be privileged in order to capture most of the regulatory interactions. Finally, for every network solution N returned by BoNeSiS, we computed

$$\exp\left(\texttt{mean}\left\{-\exp\left(-x \times a - 1\right) : (x, a) \in \mathcal{V}(N)\right\}\right),$$

where $\mathcal{V}(N)$ is the set of pairs (value, weight) associated with each topological measure

$$\mathcal{V}(N) := \{(\texttt{DS}(N), a_{\text{DS}}), (\texttt{CL}(N), a_{\text{CL}}), (\texttt{Centr}(N), a_{\text{Centr}}), (\texttt{GT}(N), a_{\text{GT}})\}.$$

The final network was the one which maximized this quantity.

B Summary of Tools in Network Identification

In order to illustrate the reproducible and automated aspect of our pipeline, we display below the name of software tools and databases (and their licence status) involved in the pipeline implemented in the associated code repository[6] as shown in Fig. 1 (Table 3).

Table 3. Software tools and databases in the network identification pipeline, as of June 13th, 2022.
[1] https://github.com/bioasp/bonesis/blob/master/README.md.
[2] https://clue.io/terms.
[3] https://www.ncbi.nlm.nih.gov/home/about/policies/.
[4] https://www.disgenet.org/terms-of-use/
[5] https://string-db.org/cgi/access?footer_active_subpage=licensing.

Tool	Version	License status
BoNeSiS [15]	64e8817 (commit)	Academic purposes only[1]
CLUE (LINCS L1000 API)	1.2	Academic purposes only[2]
Database	Version	License status
MedGen (Disease Concept IDs) [21]	N/A	Public domain[3]
LINCS L1000 [65]	Beta	Academic purposes only[2]
DisGeNet [57]	7.0	CC BY-NC-SA 4.0[4]
STRING [66]	11.5	CC BY 4.0[5]

[6] https://github.com/clreda/PrioritizationMasterRegulators.

C   Robustness on a Larger Set of 50 Solutions

Since the enumeration of solutions is still computationally expensive and time-consuming, we focused on a collection of 25 network solutions. However, in order to assess the robustness of our inference procedure, we enumerated an additional set of 25 solutions, and reproduced the two plots shown in the main paper. Note that these 25 solutions were different from the first 25 ones, yielding a set of 50 unique solutions (in terms of gene regulatory functions). The selection of the optimal model run on these 50 models returned the same network shown in the main paper. Table 1 and Fig. 3 allow us to conclude similarly to the main paper, that is, the networks obtained just before the network selection step are mostly functionally and topologically similar (Table 4 and Fig. 8).

Table 4. Distribution statistics on the number of *unique* regulatory functions (RFs) across solutions per gene, and on the value of the general topological parameter (GTP) used for network selection in step (C) of the inference procedure. All values are rounded up to the 3rd decimal place. Applied on the set of 50 solutions.

	Min	25th quantile	Median	Mean	75th quantile	Max
# RFs	1	1	2	2.635	3	17
GTP	0.794	0.796	0.797	0.797	0.799	0.802

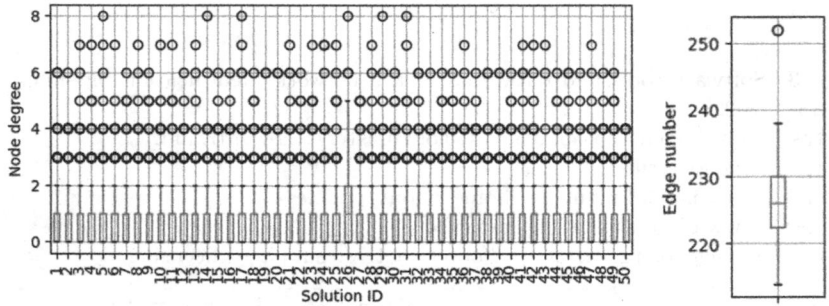

Fig. 8. *Left-hand plot*: Boxplots of node total degrees (ingoing and outgoing degree) per solution. The green lines represent median values. *Right-hand plot*: Boxplot of the number of edges across solutions (each solution comprise 232 nodes). The green line represent the median value. Applied on the set of 50 solutions. Note that the first 25 boxplots match the plot in Fig. 3. (Color figure online)

We also plotted the counterpart to Fig. 3 with *weighted* values: that is, instead of summing the edges in a solution, considering the sum of edges weighted by the STRING combined score associated with the corresponding undirected edge (divided by 1, 000). Likewise, we consider the sum of STRING combined scores across all edges involving a given node instead of the degree of this node. Resulting plots are displayed in Fig. 9. These plots aim at visualizing the differences in topology weighted by the edge-wise "evidence score" provided by the STRING database. Similar conclusions as before can be drawn from these plots, further hinting at the preservation of the network topology across inferred networks in step (C) of the network identification procedure.

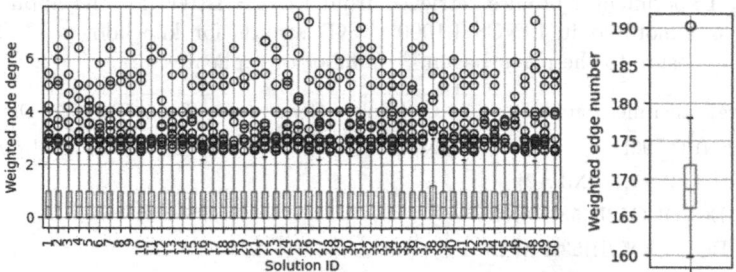

Fig. 9. *Left-hand plot*: Boxplots of node total *weighted* degrees (ingoing and outgoing degree) per solution. *Right-hand plot*: Boxplot of the weighted sum of STRING combined scores associated with edges across solutions.

Finally, we also plot the histogram of the occurrence of regulatory interactions across all 50 solutions in Fig. 10, in order to confirm the figures mentioned in Sect. 3 of the main text. A core set of 22 interactions present in all 50 solutions is observed, similarly to what was remarked on the subset of 25 solutions.

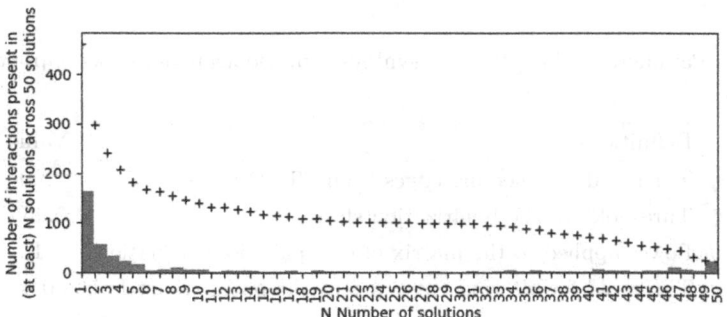

Fig. 10. Histogram of the count of regulatory interactions present in x solutions across the 50 solutions, and plot of the number of interactions present in at least x solutions across the 50 ones, for x in [[1; 50]].

D   Tables

This section of the Appendix shows supplementary tabular data about the inference of the Boolean network.

D.1   Experimental Profiles From LINCS L1000

Table 5. Experimental profiles retrieved from LINCS L1000 for the application to epilepsy, as annotated in LINCS L1000. * KD stands for knockdown. ** Time (in hours) of exposure to the perturbagen. # number of replicates

Profile brew identifier (suffix)	Cell line	Gene	Type	Time**	Dose	Nb#
KDB003_NPC_96H	NPC	PSMG1	KD*	96	$1.5\mu L$	4
Samples {X1,2.A2,X2,X3.A2}						
Treated B6_DUO52HI53LO:K16						
Control B6_DUO52HI53LO:F13						
EKW001_SHSY5Y_120H	SHSY-5Y	SOD1	KD	120	N/A	3
Samples {X1,X2,X3}						
Treated F1B3_DUO52HI53LO:J20						
Control F1B3_DUO52HI53LO:I05						
Treated F1B3_DUO52HI53LO:H17		SYT1	KD	120	N/A	3
Treated F1B3_DUO52HI53LO:I19		CACNA1C	KD	120	N/A	3
Treated F1B3_DUO52HI53LO:A03		CDC42	KD	120	N/A	3

D.2   Parameters

Table 6. Parameter values for the synthesis of Boolean networks (application to epilepsy).

	Definition	Value
η	Threshold for selecting edges from STRING [66]	400
ζ	Threshold for the binarization step	0.265
β	Power applied to the matrix of genepairwise correlations	1
τ	Threshold for filtering out edges in the putative network	0.4

E   Implementation of the Influence Maximization Algorithm

This section deals with supplementary data about the implementation of the influence maximization procedure.

Table 7. Threshold values (lower bounds on scores) for the retrieval of disease-associated genes from DisGeNet [57]. The full definitions of these indices are reported at this page [20]. EI: Evidence Index. DSI: Disease Specificity Index. DPI: Disease Pleiotropy Index.

	Score	EI	DSI	DPI	Source
Value	0	0	0.25	0	CURATED

One might wonder about what the rationale behind the choice of the minimum change in reachable attractors (by considering the maximum similarity between attractors) is. This choice leads to the good property that genes with a spread value equal to 0 can be filtered out (as mentioned in the main text), as there exists at least one attractor state which can be reached from the initial state with *and* without the perturbation of the considered gene.

E.1   Iteration of Attractor States

In order to enumerate attractors under perturbations, we used PyMaBoSS [64]. We ran PyMaBoSS with $1,000$ trajectories, for reachable attractors within 50 time steps, and parameters time_tick = 1, use_physrandgen = 0. Unfortunately, this method does not guarantee the similarity of attractors from one iteration to another, but our tests showed that, although there is some noticeable change in the resulting spread values, it does not affect the final ranking on genes. We never had to deal with the case where no attractor state is retrieved with these parameter values.

E.2   Choice of Initial States

In our application, we considered the integration of a disease-specific context by considering 48 hippocampi normalized transcriptional profiles of humans affected with Temporal Lobe Epilepsy (TLE) [46] (EMTAB 3123 on ArrayExpress). The main idea is that we specifically target genes which regulatory influence is high in a transcriptional context for epilepsy. We restricted these epileptic profiles to genes present in the network, and binarized the profiles according to the binarization procedure described in the first section, with corresponding threshold ζ equal to 0.5, so that all genes have a determined binary expression state.

F   Additional Results

This section shows additional results related to the inference of the Boolean network and the spread values.

Table 8. Regulatory interactions present in all of the 25 solutions. * strongest evidence source from the STRING database. "Association in databases" means associated in curated pathway databases.

Regulator	Regulated	Sign	Evidence source*
RBFOX1	PEG3	Inhibitory	Coexpression
SLITRK3	IL1RAP	Inhibitory	Text-mining
TSPAN7	AFF2	Activatory	Coexpression
UQCRQ	TIMM17A	Inhibitory	Coexpression
CENPJ	ANAPC1	Activatory	Coexpression
SYT13	MLLT11	Inhibitory	Coexpression
GUCY1B3	AHNAK2	Activatory	Text-mining
PLEKHG3	CCDC68	Inhibitory	Text-mining
MLLT11	TTC3	Activatory	Text-mining
SULT4A1	KIAA0232	Inhibitory	Text-mining
GRIN1	CIT	Activatory	Interaction
GPI	ANXA6	Inhibitory	Text-mining
RBFOX1	ZMAT4	Activatory	Coexpression
FAR2	FAM49A	Inhibitory	Text-mining
RAB3A	REEP2	Activatory	Coexpression
CAMK2B	AGGF1	Inhibitory	Association in databases
GAP43	ELAVL2	Inhibitory	Coexpression
ADAM22	EPB41L3	Inhibitory	Coexpression
CDC42	ARHGAP44	Activatory	Interaction
GNB5	PAK1	Inhibitory	Association in databases
STMN2	PSMG1	Inhibitory	Coexpression
AMPH	AAGAB	Activatory	Interaction
GNB5	PRKCE	Inhibitory	Association in databases
ATP1B3	FXYD7	Inhibitory	Association in databases
ATP1A3	PANK2	Activatory	Text-mining

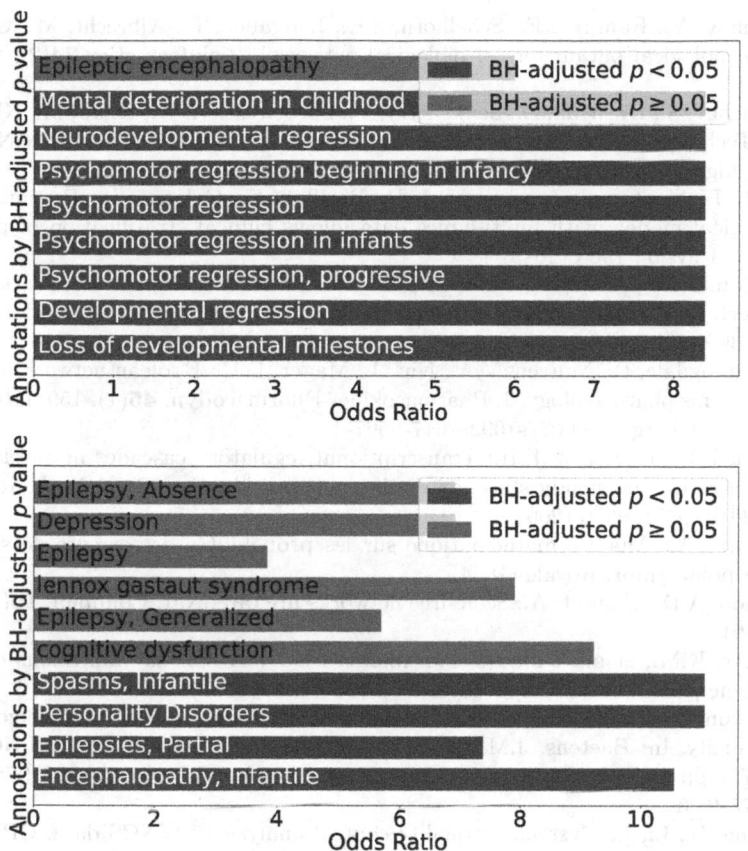

Fig. 11. Enrichment results from the ORA analysis on the filtered list of genes based on spread values from the DisGeNet annotations [57] (*top*) and GLAD4U [34] (*bottom*). The top-10 annotations (in increasing order of BH-adjusted p-value) are reported. All of these adjusted p-values are lower than 20%.

References

1. Aghamiri, S.S., Singh, V., Naldi, A., Helikar, T., Soliman, S., Niarakis, A.: Automated inference of Boolean models from molecular interaction maps using CaSQ. Bioinformatics **36**(16), 4473–4482 (2020)
2. Al-Eitan, L.N., et al.: Genetic polymorphisms of CYP3A5, CHRM2, and ZNF498 and their association with epilepsy susceptibility: a pharmacogenetic and case-control study. Pharmacogenomics Personalized Med. **12**, 225 (2019)
3. Appenzeller, S., et al.: De novo mutations in synaptic transmission genes including DNM1 cause epileptic encephalopathies. Am. J. Hum. Genet. **95**(4), 360–370 (2014)

4. Assenov, Y., Ramírez, F., Schelhorn, S.E., Lengauer, T., Albrecht, M.: Computing topological parameters of biological networks. Bioinformatics **24**(2), 282–284 (2008)
5. Babichev, S., Durnyak, B., Senkivskyy, V., Sorochynskyi, O., Kliap, M., Khamula, O.: Technique of gene regulatory networks reconstruction based on ARACNE inference algorithm. In: IDDM, pp. 195–207 (2019)
6. Béal, J., Montagud, A., Traynard, P., Barillot, E., Calzone, L.: Personalization of logical models with multi-omics data allows clinical stratification of patients. Front. Physiol. 1965 (2019)
7. Benjamini, Y., Hochberg, Y.: Controlling the false discovery rate: a practical and powerful approach to multiple testing. J. Roy. Stat. Soc.: Ser. B (Methodol.) **57**(1), 289–300 (1995)
8. Bloomingdale, P., Nguyen, V.A., Niu, J., Mager, D.E.: Boolean network modeling in systems pharmacology. J. Pharmacokinet Pharmacodyn. **45**(1), 159–180 (2018). https://doi.org/10.1007/s10928-017-9567-4
9. Bolouri, H., Davidson, E.H.: Transcriptional regulatory cascades in development: initial rates, not steady state, determine network kinetics. Proc. Natl. Acad. Sci. **100**(16), 9371–9376 (2003)
10. Bravais, A.: Analyse mathématique sur les probabilités des erreurs de situation d'un point. Impr. Royale (1844)
11. Broido, A.D., Clauset, A.: Scale-free networks are rare. Nat. Commun. **10**(1), 1–10 (2019)
12. Butler, K.M., et al.: De novo and inherited SCN8A epilepsy mutations detected by gene panel analysis. Epilepsy Res. **129**, 17–25 (2017)
13. Chatain, T., Haar, S., Paulevé, L.: Boolean networks: beyond generalized asynchronicity. In: Baetens, J.M., Kutrib, M. (eds.) AUTOMATA 2018. LNCS, vol. 10875, pp. 29–42. Springer, Cham (2018). https://doi.org/10.1007/978-3-319-92675-9_3
14. Cheng, L., Li, L.: Systematic quality control analysis of LINCS data. CPT Pharmacometrics Syst. Pharmacol. **5**(11), 588–598 (2016)
15. Chevalier, S., Froidevaux, C., Paulevé, L., Zinovyev, A.: Synthesis of Boolean networks from biological dynamical constraints using answer-set programming. In: 2019 IEEE 31st International Conference on Tools with Artificial Intelligence (ICTAI), pp. 34–41. IEEE (2019)
16. Choi, S.S., Cha, S.H., Tappert, C.C.: A survey of binary similarity and distance measures. J. Syst. Cybern. Inform. **8**(1), 43–48 (2010)
17. Cormen, T.H., Leiserson, C.E., Rivest, R.L., Stein, C.: Introduction to Algorithms. MIT Press, Cambridge (2009)
18. Cushion, T.D., et al.: De novo mutations in the beta-tubulin gene TUBB2A cause simplified gyral patterning and infantile-onset epilepsy. Am. J. Hum. Genet. **94**(4), 634–641 (2014)
19. Delahaye-Duriez, A., et al.: Rare and common epilepsies converge on a shared gene regulatory network providing opportunities for novel antiepileptic drug discovery. Genome Biol. **17**(1), 1–18 (2016)
20. DisGeNet: Faq: Original data sources (2022). https://www.disgenet.org/. Accessed 4 May 2022
21. Doğan, R.I., Leaman, R., Lu, Z.: NCBI disease corpus: a resource for disease name recognition and concept normalization. J. Biomed. Inform. **47**, 1–10 (2014)
22. Dunn, S.J., Li, M.A., Carbognin, E., Smith, A., Martello, G.: A common molecular logic determines embryonic stem cell self-renewal and reprogramming. EMBO J. **38**(1), e100003 (2019)

23. Dunn, S.-J., Yordanov, B.: Automated reasoning for the synthesis and analysis of biological programs. In: Liò, P., Zuliani, P. (eds.) Automated Reasoning for Systems Biology and Medicine. CB, vol. 30, pp. 37–62. Springer, Cham (2019). https://doi.org/10.1007/978-3-030-17297-8_2
24. Eves, E.M., Tucker, M.S., Roback, J.D., Downen, M., Rosner, M.R., Wainer, B.H.: Immortal rat hippocampal cell lines exhibit neuronal and glial lineages and neurotrophin gene expression. Proc. Natl. Acad. Sci. **89**(10), 4373–4377 (1992)
25. Finak, G., et al.: Mast: a flexible statistical framework for assessing transcriptional changes and characterizing heterogeneity in single-cell RNA sequencing data. Genome Biol. **16**(1), 1–13 (2015)
26. Gebser, M., Kaminski, R., Kaufmann, B., Ostrowski, M., Schaub, T., Wanko, P.: Theory solving made easy with clingo 5. In: Technical Communications of the 32nd International Conference on Logic Programming (ICLP 2016). Schloss Dagstuhl-Leibniz-Zentrum fuer Informatik (2016)
27. Gibbs, D.L., Shmulevich, I.: Solving the influence maximization problem reveals regulatory organization of the yeast cell cycle. PLoS Comput. Biol. **13**(6), e1005591 (2017)
28. González, F.L., et al.: Drug-resistant epilepsy: definition and treatment alternatives. Neurología (English Edition) **30**(7), 439–446 (2015)
29. Griffin, A., et al.: Phenotypic analysis of catastrophic childhood epilepsy genes. Commun. Biol. **4**(1), 1–13 (2021)
30. Han, C.L., et al.: Electrical stimulation of hippocampus for the treatment of refractory temporal lobe epilepsy. Brain Res. Bull. **109**, 13–21 (2014)
31. Harrington, E.C.: The desirability function. Ind. Qual. Control **21**(10), 494–498 (1965)
32. Huang, Y., et al.: A framework for identification of on-and off-target transcriptional responses to drug treatment. Sci. Rep. **9**(1), 1–9 (2019)
33. Johnson, M.R., et al.: Systems genetics identifies Sestrin 3 as a regulator of a proconvulsant gene network in human epileptic hippocampus. Nat. Commun. **6**(1), 1–11 (2015)
34. Jourquin, J., Duncan, D., Shi, Z., Zhang, B.: GLAD4U: deriving and prioritizing gene lists from PubMed literature. BMC Genomics **13**(8), 1–12 (2012)
35. Kalume, F., et al.: Sudden unexpected death in a mouse model of Dravet syndrome. J. Clin. Investig. **123**(4), 1798–1808 (2013)
36. Kauffman, S.A.: Metabolic stability and epigenesis in randomly constructed genetic nets. J. Theor. Biol. **22**(3), 437–467 (1969)
37. Kempe, D., Kleinberg, J., Tardos, É.: Maximizing the spread of influence through a social network. In: Proceedings of the Ninth ACM SIGKDD International Conference on Knowledge Discovery and Data Mining, pp. 137–146 (2003)
38. Kuruba, R., Hattiangady, B., Shetty, A.K.: Hippocampal neurogenesis and neural stem cells in temporal lobe epilepsy. Epilepsy Behav. **14**(1), 65–73 (2009)
39. Lal, D., et al.: Rare exonic deletions of the RBFOX1 gene increase risk of idiopathic generalized epilepsy. Epilepsia **54**(2), 265–271 (2013)
40. Lek, M., et al.: Analysis of protein-coding genetic variation in 60,706 humans. Nature **536**(7616), 285–291 (2016)
41. Liao, Y., Wang, J., Jaehnig, E.J., Shi, Z., Zhang, B.: WebGestalt 2019: gene set analysis toolkit with revamped UIs and APIs. Nucleic Acids Res. **47**(W1), W199–W205 (2019)
42. Lim, N., Pavlidis, P.: Evaluation of connectivity map shows limited reproducibility in drug repositioning. Sci. Rep. **11**(1), 1–14 (2021)

43. Liu, Y.Y., Slotine, J.J., Barabási, A.L.: Controllability of complex networks. Nature **473**(7346), 167–173 (2011)
44. Liu, Y.Y., Slotine, J.J., Barabási, A.L.: Control centrality and hierarchical structure in complex networks (2012)
45. Mattick, J.S., Taft, R.J., Faulkner, G.J.: A global view of genomic information-moving beyond the gene and the master regulator. Trends Genet. **26**(1), 21–28 (2010)
46. Mirza, N., et al.: Genetic regulation of gene expression in the epileptic human hippocampus. Hum. Mol. Genet. **26**(9), 1759–1769 (2017)
47. Montagud, A., et al.: Patient-specific Boolean models of signalling networks guide personalised treatments. Elife **11**, e72626 (2022)
48. Mudunuri, U., Che, A., Yi, M., Stephens, R.M.: bioDBnet: the biological database network. Bioinformatics **25**(4), 555–556 (2009)
49. Myers, C.T., et al.: De novo mutations in SLC1A2 and CACNA1A are important causes of epileptic encephalopathies. Am. J. Hum. Genet. **99**(2), 287–298 (2016)
50. Nicolle, R., Radvanyi, F., Elati, M.: CoRegNet: reconstruction and integrated analysis of co-regulatory networks. Bioinformatics **31**(18), 3066–3068 (2015)
51. Ogren, J.A., et al.: Three-dimensional surface maps link local atrophy and fast ripples in human epileptic hippocampus. Ann. Neurol.: Official J. Am. Neurol. Assoc. Child Neurol. Soc. **66**(6), 783–791 (2009)
52. Ohba, C., et al.: Grin 1 mutations cause encephalopathy with infantile-onset epilepsy, and hyperkinetic and stereotyped movement disorders. Epilepsia **56**(6), 841–848 (2015)
53. Ostrowski, M., Paulevé, L., Schaub, T., Siegel, A., Guziolowski, C.: Boolean network identification from perturbation time series data combining dynamics abstraction and logic programming. Biosystems **149**, 139–153 (2016)
54. Paulevé, L., Kolčák, J., Chatain, T., Haar, S.: Reconciling qualitative, abstract, and scalable modeling of biological networks. Nat. Commun. **11**(1), 1–7 (2020)
55. Perrault, P., Healey, J., Wen, Z., Valko, M.: Budgeted online influence maximization. In: International Conference on Machine Learning, pp. 7620–7631. PMLR (2020)
56. Petrovski, S., Wang, Q., Heinzen, E.L., Allen, A.S., Goldstein, D.B.: Genic intolerance to functional variation and the interpretation of personal genomes. PLoS Genet. **9**(8), e1003709 (2013)
57. Piñero, J., et al.: The DisGeNET knowledge platform for disease genomics: 2019 update. Nucleic Acids Res. **48**(D1), D845–D855 (2020)
58. Poke, G., et al.: The epileptology of GNB5 encephalopathy. Epilepsia **60**(11), e121–e127 (2019)
59. Shannon, P., et al.: Cytoscape: a software environment for integrated models of biomolecular interaction networks. Genome Res. **13**(11), 2498–2504 (2003)
60. Sharma, A., et al.: Controllability in an islet specific regulatory network identifies the transcriptional factor NFATC4, which regulates type 2 diabetes associated genes. NPJ Syst. Biol. Appl. **4**(1), 1–11 (2018)
61. Srivastava, P.K., et al.: A systems-level framework for drug discovery identifies Csf1R as an anti-epileptic drug target. Nat. Commun. **9**(1), 1–15 (2018)
62. Srivastava, P.K., et al.: Genome-wide analysis of differential RNA editing in epilepsy. Genome Res. **27**(3), 440–450 (2017)
63. Stamberger, H., et al.: STXBP1 encephalopathy: a neurodevelopmental disorder including epilepsy. Neurology **86**(10), 954–962 (2016)
64. Stoll, G., et al.: MaBoSS 2.0: an environment for stochastic Boolean modeling. Bioinformatics **33**(14), 2226–2228 (2017)

65. Subramanian, A., et al.: A next generation connectivity map: L1000 platform and the first 1,000,000 profiles. Cell **171**(6), 1437–1452 (2017)
66. Szklarczyk, D., et al.: The STRING database in 2021: customizable protein-protein networks, and functional characterization of user-uploaded gene/measurement sets. Nucleic Acids Res. **49**(D1), D605–D612 (2021)
67. Thomas, R.: Boolean formalization of genetic control circuits. J. Theor. Biol. **42**(3), 563–585 (1973)
68. Thomas, R., Thieffry, D., Kaufman, M.: Dynamical behaviour of biological regulatory networks—I. Biological role of feedback loops and practical use of the concept of the loop-characteristic state. Bull. Math. Biol. **57**(2), 247–276 (1995)
69. Vaginay, A., Boukhobza, T., Smaïl-Tabbone, M.: Automatic synthesis of Boolean networks from biological knowledge and data. In: Dorronsoro, B., Amodeo, L., Pavone, M., Ruiz, P. (eds.) OLA 2021. CCIS, vol. 1443, pp. 156–170. Springer, Cham (2021). https://doi.org/10.1007/978-3-030-85672-4_12
70. Von Mering, C., et al.: STRING: known and predicted protein-protein associations, integrated and transferred across organisms. Nucleic Acids Res. **33**(Suppl_1), D433–D437 (2005)
71. Wang, X., Goldstein, D.B.: Enhancer domains predict gene pathogenicity and inform gene discovery in complex disease. Am. J. Hum. Genet. **106**(2), 215–233 (2020)
72. Wery, M., Dameron, O., Nicolas, J., Remy, E., Siegel, A.: Formalizing and enriching phenotype signatures using Boolean networks. J. Theor. Biol. **467**, 66–79 (2019)
73. World Health Organization (WHO): Epilepsy (2022). https://www.who.int/news-room/fact-sheets/detail/epilepsy. Accessed 29 Apr 2022
74. Wu, L., Li, M., Wang, J., Wu, F.X.: CytoCtrlAnalyser: a cytoscape app for biomolecular network controllability analysis. Bioinformatics **34**(8), 1428–1430 (2018)
75. Zerrouk, N., Miagoux, Q., Dispot, A., Elati, M., Niarakis, A.: Identification of putative master regulators in rheumatoid arthritis synovial fibroblasts using gene expression data and network inference. Sci. Rep. **10**(1), 1–13 (2020)
76. Zhao, Y., Wong, L., Goh, W.W.B.: How to do quantile normalization correctly for gene expression data analyses. Sci. Rep. **10**(1), 1–11 (2020)
77. Ziegler, A., Colin, E., Goudenège, D., Bonneau, D.: A snapshot of some pLI score pitfalls. Hum. Mutat. **40**(7), 839–841 (2019)

Variable Stabilisation in Boolean Monotonic Model Pools

Samuel Pastva[✉]

Faculty of Informatics, Masaryk University, Brno, Czech Republic
xpastva@fi.muni.cz

Abstract. One of the central issues in logical modelling is whether a certain property of the model emerges due to its topological structure (i.e. its influence graph), or due to its dynamical structure (i.e. its logical update functions). In this paper, we practically evaluate a previously proposed formal instrument for studying this question: *monotonic model pools* and their associated skeleton Boolean networks. Specifically, we propose a simplified over-approximation theorem for skeleton networks and study the emergence of variable stability in these systems. Additionally, we consider the notion of minimal stabilising interventions and show how to compute such interventions symbolically. We survey the practicality of this methodology on 100+ real-world Boolean networks.

1 Introduction

The structure of gene regulation circuits and other general biochemical models is often encoded in the form of influence graphs. In an influence graph, nodes represent biochemical entities (RNA, proteins, other small molecules) and edges represent the possible dependencies between said entities. Additionally, these dependencies are often augmented with knowledge about monotonicity (activation, inhibition), as this is a common property of biochemical interactions [6].

This abstract structure may then serve as a blueprint for constructing continuous (e.g. differential equations) or logical (e.g. Boolean networks) dynamical models. However, an important question is linked to this process: *Which model properties can be attributed to the choice of dynamical rules and which are enforced (or prohibited) by the initial influence graph?*

In this paper, we address this question from the perspective of logical models, specifically asynchronous Boolean networks [9]. Here, relationship between Boolean network dynamics and its influence graph has been initially explored in the form of *Thomas conjectures* [25], later expanded for example in [10,18–20,23,26]. These works formulate necessary conditions on the influence graph for the emergence of various long-term phenomena, for example multi-stability or oscillatory behaviour. As such, they are often applied in the context of static analysis (see [17] for a survey).

Recently, a different approach towards studying this relationship has been proposed in the form of *monotonic model pools* [21], and in particular their asynchronous graph representations [22]. Here, a monotonic model pool is a collection

I. Petre and A. Păun (Eds.): CMSB 2022, LNBI 13447, pp. 122–137, 2022.
https://doi.org/10.1007/978-3-031-15034-0_6

of *all* Boolean networks whose dynamics are aligned with a prescribed signed influence graph. Instead of giving necessary or sufficient conditions on the structure of the influence graph, [22] then uses a so called *skeleton* state-transition system to collectively study the behaviour of all models within a monotonic pool.

The work on monotonic model pools and their skeletons originates in a similar approach which is tied to ordinary differential equations [7,12]. So far, the results presented as part of [21,22] are largely theoretical. In this paper, our aim is to assess this method from a practical perspective as a tool for inferring properties of large Boolean models.

To achieve this goal, we propose a slightly more strict variant of the monotonic model pool, in which we prohibit networks employing constant functions (with the exception of variables that are explicitly declared as constant in the influence graph). Arguably, such restriction is in many cases more practical, as it enforces at least some mutability for the declared variables.

We then show that the relationship between the pool and its skeleton (as proposed in [22]) can be greatly simplified for this strict pool variant. In particular, [22] defines the notion of *quotient graphs* that describe equivalence classes of state-transition systems of various networks within the monotonic pool. These are then used to formalise the relationship between a monotonic pool and its skeleton graph. Meanwhile, strict monotonic pool relates directly to the state-transition graphs of the Boolean networks within the pool.

To practically test the usefulness of skeleton networks, we propose the notion of *observably stable* variable. Such variable is guaranteed to eventually become stable for *some* initial state in *every* Boolean network in the monotonic pool. That is, regardless of which "true" Boolean network is encountered in the wild, one should always be able to observe instances where said variable stabilises.

This property represents a practically testable condition that can be used to further narrow down or validate system characteristics. For example, examining the initial conditions leading to variable stabilisation may suggest biological experiments that validate the structure of the influence graph. Alternatively, absence of observable stabilisation may be desired for some variables, e.g. ones participating in a sustained oscillation. Analysis of the skeleton network may then reveal regulatory structures which enforce the undesired stabilisation.

Additionally, we consider the notion of a *stabilising intervention*. Such intervention disrupts the dynamics of the specified variables with the goal of making all variables within our system observably stable. Variables that appear within a *minimal* stabilising intervention could be then considered, for example, as candidates for various more fine-grained control strategies [3,14,24] once more detailed knowledge about the network's dynamics is available.

We implement basic symbolic algorithms (using binary decision diagrams [1,4]) for detecting observably stable variables and stabilising interventions in skeletons of strict monotonic model pools. We compute these metrics for a large selection of signed influence graphs obtained from real-world Boolean networks, discussing how our results relate to existing knowledge that can be otherwise obtained from the influence graph using static analysis.

2   Preliminaries

Before we proceed with the main results of the paper, let us first formally introduce Boolean networks, influence graphs, monotonic model pools and other existing definitions that we use later in our reasoning.

Notation. In the following, we use $[1, n]$ to denote the set $\{1, \ldots, n\}$ and $\mathbb{B}$ to denote the set $\{0, 1\}$. We generally consider 0 and 1 to be interchangeable with *false* and *true*, respectively, and given a Boolean variable x, $\bar{x}$ denotes its negation (we may still use the symbol $\neg$ for larger formulae though). Given a vector $x \in \mathbb{B}^n$, we write x_i to denote its i-th element. Finally, for an $i \in [1, n]$ and a Boolean value $b \in \mathbb{B}$, we write $x[i \mapsto b]$ to denote a copy of x with x_i set to b.

Boolean Network. Intuitively, a Boolean network (BN) describes a system of n Boolean variables that evolves in discrete steps using Boolean update functions.

Definition 1. *Let n be the number of variables. A Boolean network is a collection $F = \{F_1, \ldots, F_n\}$, where each $F_i : \mathbb{B}^n \to \mathbb{B}$ is a Boolean update function.*

We typically refer to the set $\mathbb{B}^n$ as the *state space* of F, and to its elements as the *states*. Furthermore, we write $dep(F_i) \subseteq [1, n]$ to denote the subset of variables that actually influence the output of F_i (also called *regulators*):

$$j \in dep(F_i) \Leftrightarrow \exists x \in \mathbb{B}^n. \; F_i(x[j \mapsto 0]) \neq F_i(x[j \mapsto 1]) \tag{1}$$

Subsequently, we can also distinguish the set $dep^{\oplus}(F_i) \subseteq dep(F_i)$ that denotes the *positive* dependencies of F_i. Intuitively, j is a positive dependency (*activator*) if an *increase* in j cannot lead to a *decrease* in the output of F_i:

$$j \in dep^{\oplus}(F_i) \Leftrightarrow j \in dep(F_i) \wedge \forall x \in \mathbb{B}^n. \; F_i(x[j \mapsto 0]) \leq F_i(x[j \mapsto 1]) \tag{2}$$

Symmetrically, we define negative dependencies (*inhibitors*) $dep^{\ominus}(F_i)$ using the $\geq$ operator instead of $\leq$. Collectively, we call the function F_i locally monotonic when $dep(F_i) = dep^{\oplus}(F_i) \cup dep^{\ominus}(F_i)$. That is, when every dependency of F_i is either an activator or an inhibitor. In the following, we are only concerned with BNs where every F_i is locally monotonic. Note that this assumption covers the vast majority of biologically relevant cases (as we show in our evaluation).

Additionally, when $dep(F_i) = \emptyset$, we say that variable i is a *constant*, as we necessarily have either $F_i(x) = true$ or $F_i(x) = false$. When $dep(F_i) = \{i\}$, we call i the *input* of F, where i is either a *stable* input when $F_i(x) = x_i$ or an *unstable* input when $F_i(x) = \bar{x_i}$.

In the following, we often consider only Boolean networks *without* constant variables. This mainly serves to reduce verbosity and ease presentation of some technical arguments. Note that for practical purposes, this assumption does not pose substantial drawbacks. If a system contains constant variables, we can typically eliminate such variables without disturbing the long-term behaviour of the system. The elimination procedure simply substitutes the occurrences of the constant variable for the actual constant value.

Boolean Network Dynamics. To give meaning to the behaviour of a Boolean network, we must associate with the Boolean network some form of dynamics. There are different ways of defining the dynamics of a Boolean network. Here, we consider the *asynchronous* case which is generally assumed to properly approximate the wide range of timescales that govern the behaviour of a real biological system. Using this assumption, we can define the state-transition graph of F:

Definition 2. *Let F be a BN. A state-transition graph $\mathrm{STG}(F) = (V, E)$ is a directed graph where $V = \mathbb{B}^n$ and $E \subseteq V \times V$ is defined as follows:*

$$(x, y) \in E \Leftrightarrow x \neq y \wedge \exists i \in [1, n].\ y = x[i \mapsto F_i(x)]$$

Per convention, we can write $x \to y$, $x \to^+ y$, and $x \to^* y$ whenever we have $(x, y) \in E$, $(x, y) \in E^+$ (the transitive closure of E), and $(x, y) \in E^*$ (the reflexive and transitive closure of E), respectively.

Attractor. Within $\mathrm{STG}(F)$, we typically recognise the *attractors* as the subsets of states that determine the long-term behaviour of F. Formally, an attractor $A \subseteq \mathbb{B}^n$ of F is a *terminal* (also *bottom*) strongly connected component of $\mathrm{STG}(F)$. That is, for any $x, y \in A$, we have $x \to^* y$ (A is strongly connected), and for any $x \in A$ and $y \in \mathbb{B}^n$, $x \to^* y$ implies $y \in A$ (A cannot be escaped).

Trap Set. Additionally, we can generalize the notion of attractor by disregarding the strong connectedness, obtaining a *trap set* T [11]. Formally, we say that $T \subseteq \mathbb{B}^n$ is a trap set when T cannot be escaped, i.e. for any $x \in T$ and $y \in \mathbb{B}^n$ such that $x \to^* y$, we have $y \in T$. Clearly, every non-empty trap set contains at least one attractor and attractors are exactly the network's minimal trap sets.

Signed Influence Graph. In many practical instances where we cannot provide a fully specified BN, an *influence graph* suggesting possible dependencies between variables may be available instead. When this graph also incorporates information about interaction monotonicity, we talk about *signed influence graphs*:

Definition 3. *A signed influence graph (SIG) over n variables, $I = (V, E, L)$, is a tuple where $V = [1, n]$, $E \subseteq V \times V$ is an arbitrary edge relation, and $L : E \to \{\oplus, \ominus\}$ is a labelling that assigns a monotonicity to each graph edge.*

With a slight abuse of notation, we write $dep^\oplus(i)$ as the set of positive dependencies of variable i within I (formally, $dep^\oplus(i) = \{\ j \mid (j, i) \in E \wedge L(j, i) = \oplus\ \}$), and analogously $dep^\ominus(i)$ as the set of all negative dependencies of i. We can write $dep(i) = dep^\oplus(i) \cup dep^\ominus(i)$ to denote *all* dependencies of variable i. Analogously to the case of BNs, when $dep(i) = \emptyset$, we call the variable i *constant* and when $dep(i) = \{i\}$, we consider it to be an *input* of I (with $L(i, i) = \oplus$ being a stable input and $L(i, i) = \ominus$ being an unstable input).

We then say that a locally monotonic BN F is *consistent* with a signed influence graph I, written $F \vDash I$, when for every variable $i \in [1, n]$, we have that $dep^\oplus(F_i) \subseteq dep^\oplus(i)$ and $dep^\ominus(F_i) \subseteq dep^\ominus(i)$. Note that while F_i must adhere to I, it can still disregard some (or even all) of the dependencies declared in I. For example, every constant in I is a constant in F, but every input in I is either an input or a constant in F. In general, when i is a constant in F, but not in I, we say that i is a *hidden constant*.

Monotonic Model Pool. When studying the behaviour of biological networks, we may be interested in determining which model properties emerge due to the general structure of the network's influence graph and which properties are introduced due to the particular choice of update functions. To study this question, [21] proposes the notion of *monotonic model pools*:

Definition 4. *Let I be a signed influence graph. We then write $M(I)$ to denote the set of all Boolean networks consistent with I: $M(I) = \{F \mid F \vDash I\}$. This set $M(I)$ is called the* monotonic model pool *of I.*

Here, note that regardless of I, a monotonic pool $M(I)$ also contains many networks with hidden constants. This may be often undesirable in practical applications. For example, arbitrary steady state can be realized within $M(I)$ by simply considering every variable to be a constant with the desired basal value. However, we would hardly consider such "solution" as useful.

Monotonic Model Pool Skeleton. To study the behaviour of the Boolean networks within a monotonic model pool $M(I)$, [22] proposes a Boolean network that in some formal sense summarises the possible networks within $M(I)$:

Definition 5. *Let I be a signed influence graph. Then $\Gamma(I) = \{\Gamma_1, \ldots, \Gamma_n\}$ is the* skeleton network *of $M(I)$, where every $\Gamma_i : \mathbb{B}^n \to \mathbb{B}$ is the following:*

$$\Gamma_i(x) = x_i \Leftrightarrow \Big[\bigvee_{j \in dep^{\oplus}(i)} x_i \Leftrightarrow x_j \vee \bigvee_{j \in dep^{\ominus}(i)} x_i \Leftrightarrow x_j \Big]$$

The state-transition graph $\mathrm{STG}(\Gamma(I))$ is then called the skeleton *of I.*

In [22], the authors show several interesting propositions relating the STGs of the networks within $M(I)$ to the skeleton $\mathrm{STG}(\Gamma(I))$ through the notion of *quotient graphs*. We address this relationship later as part of the *Methods* section, using a simpler approach that avoids the quotient graphs entirely.

For now, let us observe that the skeleton $\Gamma(I)$ is a BN that is not necessarily consistent with I (i.e. $\Gamma(I) \nvDash I$). The main reason is that within $\Gamma(I)$, we always have $i \in dep(\Gamma_i)$, regardless of I. Also note that any constant in I becomes an unstable input of $\Gamma(I)$ (since $(x_i \Leftrightarrow true) \equiv \overline{x_i}$). Furthermore, any stable input of I is also a stable input of $\Gamma(I)$ (since $(x_i \Leftrightarrow (x_i \Leftrightarrow x_i)) \equiv x_i$) and any unstable input of I is also an unstable input of $\Gamma(I)$ (since $(x_i \Leftrightarrow (x_i \Leftrightarrow x_i)) \equiv \overline{x_i}$).

3   Methods

We are now ready to discuss the main theoretical results of this paper. Specifically, we propose monotonic model pools without hidden constants as a means of reducing the amount of possible behaviour within a classical monotonic model pool. We then characterise the relationship between such model pools and their skeleton graphs using the commonly understood notion of over-approximation.

Leveraging this relationship, we propose a notion of *observably stable variable* and *stabilising intervention*. These guarantee that every member of the strict monotonic pool contains initial conditions leading to stabilisation of some (or all) network variables. We describe how these can be computed symbolically using the skeleton network (Definition 5).

3.1   Theory

Strict Monotonic Model Pool. As we already suggested, a monotonic pool $M(I)$ incorporates many networks that may not be desirable with respect to the provided I. We thus propose a slightly modified notion of monotonic model pool that excludes such pathological cases:

Definition 6. *We write $M_{[!]}(I)$ to denote the subset of $M(I)$ consisting of Boolean networks with no hidden constants (with respect to I). We refer to $M_{[!]}(I)$ as the* strict *monotonic model pool of I.*

This new assumption ensures that only variables declared as constants in I can be constants in the networks within $M_{[!]}(I)$. Note that such $M_{[!]}(I)$ still admits networks with stable inputs. These variables are "effectively constant", but have no basal value towards which they tend. In other words, a variable which is *not* an input in I *can* still become an input in some network within $M_{[!]}(I)$. However, this is clearly restricted to variables with positive self-regulation in I (as opposed to classical $M(I)$, where any variable could become a constant).

Over-Approximation of a Strict Monotonic Pool. Now, we aim to establish a formal relationship between the skeleton state-transition graph $\mathrm{STG}(\Gamma(I))$ and the networks within a strict monotonic model pool $M_{[!]}(I)$. For this, let us first observe the following:

Lemma 1. *Let $f : \mathbb{B}^n \to \mathbb{B}$ be a non-constant locally monotonic function and let us define the following abbreviations based on f:*

$$\mathrm{AND}_f(x) = \bigwedge_{i \in dep^\oplus(f)} x_i \wedge \bigwedge_{i \in dep^\ominus(f)} \overline{x_i} \;\; and \;\; \mathrm{OR}_f(x) = \bigvee_{i \in dep^\oplus(f)} x_i \vee \bigvee_{i \in dep^\ominus(f)} \overline{x_i}$$

We then claim that for every $x \in \mathbb{B}^n$, $\mathrm{AND}_f(x) \le f(x) \le \mathrm{OR}_f(x)$.

Proof. Let us focus on the first inequality, i.e. $\text{AND}_f(x) \leq f(x)$. Clearly, the inequality trivially holds whenever $\text{AND}_f(x) = 0$. We thus consider a fixed but arbitrary $x^\top$ where $\text{AND}_f(x^\top) = 1$. For contradiction, we assume that $f(x^\top) = 0$, resulting in $\text{AND}_f(x^\top) > f(x^\top)$. Such $x^\top$ must clearly have $x_i^\top = 1$ for every $i \in dep^\oplus(f)$ and $x_j^\top = 0$ for every $j \in dep^\ominus(f)$ (to achieve $\text{AND}(x^\top) = 1$). Now, because f is not constant, there must be some $y \in \mathbb{B}^n$ such that $f(y) = 1$ and y only differs from $x^\top$ in the components within $dep(f) = \{d_1, \ldots, d_k\}$.

Let us consider a sequence $z^{(0)}, \ldots, z^{(k)}$ where $z^{(0)} = x^\top$ and for every other $z^{(i)}$, we have $z^{(i)} = z^{(i-1)}[d_i \mapsto y_i]$. Intuitively, in this sequence, we flip the relevant bits of $x^\top$ one-by-one to match y (consequently, $z^{(k)} = y$).

Let $i > 0$ be the smallest index for which $f(z^{(i-1)}) < f(z^{(i)})$ (such i must exist because $f(x^\top) < f(y)$). If $i \in dep^\oplus(f)$, then we must have $z_i^{(i-1)} = 1$ (since $x_i^\top = 1$) and $z_i^{(i)} = 0$. However, this contradicts the monotonicity of f, as we must have $f(x[i \mapsto 0]) \leq f(x[i \mapsto 1])$ for every $x \in \mathbb{B}^n$ (recall Eq. 2). Symmetrically, when $i \in dep^\ominus(f)$, we must have $z_i^{(i-1)} = 0$ and $z_i^{(i)} = 1$ which contradicts the assumption that $f(x[i \mapsto 0]) \geq f(x[i \mapsto 1])$.

Consequently, there can be no such $x^\top$ where $\text{AND}_f(x^\top) = 1$ but $f(x^\top) = 0$. The other inequality is analogous: we consider an input $x^\perp$ where $\text{OR}_f(x^\perp) = 0$ but $f(x^\perp) = 1$ and equally show that this violates the monotonicity of f. $\square$

Using this simple lemma, we can argue the following:

Theorem 1. *Let I be a signed influence graph and let $M_{[!]}(I)$ be its strict monotonic model pool (i.e. without hidden constants). For every $F \in M_{[!]}(I)$, we have that the edge relation of $\text{STG}(F)$ is a subset of the edge relation of $\text{STG}(\Gamma(I))$.*

Proof. Let F be an arbitrary but fixed BN from $M_{[!]}(I)$. Our goal is to show that for any two $x, y \in \mathbb{B}^n$ such that $y = x[i \mapsto \overline{x_i}]$ for some $i \in [1, n]$ (x and y only differ in component i), $F_i(x) = \overline{x_i}$ implies $\Gamma_i(x) = \overline{x_i}$. In other words of Definition 2, if there is a transition $x \to y$ in $\text{STG}(F)$ (F_i flips the value of i in state x), then also $x \to y$ in $\text{STG}(\Gamma(I))$ (Γ_i also flips the value of i in x).

First, observe that if F_i is a constant function, then i must be a constant in I and consequently an unstable input in $\Gamma(I)$. This means that regardless of x, there is a transition $x \to x[i \mapsto \overline{x_i}]$ and the implication is trivially satisfied.

To show this implication for general F_i, first assume that $x_i = 0$, meaning that $F_i(x) = 1$. Under this assumption, we can simplify the function Γ_i (see Definition 5) to $\bigvee_{j \in dep^\oplus(i)} x_j \vee \bigvee_{j \in dep^\ominus(i)} \overline{x_j}$ (by observing that $0 \nleftrightarrow x \equiv x$ and $0 \leftrightarrow x \equiv \overline{x}$). This is exactly the function OR_{F_i} from Lemma 1, and we thus have $F_i(x) \leq \text{OR}_{F_i}(x)$. Since, $F_i(x) = 1$, we get that $\text{OR}_{F_i}(x) = 1$ and by extension $\Gamma_i(x) = 1$, satisfying the desired implication.

The case of $x_i = 1$ is then analogous. Clearly, $F_i(x) = 0$ and Γ_i can be simplified to $\neg \left[\bigvee_{j \in dep^\oplus(i)} \overline{x_j} \vee \bigvee_{j \in dep^\ominus(i)} x_j \right]$, which is equivalent to $\bigwedge_{j \in dep^\oplus(i)} x_j \wedge \bigwedge_{j \in dep^\ominus(i)} \overline{x_j}$ (function AND_{F_i} from Lemma 1). Because $\text{AND}_{F_i}(x) \leq F_i(x)$ and $F_i(x) = 0$, we have that $\Gamma_i(x) = 0$ as well, and the implication is satisfied for both $x_1 = 0$ and $x_i = 1$. $\square$

Intuitively, Theorem 1 states that $\text{STG}(\Gamma(I))$ is an *over-approximation* of $\text{STG}(F)$ for any $F \in M_{[!]}(I)$. Conceptually, this is a similar result to Theorem 5 of [22]. However, we avoid the notion of quotient graphs entirely by focusing on strict monotonic pools only.

Subjectively, this makes the result much easier to relate with other formal methods that derive their correctness form the notion of graph over- and under-approximation, as we are directly comparing the state-transition graphs of $\Gamma(I)$ and the networks within $M_{[!]}(I)$. Theorem 1 then allows us to derive several key observations (akin to Corollary 7 of [22]):

Corollary 1. *Let F be an arbitrary network in $M_{[!]}(I)$.*

- *When $x \not\rightarrow y$ in $\text{STG}(\Gamma(I))$, then $x \not\rightarrow y$ in $\text{STG}(F)$.*
- *When $X \subseteq \mathbb{B}^n$ is SCC-closed in $\text{STG}(\Gamma(I))$, X is SCC-closed in $\text{STG}(F)$.*
- *When $X \subseteq \mathbb{B}^n$ is a trap set in $\text{STG}(\Gamma(I))$, X is a trap set in $\text{STG}(F)$.*
- *If $x \in \mathbb{B}^n$ is not a part of any cycle in $\text{STG}(\Gamma(I))$, then it is not a part of any cycle in $\text{STG}(F)$.*

Of course, Lemma 1 suggests that the degree of over-approximation within $\text{STG}(\Gamma(I))$ is quite high, and it is indeed easy to find examples of useful properties that do not transfer to skeleton graphs. For example, it is well know that given an acyclic influence graph I, any Boolean network F consistent with I has an acyclic $\text{STG}(F)$ [17]. However, it is easy to find cases where $\text{STG}(\Gamma(I))$ is not acyclic even though I is. It may thus seem questionable how useful Theorem 1 really is in practice. We attempt to address this empirically as part of *Results*.

Variable Stabilisation in Model Pools. To demonstrate the practical merit of (strict) monotonic model pools, we propose the problem of variable stabilisation. In biological applications, sustained activity (or inactivity) of a particular variable within the network is often associated with a known biological phenotype. By studying which variables eventually become stable in the long-term, we get an indication as to what phenotypes can be expressed by a particular network.

Definition 7. *Let F be a BN of n variables. We call a variable $i \in [1, n]$ observably stable if there exists an attractor $A \subseteq \mathbb{B}^n$ such that $\forall x, y \in A. x_i = y_i$.*

Intuitively, observable stability guarantees that there is *some* long-term outcome where the system no longer updates the chosen variable. However, this does not *guarantee* that this outcome is always achieved, nor that it is achievable from an arbitrary initial state. Achieving these stronger properties across all networks in a monotonic model pool is very unlikely due to the high variability of the possible update functions.

Nevertheless, when considering a monotonic model pool $M_{[!]}(I)$, a natural question is whether certain variables are observably stable regardless of the choice of $F \in M_{[!]}(I)$. To study this phenomenon, we can expand Corollary 1:

Corollary 2. *Let I be a SIG. If $\text{STG}(\Gamma(I))$ contains a trap set T such that for some i, $\forall x, y \in T. x_i = y_i$, then i is observably stable in every $F \in M_{[!]}(I)$.*

This corollary follows from the previously shown facts that every trap set in $\mathrm{STG}(\Gamma(I))$ is a trap set in $\mathrm{STG}(F)$ and that every trap set contains at least one attractor. Of course, most networks will only contain a small subset of observably stable variables. We may thus also consider a notion of *intervention* that aims to increase this number:

Definition 8. *Given a signed influence graph I and a set of variables $\mathcal{V} \subseteq [1, n]$ (called* intervention*), we write $I_{\mathcal{V}}$ to denote a copy of I where every $i \in \mathcal{V}$ is a stable input. Formally, $(j, i) \in E(I_{\mathcal{V}})$ when $(j, i) \in E(I)$ and $i \notin \mathcal{V}$, plus $(i, i) \in E(I_{\mathcal{V}})$ for every $i \in \mathcal{V}$ (with $L(i, i) = \oplus$). Sign labels of other existing influence edges are preserved.*

Intuitively, such intervention $\mathcal{V}$ simply erases influences targeting variables from $\mathcal{V}$ and turns these variables into stable inputs. In particular, the resulting influence graph $I_{\mathcal{V}}$ may admit more observably stable variables than the original I:

Definition 9. *We say that $\mathcal{V}$ is a* stabilising intervention *of I when all variables are observably stable in every BN within the monotonic pool $M_{[!]}(I_{\mathcal{V}})$.*

In particular, we may be interested in *minimal* stabilising interventions, as these guarantee that every variable is observably stable, but the least amount of variables need to be externally modified. In a practical application, we may further restrict our search to a minimal stabilising intervention of variables for which an intervention is actually biologically feasible. Here, we consider the general case (intervention can be applied to any variable), but a modification focusing on a smaller subset of variables is trivial.

Note that conceptually, this follows the notion of *permanent* perturbations as considered in the more fine-grained control approaches to Boolean networks (e.g. [3,24]). However, we do not prescribe fixed values for the perturbed variables. Again, due to over-approximation, it is hard to ensure that other control strategies (e.g. one-step, temporary or sequential) can achieve the desired result for all networks within a monotonic pool.

3.2   Algorithms and Implementation

Since every skeleton graph can be seen as an asynchronous Boolean network, it can be analysed using the symbolic framework of the tool AEON [1]. Here, the subsets of network states are encoded into Boolean formulae: a state is a member of a set if and only if it is a satisfying valuation of the corresponding formula. Such formula is then stored in the form of a binary decision diagram (BDD) [4], in which it is encoded as a directed acyclic graph that is often much smaller than the formula itself. In this paradigm, set operations can be represented by logical operations on the corresponding formulae ($\cup \equiv \vee$, $\cap \equiv \wedge$, etc.), which correspond to known graph algorithms on the BDDs.

Algorithms for exploration of a Boolean network F then utilise the set operation $\mathrm{VARPOST}_i(X)$. Given a set of states $X \subseteq \mathbb{B}^n$, this operation produces the set of successors reachable by modifying variable $i \in [1, n]$. Formally,

$\text{VARPOST}_i(X) = \{y \mid \exists x \in X.\ x \rightarrow y \wedge y = x[i \leftarrow \overline{x_i}]\}$. Analogously, we can define the operation $\text{VARPRE}_i(X)$ that computes the predecessors with respect to the i-th variable.

Observably Stable Variables. Using this formalism, we can define the $\text{TRAP}(X)$ operation which computes the maximal subset of X that is still a trap set (Algorithm 1). In each step, the algorithm identifies a subset of X that can reach a state outside of X (the set step) and eliminates it from X. Once no further states can be eliminated, X is a trap set and the procedure terminates.

Note that this algorithm uses the idea of saturation [5] when considering transitions using different variables. Specifically, instead of considering successors following all transitions in one step, the individual variables are modified gradually, restarting (using *break*) whenever a modification is found. The resulting procedure has been shown to produce more compact encoding of intermediate sets due to the symmetries that typically appear in asynchronous systems.

Input: An $\text{STG}(F)$ of a BN F and an initial set X.
Output: Largest subset of X that is a trap set.
do
    done $\leftarrow$ *true*;
    for $i \in [1, n]$ do
        step $\leftarrow X \cap \text{VARPRE}_i(\overline{X})$;
        if step $\neq \emptyset$ then
            done $\leftarrow$ *false*;
            $X \leftarrow X \setminus$ step;
            break;
        end
    end
while $\overline{\text{done}}$;
return X;

Algorithm 1: Symbolic trap space computation with saturation.

Using this algorithm and Corollary 2, it is easy to see that a variable is observably stable for every $F \in M_{[\mathbb{I}]}(I)$ when $\text{TRAP}(\{x \in \mathbb{B}^n \mid x_i = true\})$ or $\text{TRAP}(\{x \in \mathbb{B}^n \mid x_i = false\})$ is not empty.

Minimal Stabilising Interventions. In theory, we could iterate through all possible interventions of a particular influence graph (starting with the smaller ones) and detect the cases which ensure that every variable is observably stable using the algorithm proposed above. However, such method quickly becomes unfeasible for networks requiring larger interventions.

As an alternative, the symbolic framework in AEON provides a way to specify m logical *parameters* that can appear in the individual update functions and thus influence the behaviour of the network. By fixing a particular parameter valuation $p \in \mathbb{B}^m$, we obtain a standard Boolean network. However, due to this symbolic representation, we can often perform operations on all the $\mathbb{B}^m$ possible networks as a whole. For example, instead of a subset of states $X \subseteq \mathbb{B}^n$, we can consider a relation $Y \subseteq \mathbb{B}^n \times \mathbb{B}^m$ that can contain different states for different parametrisations. We can also extend the VARPOST_i and VARPRE_i operations such that they operate using these relations.

Intuitively, we use this mechanism to simultaneously consider different possible interventions during one computation. Formally, we adapt the approach proposed within [3]. Here, every update function F_i is modified in the following way, using a fresh parameter p_i:

$$F_i'(x) = (p_i \Rightarrow x_i) \wedge (\overline{p_i} \Rightarrow F_i(x)) \tag{3}$$

Depending on p_i, such function F_i' is then either constant, or equivalent to F_i. In other words, parameter p_i triggers intervention of variable i. We thus call the introduced p_i *intervention parameters*. Subsequently, we can define operation $\text{TRAPINT}(X)$ based on $\text{TRAP}(X)$ that instead of the actual trap set computes the subset of $\mathbb{B}^m$ for which X contains a non-empty trap set. Due to the encoding that considers all possible interventions simultaneously, the pseudocode is almost identical, with the only exception being that the information regarding network states is eliminated in the end, leaving only a set of parametrisations.

Consequently, we can obtain the set of all possible stabilising interventions by computing:

$$\bigcap_{i \in [1,n]} \bigcup_{b \in \mathbb{B}} \text{TRAPINT}(\{ (x,y) \in \mathbb{B}^n \times \mathbb{B}^m \mid x_i = b \}) \tag{4}$$

Once this set is know, we use a simple branch-and-bound search of the set elements to identify the minimal valuation of the intervention parameters that still results in variable stabilisation. For cases where the minimal intervention is not trivial (e.g. five or more variables), this approach is generally more efficient than the previously discussed naive method of testing individual perturbations in isolation. This is because even though the explicit search still has to be performed by the branch-and-bound algorithm, it is only performed on the pruned set of interventions that are guaranteed to be valid.

4   Results

To asses the practical utility of (strict) monotonic model pools, we rely on the repository of benchmark models[1] that are available alongside the tool AEON [1]. This repository contains 145 real-world Boolean models that are available across various biological databases and tool repositories [8,13,15,16]. The code and other data relevant for reproducibility of the presented results is available at https://doi.org/10.5281/zenodo.6659615.

---

[1] https://github.com/sybila/biodivine-boolean-models.

Data Filtration and Preprocessing. Not all studied models incorporate a proper signed influence graph. Some older models completely lack this information. Meanwhile, some Boolean networks contain functions that are not consistent with their published influence graph (i.e. $F \not\models I$). Finally, some networks simply contain update functions that are not locally monotonic.

We thus implement the following preprocessing step: For each network, the published influence graph I is erased and a new graph I' is inferred to exactly match the published update functions of the network (this can be trivially implemented by inspecting the function's truth table). This step also filters out networks with functions that are not locally monotonic. This eliminates 15 out of the 145 networks (i.e. only approx. 10%).

A second preprocessing step is concerned with inputs and constants. As already suggested, the difference between a constant and a stable input with regards to long-term behaviour is mostly technical. The actual format of the published model thus largely depends on the author's preference. To avoid discrepancies in this regard, every constant variable is transformed into a stable input (existing stable or unstable inputs are left unchanged).

Finally, the resulting influence graph is used to generate two skeleton Boolean networks: One *basic* skeleton as outlined in Definition 5, and a second skeleton with *intervention parameters* as outlined within the *Methods* section. Both of these networks can be then loaded into the tool AEON for further analysis.

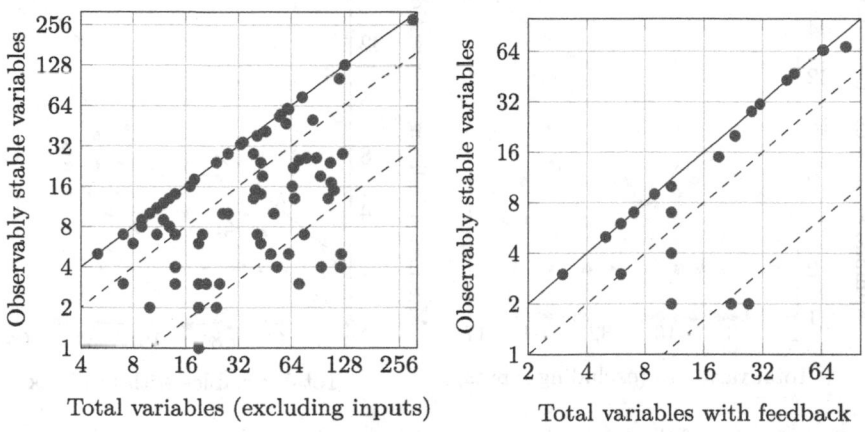

Fig. 1. Left: Number of observably stable variables with respect to the network size (excluding stable inputs). Right: Number of observably stable variables within non-trivial feedback cycles. All axis are in logarithmic scale. Dashed diagonal lines show points below 50% and 10% of the x-axis (variable count).

Observably Stable Variables. The first experiment we perform is detection of observably stable variables in the basic monotonic pools of each network. Here, the analysis can be easily performed using Algorithm 1 for all networks within a few seconds. Since stable inputs are inherently observably stable in any network, they are not included in the resulting dataset.

We find that there are 37 networks with no detected observably stable variables. The remaining 93 networks contain at least one observably stable variable. Interestingly, there are 26 networks where every variable is observably stable. Of these 26 networks, 17 are acyclic. The ratio of observably stable variables with respect to the network size is then as plotted in Fig. 1 (left).

Another important aspect influencing variable stability are feedback cycles within the influence graph. Variables that are not part of such a cycle can be often easily shown to stabilise in any monotonic model. We may thus consider limiting our inquiry only to variables that are members of at least one non-trivial (i.e. not self-loop) feedback cycle.

To address this question, Fig. 1 (right) depicts a plot that shows the relative number of observably stable variables that are members of at least one feedback cycle. Note that this excludes previously mentioned 17 networks that are completely acyclic, as well as 41 other networks where no variable on a feedback cycle was found to be observably stable.

This still leaves 13 networks with non-trivial cycles where all variables are observably stable. Additionally, 24 other networks have at least one observably stable variable within a non-trivial feedback cycle.

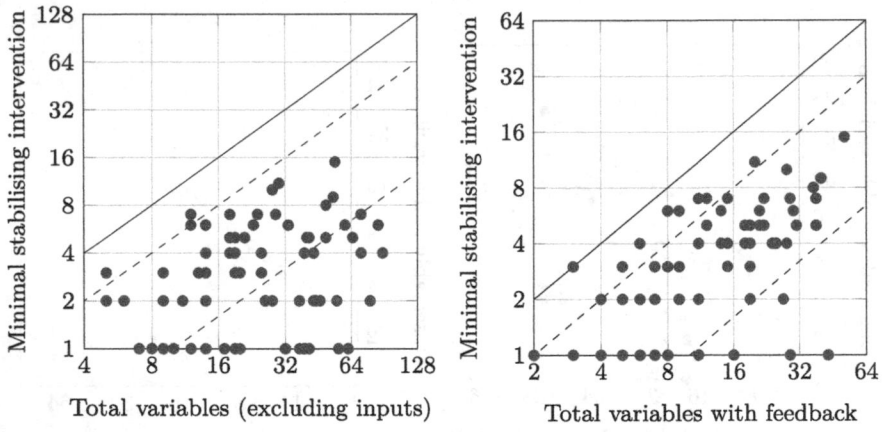

Fig. 2. Left: Size of the minimal stabilising intervention with respect to the network size (excluding stable inputs). Right: Size of the minimal stabilising intervention with respect to the number of network variables on non-trivial feedback cycles. All axis are in logarithmic scale. Dashed diagonal lines show points below 50% and 10% of the x-axis (variable count).

Stabilising Interventions. For our second experiment, we consider the notion of stabilising interventions. First, a monotonic pool with intervention parameters is generated for each network. Afterwards, we use the trap set method with intervention parameters to infer the possible set of interventions leading to stabilisation. We then identify the smallest intervention leading to observable stabilisation of *all* network variables.

Overall, this computation is much more demanding, and as a result, there were 21 networks where a timeout of 24 h was reached before a solution was computed. However, it should be noted that in all cases, the memory usage was still very reasonable and it is conceivable that a higher timeout could solve even these problem instances. Also, in all cases the timeout occurred in the symbolic trap set computation, not in the branch-and-bound search algorithm.

Out of the 109 networks, 26 networks already have all variables observably stable, leaving 83 networks for which non-trivial stabilising intervention was discovered via this method. Out of these 83 networks, 16 (19%) only require intervention of size one and 15 (18%) intervention of size two. This is in line with existing knowledge about Boolean networks being typically controllable by a small number of variables [2]. Nevertheless, we still observe that 6/83 networks require an intervention of 10 or more variables. In Fig. 2, we show how the size of the minimal intervention relates to the network size and to the number of variables on feedback cycles.

5   Discussion

In this paper, we considered the notion of monotonic model pools and their corresponding skeleton state-transition graphs to uncover high-level properties of logical models based on the structure of their influence graph. Consequently, we proposed a simplified over-approximation theorem relating the skeleton state-transition graph to the *strict* monotonic pool without unnecessary constant variables. Considering a symbolic encoding of the skeleton network, we then analysed the prospects of long-term stabilisation in individual network variables. We also attempted to uncover minimal disruptions to the network structure that guarantee the possibility of long-term stabilisation.

From our survey of monotonic pools of a large collection of real-world influence graphs, we can draw several relevant conclusions: First, observably stable variables are common in most networks (93/130). However, upon closer inspection, they are largely delegated to "mediator" variables that are otherwise not part of any feedback cycle. If we consider only variables with non-trivial feedback, the ratio is substantially lower (37/113; excluding models without feedback cycles). If we instead focus on the interventions leading to variable stabilisation, we find that most networks only require small interventions (3 or fewer variables), which is in line with our expectations. Nevertheless, there is also a non-negligible portion of outliers that require an intervention of size 10 or more.

Another important observation is that using the symbolic encoding, we were able to analyse variable stability in all monotonic pools and detect minimal interventions in 109 out of 130 pools (within one day of compute time). This suggests that the pool skeletons can indeed be efficiently analysed using existing algorithmic techniques and can be of practical importance in logical modelling.

However, it should be noted that the obtained results appear to be largely comparable to the existing Thomas conjectures and the output of other static analysis tools. Specifically, we find that the observably stable variables are those that do not appear on any feedback cycle with an odd number of negative regulations (which is know necessary condition of variable instability). We also observe that all the discovered minimal stabilising interventions cause the elimination of such cycles from the influence graph entirely.

Nevertheless, we believe that the results of the method are still valuable in that they not only identify the structural properties of the model (observably stable variables, stabilising interventions) but give a full characterisation of the trap sets that support these claims. These could then be beneficial when performing subsequent analysis of individual networks within the monotonic pool. For example, attractor analysis of any network in the pool can be restricted only to the trap sets of the skeleton network, which are often easier to compute.

Finally, the proof of Theorem 1 suggests that a similar over-approximation result could be obtained not only for general influence graphs, but for any logical model that incorporates *some* unknown dynamics. As future work, in such a hybrid approach, the skeleton network could then provide more precise overview of the studied system, as the over-approximation would be limited to the truly unknown interactions within the network (as opposed to the whole interaction graph in this case).

References

1. Beneš, N., Brim, L., Kadlecaj, J., Pastva, S., Šafránek, D.: AEON: attractor bifurcation analysis of parametrised Boolean networks. In: Lahiri, S.K., Wang, C. (eds.) CAV 2020. LNCS, vol. 12224, pp. 569–581. Springer, Cham (2020). https://doi.org/10.1007/978-3-030-53288-8_28
2. Borriello, E., Daniels, B.C.: The basis of easy controllability in Boolean networks. Nat. Commun. **12**(1), 1–15 (2021)
3. Brim, L., Pastva, S., Šafránek, D., Šmijáková, E.: Robust control of partially specified Boolean networks. arXiv preprint arXiv:2202.13440 (2022)
4. Bryant, R.E.: Graph-based algorithms for Boolean function manipulation. Comput. IEEE Trans. **100**(8), 677–691 (1986)
5. Ciardo, G., Marmorstein, R., Siminiceanu, R.: The saturation algorithm for symbolic state-space exploration. Int. J. Softw. Tools Technol. Transfer **8**(1), 4–25 (2006)
6. De Jong, H.: Modeling and simulation of genetic regulatory systems: a literature review. J. Comput. Biol. **9**(1), 67–103 (2002)
7. Eisenack, K., Petschel-Held, G.: Graph theoretical analysis of qualitative models in sustainability science. In: Working Papers of 16th Workshop on Qualitative Reasoning, pp. 53–60 (2002)
8. Helikar, T., et al.: The cell collective: toward an open and collaborative approach to systems biology. BMC Syst. Biol. **6**(1), 1–14 (2012)
9. Kauffman, S.A.: Metabolic stability and epigenesis in randomly constructed genetic nets. J. Theor. Biol. **22**(3), 437–467 (1969)
10. Kaufman, M., Soulé, C., Thomas, R.: A new necessary condition on interaction graphs for multistationarity. J. Theor. Biol. **248**(4), 675–685 (2007)

11. Klarner, H., Bockmayr, A., Siebert, H.: Computing maximal and minimal trap spaces of Boolean networks. Nat. Comput. **14**(4), 535–544 (2015). https://doi.org/10.1007/s11047-015-9520-7
12. Kuipers, B.: Commonsense reasoning about causality: deriving behavior from structure. Artif. Intell. **24**(1–3), 169–203 (1984)
13. Le Novere, N., et al.: Biomodels database: a free, centralized database of curated, published, quantitative kinetic models of biochemical and cellular systems. Nucleic Acids Res. **34**, D689–D691 (2006)
14. Mandon, H., Su, C., Haar, S., Pang, J., Paulevé, L.: Sequential reprogramming of Boolean networks made practical. In: Bortolussi, L., Sanguinetti, G. (eds.) CMSB 2019. LNCS, vol. 11773, pp. 3–19. Springer, Cham (2019). https://doi.org/10.1007/978-3-030-31304-3_1
15. Naldi, A., Berenguier, D., Fauré, A., Lopez, F., Thieffry, D., Chaouiya, C.: Logical modelling of regulatory networks with GINsim 2.3. Biosystems **97**(2), 134–139 (2009)
16. Ostaszewski, M., et al.: COVID-19 disease map, building a computational repository of SARS-CoV-2 virus-host interaction mechanisms. Sci. Data **7**(1), 1–4 (2020)
17. Paulevé, L., Richard, A.: Static analysis of Boolean networks based on interaction graphs: a survey. Electron. Notes Theor. Comput. Sci. **284**, 93–104 (2012)
18. Remy, É., Ruet, P., Thieffry, D.: Graphic requirements for multistability and attractive cycles in a Boolean dynamical framework. Adv. Appl. Math. **41**(3), 335–350 (2008)
19. Richard, A.: Negative circuits and sustained oscillations in asynchronous automata networks. Adv. Appl. Math. **44**(4), 378–392 (2010)
20. Richard, A., Comet, J.P.: Necessary conditions for multistationarity in discrete dynamical systems. Discret. Appl. Math. **155**(18), 2403–2413 (2007)
21. Schwieger, R., Siebert, H.: Graph representations of monotonic Boolean model pools. In: Feret, J., Koeppl, H. (eds.) CMSB 2017. LNCS, vol. 10545, pp. 233–248. Springer, Cham (2017). https://doi.org/10.1007/978-3-319-67471-1_14
22. Schwieger, R., Siebert, H.: Structure and behavior in Boolean monotonic model pools. Biosystems **214**, 104610 (2022)
23. Soulé, C.: Graphic requirements for multistationarity. ComPlexUs **1**(3), 123–133 (2003)
24. Su, C., Paul, S., Pang, J.: Controlling large Boolean networks with temporary and permanent perturbations. In: ter Beek, M.H., McIver, A., Oliveira, J.N. (eds.) FM 2019. LNCS, vol. 11800, pp. 707–724. Springer, Cham (2019). https://doi.org/10.1007/978-3-030-30942-8_41
25. Thomas, R.: On the relation between the logical structure of systems and their ability to generate multiple steady states or sustained oscillations. In: Numerical Methods in the Study of Critical Phenomena, pp. 180–193. Springer (1981). https://doi.org/10.1007/978-3-642-81703-8_24
26. Thomas, R., Kaufman, M.: Multistationarity, the basis of cell differentiation and memory II. Logical analysis of regulatory networks in terms of feedback circuits. Chaos Interdisc. J. Nonlinear Sci. **11**(1), 180–195 (2001)

Variable-Depth Simulation of Most Permissive Boolean Networks

Théo Roncalli and Loïc Paulevé[(✉)]

Univ. Bordeaux, CNRS, Bordeaux INP, LaBRI, UMR 5800, 33400 Talence, France
theo.roncalli@universite-paris-saclay.fr, loic.pauleve@labri.fr

Abstract. In systems biology, Boolean networks (BNs) aim at modeling the qualitative dynamics of quantitative biological systems. Contrary to their (a) synchronous interpretations, the Most Permissive (MP) interpretation guarantees capturing all the trajectories of any quantitative system compatible with the BN, without additional parameters. Notably, the MP mode has the ability to capture transitions related to the heterogeneity of time scales and concentration scales in the abstracted quantitative system and which are not captured by asynchronous modes. So far, the analysis of MPBNs has focused on Boolean dynamical properties, such as the existence of particular trajectories or attractors.

This paper addresses the sampling of trajectories from MPBNs in order to quantify the propensities of attractors reachable from a given initial BN configuration. The computation of MP transitions from a configuration is performed by iteratively discovering possible state changes. The number of iterations is referred to as the *permissive depth*, where the first depth corresponds to the asynchronous transitions. This permissive depth reflects the potential concentration and time scales heterogeneity along the abstracted quantitative process. The simulation of MPBNs is illustrated on several models from the literature, on which the depth parametrization can help to assess the robustness of predictions on attractor propensities changes triggered by model perturbations.

1 Introduction

Boolean networks (BNs) have been employed to model the temporal evolution of gene expression and protein activities in biological systems [5,7,15,16]. A BN is composed of a finite set of components having two states, either 0 (false) or 1 (true). The BN then specifies in which contexts the components can change state, such as "component 3 can switch to state 1 if and only if component 1 is in state 0 *and* component 2 is in state 1; in all other contexts, component 3 can only switch to 0". These rules can be given as one Boolean function per component, associating each possible configuration (which associates each component to a state) to a Boolean value: in our example, the function of component 3 is f_3 : $\{0,1\}^n \to \{0,1\}$ with $f_3(x) = \neg x_1 \wedge x_2$, where $\neg$ and $\wedge$ denote the logical negation and conjunction, respectively.

The *execution* of a BN relies on a given *update mode* which specifies how are computed the evolution of the states of components. In the systems biology

I. Petre and A. Păun (Eds.): CMSB 2022, LNBI 13447, pp. 138–157, 2022.
https://doi.org/10.1007/978-3-031-15034-0_7

literature, two main update modes are widely employed: the *synchronous* (or *parallel*) and *fully-asynchronous*. In synchronous, each component is updated simultaneously: the configuration $x = (x_1, x_2, x_3)$ evolves in one step to $(f_1(x), f_2(x), f_3(x))$. In fully-asynchronous, only one component can be updated at a time, possibly leading to non-deterministic transitions: the configuration x can evolve either to $(f_1(x), x_2, x_3)$, $(x_1, f_2(x), x_3)$, or $(x_1, x_2, f_3(x))$. Whatever the chosen update mode, as there is a finite number of configurations, an execution eventually reaches a set of *limit* configurations which will be visited infinitely often. These sets of configurations are called *attractors*: each execution eventually reaches and stays within one attractor. Attractors are prominent dynamical features when modeling biological processes: they represent stable behaviors and are usually associated to cellular phenotypes.

Computing the attractors *reachable* from a given initial configuration is at the core of many studies of biological processes with BNs [1,2,6,7,11]. These studies typically involve comparing the effect of a network mutation (forcing some components to have fixed value) on the sets of reachable attractors and their propensities. This later notion is often related to the number of paths leading from the initial configuration to each attractor: under some mutations, the same set of attractors may still be reachable, but the proportion of trajectories leading to them may substantially differ. This motivated the development of *simulation* algorithms for BNs in order to sample trajectories and quantify the propensities to reach attractors. These methods replace the non-determinism of asynchronous transitions by a probabilistic choice [6,14].

Whenever employed as qualitative models of quantitative systems, dynamics of BNs aim at giving a coarse-grained view of system dynamics without requiring numerous quantitative parameters. However, the Boolean (a)synchronous modes are not correct abstractions of quantitative dynamics [8]: they lead at the same time to predict spurious transitions and, importantly, preclude transitions which are actually possible when considering delays for instance. Let us illustrate this with the following BN, denoted (A) in the rest of the text:

$$f_1(x) = 1 \qquad f_2(x) = x_1 \qquad f_3(x) = (\neg x_1 \wedge x_2) \vee x_3 \qquad \text{(A)}$$

From the initial configuration where all components are 0, we write 000, there is only one possible transition: the activation of 1, leading to the configuration 100. Then, again, only one transition is possible, activating 2, thus leading to 110. There, the execution will stay infinitely on this configuration: no other state changes are possible: from 000 it is impossible to eventually activate 3, and {110} is the only reachable attractor. However, it is known that this system can actually activate 3 for a range of kinetics, as observed experimentally [13], and easily captured with quantitative models [4,12]. Indeed, consider that 1 has actually several activation levels: one intermediate $1/2$ which is sufficient to activate 2 but not enough sufficient to inhibit 3: when in $1/200$, 2 can be activated, going to $1/210$, and then 3 can change state, going to $1/211$. When 1 becomes fully active (111), 3 will self-maintain its activation. Thus, a correct Boolean analysis of the BN f above should conclude that from 000, two attractors are reachable:

{110}, when 1 goes rapidly to its maximum level, and {111} when 3 had time to activate before the full activation of 1. This example shows the limit of (a) synchronous interpretations of BNs when used as abstraction of quantitative systems: they enforce that the Boolean 0 matches with the quantitative 0, and the Boolean 1 matches with the quantitative non-zero (> 0), making impossible to capture transitions happening at different activation levels or different time scales.

The *Most Permissive* (MP) update mode of BNs [8,9] is a recently-introduced execution paradigm which enables capturing dynamics precluded by the asynchronous ones. The main idea behind the MP update mode is to systematically consider a potential delay when a component changes state, and consider any additional transitions that could occur if the changing component is in an intermediate state. It can be modeled as additional *dynamic* states "increase" ($\nearrow$) and "decrease" ($\searrow$): when a component can be activated, it will first go through the "increase" state where it can be interpreted as either 0 or 1 by the other components, until eventually reaching the Boolean 1 state. With the previous example, starting from 000, the first component is put as increasing, thus going to the MP configuration $\nearrow 00$. In this configuration, 2 can be activated because $\nearrow$ can be interpreted as 1, leading to $\nearrow\nearrow 0$. Then, 3 can be activated, as it can interpret the dynamic state of 1 as the Boolean 0, and 2 as the Boolean 1. This model the fact that the component 1 is not high enough for inhibiting 3, while 2 is high enough to activate it. Thus, in this example, there is a trajectory from 000 to 111, i.e., a configuration where all the components are active. As 111 is a fixed point of f, the MP analysis would thus conclude that two attractors are reachable from 000: {110} and {111}.

The MP update mode brings a formal abstraction property to BN dynamics with respect to quantitative models, without requiring additional parameters: essentially, MPBNs capture any behavior that is achievable by any quantitative model being compatible with the logic of the BN. This includes, for instance, models which result from introducing quantitative parameters, such as transition speed and interaction thresholds. We give here a brief informal overview of the property (see [8] for details): let us consider multivalued networks (MNs) of dimension n where components can have values in $\mathbb{M} = \{0,\ldots,m\}$ for some $m \in \mathbb{N}_{>0}$. A MN can be considered as map from multivalued configurations to the derivative of the value of the components, i.e., of the form $F : \mathbb{M}^n \to \{-1,0,1\}$. A multivalued configuration $z \in \mathbb{M}^n$ can be *binarized* by associating components with value 0 to the Boolean state 0, components with value m to the Boolean state 1, and other components to any Boolean state. A MN F is a *refinement* of a BN f whenever for any multivalued configuration z and for each component i, if component i increases (resp. decreases), i.e., $F_i(z) > 0$ (resp. $F_i(z) < 0$), there exists a binarization of this configuration such that f_i is evaluated to 1 (resp. to 0). Then, for any pair of Boolean configurations x and y, if there exists a trajectory from $m \cdot x$ to $m \cdot y$ in the asynchronous dynamics of the MN, there necessarily exists a trajectory from x to y in the MP dynamics of the BN. Moreover, to any transition computed according the MP update mode, there is refinement of the BN which realizes it. One of the major consequences of the abstraction property is that if a configuration is not reachable from another with the MP update mode, then no quantitative models being compatible with the BN can produce the trajectory. In addition, the MP mode has a lower computational complexity for computing reachability and attractor properties, enabling formal analysis of genome-scale BNs [8,10].

However, the simulation of MPBNs, i.e., the sampling of transitions following the MP update mode, has not been addressed so far. Thus, besides computing Boolean properties, such as the existence or absence of reachable attractors, there is no algorithm nor tools to approach the effect of a mutation on the propensities to reach attractors, as we mentioned above with asynchronous BNs.

In this paper, we present a first algorithm for sampling trajectories of BNs with the MP update mode, subject to additional simulation parameters for assigning probabilities to the transitions. The MP transitions enabled from a single configuration are computed iteratively, and we refer to the number of times this iteration is performed as the *depth* of the MP computation. At depth 1, the transitions match with the asynchronous update mode, but further depths bring additional behaviors. These iterations capture possibly different time scales: while component 1 is changing (depth 1), 2 can change (depth 2); then while 2 is changing, 3 can change (depth 3), etc. In our simulation algorithm, the probability of a transition can be affected by its depth and the number of components it changes simultaneously. Thus, similarly to [6,14] with the fully-asynchronous mode, our sampling of trajectories can be assimilated to a random walk in the MP dynamics, where the probability of transitions can be tuned with the simulation parameters. As we will show on case studies, the MP interpretation can lead to drastic changes in the predicted probabilities of reachable attractors, enabling assessing the robustness of prediction to the heterogeneity of time scales and concentration scales in the quantitative system captured by the discrete MP dynamics. Nevertheless, the simulation parameters and derived transition probabilities are empirical and cannot be formally related to a putative abstracted system.

2   Background

2.1   Boolean Networks and Dynamics

A *Boolean network* (BN) of dimension n is a function $f : \mathbb{B}^n \to \mathbb{B}^n$, with $\mathbb{B} = \{0, 1\}$. For each $i \in \{1, \ldots, n\}$, $f_i : \mathbb{B}^n \to \mathbb{B}$ is the *local function* of component i. The Boolean vectors $x \in \mathbb{B}^n$ are the *configurations* of f, where for each $i \in \{1, \ldots, n\}$, x_i is the state of i. Given two configurations $x, y \in \mathbb{B}^n$, the components having a different state are noted $\Delta(x, y) = \{i \in \{1, \ldots, n\} \mid x_i \neq y_i\}$.

The *influence graph* $G(f)$ of a BN f is a signed digraph whose nodes are the components and edges mark the dependencies between them in the local functions: for all $i, j \in \{1, \ldots, n\}$, there is an edge $i \xrightarrow{s} j$ in $G(f)$ with $s \in \{-1, +1\}$ if and only if there exists a configuration $x \in \mathbb{B}^n$ with $x_i = 0$ such that $s = f_j(x_1, \ldots, x_{i-1}, 1, x_{i+1}, \ldots, x_n) - f_j(x)$: the sole increasing of component i

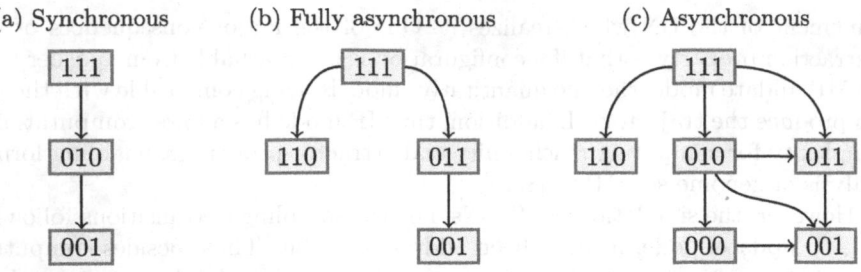

Fig. 1. Transitions of the BN (B) from the configuration 111 with different modes

causes f_j to increase ($s = +1$) or decrease ($s = -1$). Note there may exist i and j so that both $i \xrightarrow{+1} j$ and $i \xrightarrow{-1} j$ are in $G(f)$, for instance with $f_j(x) = x_i$ xor x_k. A BN f is *locally monotone* whenever there is no $i, j \in \{1, \ldots, n\}$ such that both $i \xrightarrow{+1} j$ and $i \xrightarrow{-1} j$ are in $G(f)$, i.e., each of its local function is *unate*. Computing $G(f)$ is a DP-complete problem (both in NP and coNP) [3]. For the example BN f of Eq.(A), $G(f) = \{1 \xrightarrow{+1} 2, 1 \xrightarrow{-1} 3, 2 \xrightarrow{+1} 3, 3 \xrightarrow{+1} 3\}$: it is locally monotone.

An *update mode* μ of f specifies a binary transition relation between configurations $\rightarrow_\mu \subseteq \mathbb{B}^n \times \mathbb{B}^n$. Classical update modes include the *synchronous* (or *parallel*) update mode where $x \rightarrow_s y$ iff $x \neq y$ and $y = f(x)$; the *fully-asynchronous* update mode where $x \rightarrow_a y$ iff x and y differ on only one component i and $y_i = f_i(x)$; and the (general) *asynchronous* update mode where $x \rightarrow_g y$ iff $x \neq y$ and for each component $i \in \Delta(x, y), y_i = f_i(x)$.

Given an update mode μ, a configuration $y \in \mathbb{B}^n$ is *reachable* from a configuration $x \in \mathbb{B}^n$, noted $x \rightarrow_\mu^* y$, if and only if either $x = y$ or there exists a sequence of transitions $x \rightarrow_\mu \cdots \rightarrow_\mu y$. A non-empty set of configurations $A \subseteq \mathbb{B}^n$ is an *attractor* if and only if for each pair of configurations $x, y \in A$, y is reachable from x, and there is no configuration $z \in \mathbb{B}^n \setminus A$ that is reachable by a configuration in A. Remark that attractors are the bottom strongly connected components of the digraph $(\mathbb{B}^n, \rightarrow_\mu)$. Whenever A is a singleton configuration $\{x\}$, it is said to be a *fixed point* of the dynamics. Otherwise, A is a cyclic attractor. In the case of (a) synchronous and MP update modes, the fixed points of the dynamics match exactly with the fixed points of f, i.e., the configurations $x \in \mathbb{B}^n$ such that $f(x) = x$. Finally, the *strong basin* of an attractor A is the set of configurations that can reach A and no other distinct attractor.

Figure 1 shows the transitions computed with the above defined update modes with the BN f defined as follows:

$$f_1(x) = x_1 \wedge \neg x_3 \qquad f_2(x) = x_1 \qquad f_3(x) = \neg x_1 \tag{B}$$

This model has two attractors, being fixed points 001 and 110. With the synchronous mode, only 001 is reachable from 111, whereas both are reachable with the asynchronous modes.

2.2   Sub-hypercubes and Closures

The MP update mode and the algorithm presented in this paper gravitate around the notion of sub-hypercube of $\mathbb{B}^n$ and their partial closure by f.

A Boolean *sub-hypercube* of dimension n is specified by a vector in $\{0, 1, *\}^n$, where components having value $*$ are said *free*, otherwise they are *fixed*. The number of free components in a sub-hypercube $h \in \{0, 1, *\}^n$ is denoted by $\mathrm{rank}(h) = |\{i \in \{1, \ldots, n\} \mid h_i = *\}|$. A sub-hypercube h has $2^{\mathrm{rank}(h)}$ vertices, denoted by $c(h) = \{x \in \mathbb{B}^n \mid \forall i \in \{1, \ldots, n\}, h_i \in \mathbb{B} \Rightarrow x_i = h_i\}$. A sub-hypercube $h' \in \{0, 1, *\}^n$ is *smaller* than a sub-hypercube $h \in \{0, 1, *\}^n$ whenever for each $i \in \{1, \ldots, n\}$ fixed in h ($h_i \in \mathbb{B}$), it is fixed to the value in h' ($h'_i = h_i$). Then, remark that $c(h') \subseteq c(h)$.

A sub-hypercube h is *closed* by a BN f if the result of f applied to any of its vertices is one of its vertices: $\forall x \in c(h), f(x) \in c(h)$; it is also known as a *trap space*. Given components $K \subseteq \{1, \ldots, n\}$, h is *K-closed* by f whenever, for each component $i \in K$, either i is free in h, or f_i applied on any vertices of h results in the fixed value h_i. In other words, for all configurations in the K-closed sub-hypercube h, the next states of the components $i \in K$ are in h:

$$\forall x \in c(h), \forall i \in K, h_i \neq * \Rightarrow f_i(x) = h_i. \tag{1}$$

Let us consider the BN f defined in (B). The sub-hypercube $h = 0 * *$ is closed by f, where $c(0 * *) = \{000, 010, 001, 011\}$. This trap space indicates that component 1 can never get activated once deactivated. The sub-hypercube $h = 0 * 1$ is $\{1, 2\}$-closed by f as $f_1(001) = f_1(011) = 0$. Sub-hypercubes $1 * 0$ and $* * 0$ are also $\{1, 2\}$-closed by f, where $1 * 0$ is smaller than $* * 0$. In contrast, $1 * 1$ is not $\{1, 2\}$-closed by f.

2.3   The Most Permissive Update Mode

Whenever a component changes state, e.g., increases from its minimal value 0 to its maximal value 1, the MP mode captures any behavior that could arise in the course of this change. For example, it may be that at some point, the component becomes high enough to activate one of its targets, whereas it remains not high enough to activate another one (because it has not reached its maximal value yet). These *dynamic states* can be captured by sub-hypercubes, where the changing components are free: they can be read both as 0 or 1.

Formally, the MP update mode can be defined as follows. Given a set of components $K \subseteq \{1, \ldots, n\}$, let us denote by $h^{\langle x, K \rangle}$ the *smallest* sub-hypercube of dimension n that contains x and that is K-closed by f. There is an MP transition from x to y whenever (a) y is a vertex of $h^{\langle x, K \rangle}$, and (b) the new state of all the components in K can be computed from $h^{\langle x, K \rangle}$:

$$\forall x, y \in \mathbb{B}^n, \quad x \rightarrow_{\mathrm{MP}} y \iff \exists K \subseteq \{1, \ldots, n\} : y \in c(h^{\langle x, K \rangle})$$
$$\wedge \, \forall i \in K, \exists z \in c(h^{\langle x, K \rangle}) : y_i = f_i(z) . \tag{2}$$

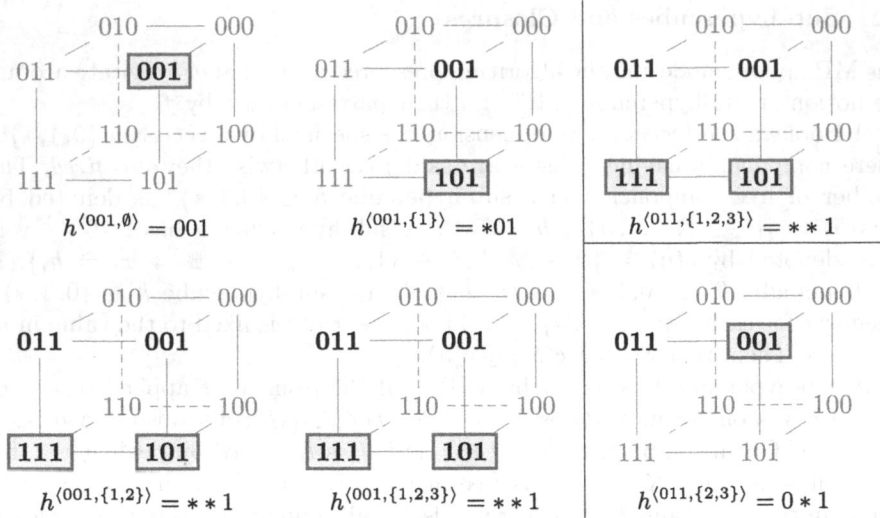

Fig. 2. Computation of sub-hypercubes with BN $f(x) = (1, x_1, (\neg x_1 \wedge x_2) \vee x_3)$

The abstraction properties and complexity results are detailed in [8]. One can remark that the transition and reachability relations are identical: y is reachable from x if and only if $x \rightarrow_{MP} y$.

As this formalism serves as the cornerstone for the simulation algorithm, let us illustrate it by using the BN (A). Figure 2 depicts the smallest sub-hypercubes with respect to a configuration x and a set K. Bold configurations belong to $h^{\langle x, K \rangle}$ and those satisfying reachability condition (b) are boxed. The left side focuses on the initial configuration 001 and shows an iterative computation of $h^{\langle 001, K \rangle}$ from $K = \emptyset$ to $K = \{1, 2, 3\}$. As component 1 can flip in configuration 001, one obtains $h^{\langle 001, \{1\} \rangle} = *01$. If component 1 is active, component 2 can flip, leading to $h^{\langle 001, \{1,2\} \rangle} = **1$. For all configurations $z \in c\left(h^{\langle 001, \{1,2\} \rangle}\right)$, component 3 stays active: $f_3(z) = 1$. Hence, one obtains $h^{\langle 001, \{1,2,3\} \rangle} = **1$. Here, some reachable configurations from x can be deduced. For instance, transitions 001 $\rightarrow_{MP}$ 101 and 001 $\rightarrow_{MP}$ 111 exist because reachability properties (a) and (b) are verified, However, 000 and 011 do not verify properties (a) and (b) respectively. Multiple sub-hypercubes may be required to capture all the MP transitions. This is illustrated in Fig. 2 on the right side with the configuration 011. The sub-hypercube $h^{\langle 011, \{1,2,3\} \rangle}$ does not capture the transition to 001 because property (b) is not satisfied, while $h^{\langle 011, \{2,3\} \rangle}$ does.

3   Simulation Algorithm

3.1   Main Principle

In its basic formal definition (2), the MP update mode defines the possible next configurations by considering all the subsets of components $K \subseteq \{1, \ldots, n\}$ and compute the associated sub-hypercube $h^{\langle x,K\rangle}$, that is the smallest K-closed sub-hypercube containing x. However, part of these subsets may be redundant. Let H be the components free in $h^{\langle x,K\rangle}$ (3). By the minimality of $h^{\langle x,K\rangle}$, it means that for each component $i \in H$, there exists a configuration $y \in c(h^{\langle x,K\rangle})$ such that $f_i(y) \neq x_i$. The set H can then be split in two: the components J for which their local function can be evaluated both to 0 and 1 from the configurations of $h^{\langle x,K\rangle}$ (4), and the components L for which their local function is always evaluated to $\neg x_i$ from any of the configurations of $h^{\langle x,K\rangle}$:

$$H = \left\{ i \in K \mid h_i^{\langle x,K\rangle} = * \right\} , \tag{3}$$

$$J = \left\{ i \in H \mid \exists y, z \in c(h^{\langle x,K\rangle}) : f_i(y) = 0 \wedge f_i(z) = 1 \right\} , \tag{4}$$

$$L = \left\{ i \in H \mid \forall y \in c(h^{\langle x,K\rangle}) : f_i(y) \neq x_i \right\} . \tag{5}$$

We have $H = J \cup L$ and $J \cap L = \emptyset$. Then, remark that in (2), for a fixed x and K, there exists an MP transition $x \rightarrow_{\mathrm{MP}} y$ if and only if $L \subseteq \Delta(x,y) \subseteq H$. Indeed, for each component $i \in \{1,\ldots,n\}$, if $y_i \neq x_i$ then necessarily $i \in H$. Moreover, if $i \in L$, the second condition of (2) imposes that $y_i = \neg x_i$.

Whenever $L = \emptyset$, it results that for any strict subset $K' \subsetneq K$, the transitions generated from the sub-hypercube $h^{\langle x,K'\rangle}$ form a (strict) subset of transitions generated from $h^{\langle x,K\rangle}$: the set of free components is a (strict) subset of H. Therefore, it is useless to explore any subset of K.

Whenever $L \neq \emptyset$, by the K-closeness property of $h^{\langle x,K\rangle}$, changing the state of a component $i \in L$ would be *irreversible* while updating only components in K. Let us illustrate it by using the previous example without the $\vee x_3$ part: $f(x) = (1, x_1, \neg x_1 \wedge x_2)$. With the initial configuration $x = 001$, one can obtain the sub-hypercube $h^{\langle 001,\{1,2,3\}\rangle} = ***$, $L = \{1\}$ and $J = \{2,3\}$. $L = \{1\}$ means that component 1 cannot return to its initial state when it has begun to flip. Moreover, by the definition of MP transitions, all the components in L are modified. Therefore, whenever L is not empty, one should consider all the sub-hypercubes $h^{\langle x,K'\rangle}$ with $(H \setminus L) \subseteq K' \subsetneq H$, to account for all these potential dependencies. Let us consider $K' = H \setminus L = \{2,3\}$ in the previous example. One can obtain $h^{\langle 001,\{2,3\}\rangle} = 00*$ and $L = \{3\}$, which allows discovering a new transition: $001 \rightarrow_{\mathrm{MP}} 000$.

This approach leads us to compute a set of sub-hypercubes $h^{\langle x,K\rangle}$ from subsets K of $\{1,\ldots,n\}$, and where each of them can be characterized by a triplet (x,H,L) to which we refer to as a *space*. Each space (x,H,L) characterizes a set of MP transitions from x where the state of all the components in L is flipped, as well as any subset of components in $H \setminus L$: the transitions $x \rightarrow_{\mathrm{MP}} y$ with $L \subseteq \Delta(x,y) \subseteq H$. In our algorithm, the set of transitions generated by a space is never enumerated explicitly, as it is exponential in $|H \setminus L|$. Instead, we count the number of transitions changing m components, from $m = |L|$ (or 1 whenever L is empty) to $|H|$, i.e., the binomial coefficient $\binom{|H \setminus L|}{m-|L|}$.

Overall, the sampling of the configuration following x can be summarized with the following steps:

1. compute a set of spaces $S = \{(x, H^1, L^1), \ldots, (x, H^q, L^q)\}$, corresponding to the sub-hypercubes $\{h^{\langle x, K^1 \rangle}, \ldots, h^{\langle x, K^q \rangle}\}$: starting with $K^1 = \{1, \ldots, n\}$, we consider the sub-hypercubes closed along $K^1 \setminus \ell$ with each $\ell \subseteq L^1$, and recursively (thus whenever $L^1 = \emptyset$, $q = 1$).
2. for each space, count the number of transitions it can generate per number of changing components;
3. generate one random number to select the space (x, H, L) and the number m of components to flip;
4. randomly select $m - |L|$ components in $H \setminus L$, let us denote by C these chosen components;
5. flip the state of the components in $L \cup C$.

The size of sets H and L is at most n. However, the number of spaces q can be exponential with n in the worst case.

Now, let us focus on the computation of $h^{\langle x, K \rangle}$, the smallest sub-hypercube containing x and K-closed by f. The main principle is to start from the sub-hypercube of rank 0 with x as the sole vertex, and iteratively free components to fulfill the K-closure property. To do so, we collect the set of components (among K) which can flip of state from at least one vertex of the sub-hypercube: for each $i \in K$, if there exists a vertex y of the sub-hypercube such that $f_i(y) \neq x_i$, then i must be free to verify the closeness property. This process is then repeated until the K-closeness property is verified, which in the worst case may require n iterations. It appears that the transitions generated by the asynchronous update mode match with the components marked as free in the first iteration only: the components $i \in \{1, \ldots, n\}$ such that $f_i(x) \neq x_i$. Thus, the additional transitions brought by the MP update mode are discovered in the later iterations only. This highlights the concept of *permissive depth* we introduce in this paper: the number of iterations in the computation of $h^{\langle x, K \rangle}$ required to discover an MP transition. A depth of 1 corresponds to the asynchronous transitions, while a depth of n corresponds to the full MP dynamics. The simulation algorithm we propose in this paper allows controlling the depth of MP transitions, for instance by following a probabilistic distribution.

3.2 Algorithm

Listings 1 and 2 detail the steps for computing the reachable spaces and sampling the next configurations from them, as sketched in the previous section. In the description of the algorithms, we assumed fixed (1) a BN f of dimension n; (2) a function **depth** to determine the depth threshold for computing the sub-hypercubes (it has to be an integer between 1 and n). For instance, it can be a constant, or a sampling from a discrete distribution; (3) a weighting vector

Listing 1. Computation of reachable spaces with the MP update mode

```
1  def can_flip(x: configuration, i: index, H: index set, v: bool):
2      # assumes f is locally monotone
3      if v == 1:
4          z = min_configuration(x, i, H)
5      else:
6          z = max_configuration(x, i, H)
7      return f[i](z) != v
8
9  def spread(x: configuration, K: index set, d: depth):
10     # returns subset of K that can flip within the given depth
11     H = {}
12     repeat d times:
13         H = H ∪ {i for i in K if can_flip(x, i, H, x[i])}
14         K = K \ H
15     return H
16
17 def irreversible(x: configuration, H: index set):
18     # returns subset of H that cannot flip back
19     return {i for i in H if not can_flip(x, i, H, 1-x[i])}
20
21 def reachable_spaces(x: configuration):
22     d = depth()
23     S = []  # map of index set -> (index set, index set)
24     K = {1, ..., n}
25     Q = {K}
26     while Q is not empty:
27         K = Q.pop()
28         H = spread(x, K, d)  # H is subset of K
29         L = irreversible(x, H) if d > 1 else {}  # L is subset of H
30         for each M non-empty subset of L:
31             J = K \ M
32             if J not in S and J not in Q:
33                 Q.push(J)
34         S[K] = (H, L)
35     return S
```

$W \in \mathbb{R}_{\geq 0}^n$: W_m is the weight of a transition modifying the state of m components simultaneously. A uniform random walk along MP transitions is thus obtained with $W = \mathbf{1}_n$ and depth=n.

The spread function computes the K-closure of the sub-hypercube starting from the configuration x and stops after d iterations, d being a given depth. Determining whether the component i can change state requires determining the existence of a vertex z of the sub-hypercube such that $f_i(z) \neq x_i$. This is an instance of the classical Boolean satisfiability (SAT) problem, which, in general, is NP-complete.

Listing 2. Sampling of the next configurations with the MP update mode

```
1  def sample_next_configuration(x: configuration)
2      S = reachable_spaces(x)
3      if |S| = 0: return x  # fixed point
4      # compute apparent rate of transitions
5      R = 0(len(S),n)  # len(S)*n zero-filled matrix
6      for i = 1 to len(S):
7          if |L| > 0:
8              R[i,|L|] = |L| * W[|L|]
9          for j = 1 to |H\ L|:
10             R[i,|L|+j] = binom(|H\L|,j) * W[|L|+j]
11     r = U[0, sum(R)[  # uniform sampling between 0 and sum(R) excluded
12     s,m = where(cumsum(R) > r)  # s = space, m = nb of components to flip
13     H, L = S[s]
14     C = L ∪ random.sample(H\L, m - |L|)
15     y = copy(x)
16     y[C] = 1 - y[C]
17     return y
```

The given algorithm makes the assumption that the BN f is locally monotone. In that case, given a sub-hypercube h, one can build in linear time from the influence graph $G(f)$ a configuration $x^{\min} \in c(h)$ so that $f_i(x^{\min}) = \min\{f_i(x) \mid x \in c(h)\}$: the idea is that for each component j free in h, if j has a positive (monotone) influence on i (i.e., $j \xrightarrow{+1} i \in G(f)$), then $x_j^{\min} = 0$; if it has a negative influence on i, then $x_j^{\min} = 1$. If j has no influence on i, its state in $x_j^{\min}$ can be arbitrary. A similar reasoning can be applied to compute a configuration $x^{\max} \in c(h)$ so that $f_i(x^{\max}) = \max\{f_i(x) \mid x \in c(h)\}$. Then, one can decide in linear time whether i can change state in x in the scope of the sub-hypercube h: if x_i is 1, then $f_i(x^{\min})$ must be 0, and if x_i is 0, then $f_i(x^{\max})$ must be 1. Without this assumption, the can_flip function should be replaced with a call to a SAT solver.

3.3 Correctness, Complexity, and Parametrization

The sampling of the next configuration is driven by two parameters: a distribution over permissive depth, and a weight $W \in \mathbb{R}^n_{\geq 0}$ for transitions depending on the number of components they flip. These parameters can affect the generated dynamics and the complexity of the sampling.

General case: full MP dynamics. Let us assume that the depth function can always return n (either it is a constant function returning n, or the returned values follow a discrete distribution where the probability of drawing n is not 0), and that $W \in \mathbb{R}^n_{>0}$. Then, any MP transition has a non-zero probability to be sampled.

Given a space (x, H, L), as computed by reachable_spaces and character-ized by a configuration x, a set $H \subseteq \{1, \ldots, n\}$ of free components, and a subset $L \subseteq H$ of irreversible components, let us denote by $\mathrm{tr}(x, H, L)$ the set of candi-date next configurations considered by sample_next_configuration:

$$\mathrm{tr}(x, H, L) = \{y \in \mathbb{B}^n \mid L \subseteq \Delta(x, y) \subseteq H\} . \tag{6}$$

We prove that the set of transitions the algorithm can generate from the com-puted spaces is equal to the full MP dynamics, except self-loops (Lemma 1), and that spaces generate disjoint sets of transitions (Lemma 2).

Lemma 1. *Given a BN f of dimension n and one of its configuration $x \in \mathbb{B}^n$, and denoting by S the set of spaces returned by reachable_spaces(x) function,*

$$\bigcup_{(x, H, L) \in S} \mathrm{tr}(x, H, L) = \{y \mid x \rightarrow_{\mathrm{MP}} y, x \neq y\} .$$

Lemma 2. *Given a BN f of dimension n and one of its configuration $x \in \mathbb{B}^n$, and denoting by S the set of spaces returned by reachable_spaces(x), for any distinct pair of spaces (x, H, L), $(x, H', L') \in S$, $\mathrm{tr}(x, H, L) \cap \mathrm{tr}(x, H', L') = \emptyset$.*

Proofs are given in Appendix A.

Worst-case complexity. Assuming locally-monotone BNs, can_flip is performed in linear time (we assume the influence graph $G(f)$ is given); spread makes in the worst case n^2 call to can_flip, resulting in a cubic time in n; whereas irreversible is quadratic in n. Function reachable_spaces can then generate an exponential number of spaces. The sampling is then linear with respect to the number of spaces generated, and linear with respect to n. In the non-monotone case, can_flip is an NP-complete problem which currently can be solved in exponential time and space with SAT solvers.

Unitary depth: asynchronous and fully-asynchronous dynamics. Let us consider the case whenever depth function always returns 1. The algorithm computes only one space $(x, H, L = \emptyset)$ with H being the set of components i such that $f_i(x) \neq x_i$, and can generate any transition to $y \neq x$ where $\Delta(x, y) \subseteq H$. This corresponds exactly to the (general) asynchronous dynamics $\rightarrow_{\mathrm{g}}$ assuming $W \in \mathbb{R}^n_{>0}$. Moreover, as only one space is computed, the complexity drops to being linear in n, without any assumption on the local-monotony of f.

Finally, whenever $W_1 > 0$ and for all $m \in \{2, \ldots, n\}$, $W_m = 0$, only transi-tions modifying one component are generated, matching exactly with the fully-asynchronous dynamics $\rightarrow_{\mathrm{a}}$ with equiprobable transitions, and with complexity similar to the previous restriction.

Listing 3. Algorithms for sampling a reachable attractor with the MP update mode

```
 1  def sample_reachable_attractor(x: configuration):
 2      stop = False
 3      while not stop:
 4          x = sample_next_configuration(x)
 5          every k iterations:  # no need to verify at each step
 6              if in_attractor(x):
 7                  stop = True
 8      A = reachable_attractors(x)
 9      return A[1]  # A contains only one element
10
11  def filter_reachable_attractors(A: sub-hypercube list, x: configuration):
12      H = spread(x, {1,...,n}, n)
13      return [a for a in A if a ⪯ x/H]  # a is smaller than the sub−hypercube
14                                          # formed by x and H
15  def sample_reachable_attractor_bis(x: configuration):
16      A = reachable_attractors(x)  # list of attractors
17      while len(A) > 1:
18          x = sample_next_configuration(x)
19          every k iterations:  # no need to verify at each step
20              A = filter_reachable_attractors(A, x)
21      return A[1]  # A contains only one element
```

3.4 Sampling Reachable Attractors

The simulation of BNs is typically employed to assess the probability of reaching the different attractors. Because determining whether a configuration belongs to a cyclic attractor is a PSPACE-complete problem with (a) synchronous update modes [8], most simulation algorithms are parametrized with a maximum number of steps to sample, without guarantee that an attractor has been reached.

In the case of the MP update mode, the attractors turn out to be exactly the smallest closed sub-hypercubes (minimal trap spaces) of f, which can be computed at a much lower cost, albeit still relying on SAT solving [8]. Thus, instead of fixing an arbitrary number of simulation steps, one can verify during the simulation whether the current configuration belongs to an attractor, and stop the sampling in that case. (sample_reachable_attractor of Listing 3).

In practice, the number of attractors reachable from a fixed initial configuration is usually small, and can be efficiently enumerated with the MP update mode, for instance using the MPBN tool [8][1]. In that case, the MP simulations can be employed to estimate the probability of reaching the different attractors depending on the depth and weight parameters. We then proceed as follows. Before simulating, we first compute the full set of attractors reachable from the initial configuration x. This set is then progressively refined during the simulation by removing attractors which are no longer included in $h^{\langle x,\{1,\dots,n\}\rangle}$. The simulation can then stop as soon as only one attractor can

---

[1] https://github.com/bnediction/mpbn.

be reached, i.e., the current configuration is in the strong basin of an attractor (`sample_reachable_attractor_bis` of Listing 3).

These two schemes assume that the full MP dynamics is sampled, or that the attractors of the sampled dynamics match with the MP attractors as follows: each MP attractor is a superset of one and only one attractor of the sampled dynamics, and each attractor of the sampled dynamics is a subset of one and only one MP attractor. This is always the case for BNs having no cyclic attractors.

4   Evaluation

Using a prototype written in Python[2], we demonstrate the applicability of MP simulation on several Boolean models from the literature comprising about thirty components. After an illustration on toy examples, we study the effect of the parametrization on the assessment of the robustness of mutations impacting the propensities of reachable attractors in those models.

Our MP simulation algorithm has two parameters: the permissive depth and the weight of transitions W depending on the number of binary state changes. For the permissive depth, we will consider the constant n (full MP dynamics), constant 1 (asynchronous dynamics), and random sampling from a discrete exponentially decreasing distribution: depth d has probability $1/(2^d.M)$ with $M = \sum_{i=1}^{n} 1/2^i$ the normalization factor. As depth n has a non-zero probability, this parametrization also enables the full MP dynamics, although largely prioritize transitions from low permissive depths. Regarding W, we will consider the uniform weight $\mathbf{1}_n$ (random walk), and one-change only $\mathbf{10}_{n-1}$.

4.1   Toy Examples

Let us first consider the bi-stable example (A) from the configuration 000. This BN has two fixed points: 110 and 111. As explained in Sect. 1, the (a) synchronous dynamics from 000 predicts only one trajectory: $000 \rightarrow_g 100 \rightarrow_g 110$. Thus, only one attractor is reachable (100% propensity). The MP dynamics uncovers another reachable attractor (111), as depicted in Fig. 3(left). Considering all the transitions, the MP simulation would conclude that the reachability of these two attractors are equiprobable: from the 4 initial transitions, 2 lead to the strong basin of 110 and two to the strong basin of 111. Whenever the depth is a random variable with an exponentially decreasing distribution, reaching 111 requires drawing a depth of 3 (probability of 1/7), leading to reaching 111 with probability 1/14, and 110 with probability 13/14. This indicates the sensitivity of the reachability of 111 to the permissive depth, and thus to the time scales of the underlying quantitative model.

Let us now consider the example (B) where the two attractors reachable with the asynchronous mode (Fig. 1(c)) are identical to ones reachable with MP mode (Fig. 3(right)). In fully-asynchronous mode, they have an equal propensity,

---

[2] https://github.com/bnediction/mpbn-sim.

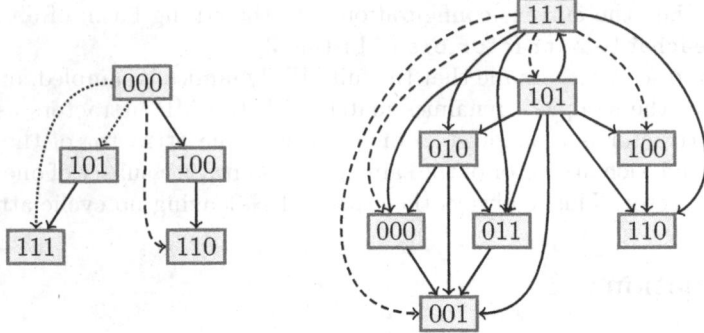

Fig. 3. MP transitions of BN (A) from configuration 000 (left) and of BN (B) from configuration 111 (right). Plain transitions have permissive depth 1, dashed depth 2, and dotted depth 3.

however in the (general) asynchronous mode, 001 propensity is twice the one of 110. Whereas the MP dynamics uncover additional trajectories and reachable configurations, in this particular example, the propensities of attractors is the same as with the general asynchronous case.

An important aspect of MP dynamics is that it is transitive: any state reachable in a sequence of transitions is reachable in one transition. Thus, the size of the (reachable) strong basins of attractors can contribute significantly to their propensities. In the above two examples, the propensities of the attractors in MP uniform random walk is entirely determined by the size of their basins.

It should be also stressed that the parameters for determining the transition probabilities are usually arbitrary. Thus, analyzing the impact of parameters on the probability of reaching the different attractors by a random walk in the MP dynamics may only give an empirical insight on the Boolean dynamics, and cannot be formally transfered to an associated quantitative model.

4.2   Models from Literature with Different Mutation Conditions

From the literature, we selected BNs modeling cell fate decision processes: the reduced cell death receptor model [1] with 14 components; the tumor invasion model [2] with 32 components; and the bladder model [11] with 35 components. These models have been designed with the fully-asynchronous update mode, and have been evaluated with respect to their ability to predict the changes of the attractors propensities subject to different mutation conditions (modeled by forcing some components to some state). For each model, we performed 10,000 simulations from the relevant initial configurations and for several mutation conditions, with different simulation parameters. With our prototype, the computational cost of the permissive depth is substantial, the simulations being between 3 times to 50 times slower than with depth 1 only, depending on the number of reachable spaces computed at each simulation step. However, no particular optimization has been implemented.

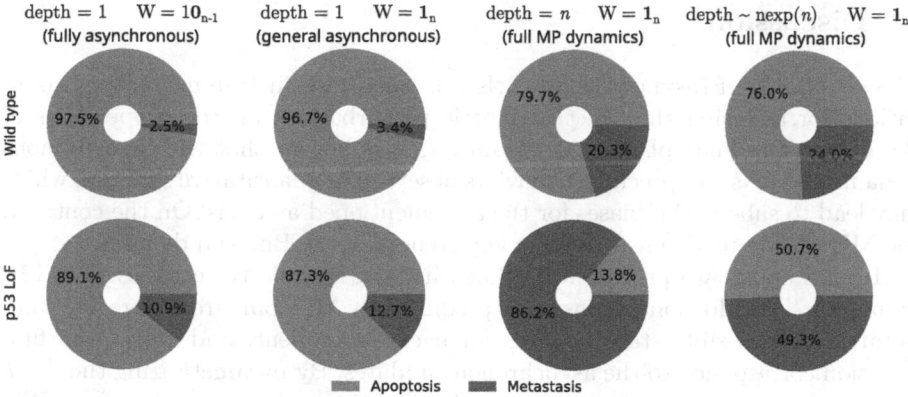

Fig. 4. Estimated propensities of reachable attractors from a unique configuration of the tumor invasion model in two mutation conditions (no mutation; p53 forced to 0), and different simulation parameters.

Figure 4 gives an example of the estimated propensities of attractors reachable from a unique initial configuration of the tumor invasion model with different mutation conditions and parameters. On the one hand, we can observe that the permissive depth can change drastically the proportions of reachable attractors: there are much more paths to the Metastasis attractor when considering potential time scale heterogeneity. The difference can be reduced by still considering the full MP dynamics, but giving exponentially decreasing probability to transitions with high permissive depth (computing transitions at depth at least D is done with probability $\sum_{d=D}^{n} (2^d \cdot \sum_{i=1}^{n} 2^{-i})^{-1}$, with n the number of nodes in the model, 32 in this case). On the other hand, the qualitative effect of the p53 mutation is similar whenever analyzed only with the asynchronous simulations (permissive depth = 1), or in the most permissive cases: the propensity of the Metastasis attractor is larger in the mutant.

From a global analysis on the 3 models[3], we observe that the absolute value of the propensities of the reachable attractors changes substantially when considering MP dynamics, in particular with depth fixed to n and uniform transition rates. Variable depth enables reducing the difference with the asynchronous dynamics while giving access to the full MP dynamics. Regarding the effect of mutations, it appears that, in most cases, the propensities of reachable attractors are affected in qualitatively the same direction, with various MP simulation parameters. This may indicate that these perturbations should be robust to heterogeneous time scales in the quantitative models captured by the BN abstraction. On the contrary, perturbations having a different qualitative effect on attractor propensities when considering permissive depth may be sensitive to the variety of time scales of the actual system, which could not be captured with usual (a) synchronous modes.

---

[3] See https://doi.org/10.5281/zenodo.6725844 for supplementary information.

5 Discussion

The simulation of (asynchronous) BNs is a usual task in systems biology applications for assessing the effect of genetic perturbations on the propensities of the different cellular phenotypes. However, it is known that the asynchronous dynamics of BNs can preclude behaviors observed in quantitative systems, which may lead to substantial biases for the aforementioned analysis. On the contrary, the MP update mode ensures the completeness of the Boolean dynamics.

In this paper, we provide a first algorithm to sample trajectories from MP dynamics. The additional transitions predicted by MP come from iterative computations of possible state changes for each component, and where the first iteration corresponds to the asynchronous updates. By parameterizing the *depth* of this computation, we obtain a generalization of transition sampling from BNs with MP, (general) asynchronous, and fully-asynchronous dynamics. The main bottleneck of the provided algorithm is its potential exponential blow-up when considering all reachable spaces. Further work will address its efficient implementations and approximations.

As illustrated on different models from the literature, the simulation with the MP mode can largely affect predictions, both in terms of reachable attractors, and in terms of their propensities. Because of the transitive properties of MP dynamics (there is a direct transition to any reachable configuration), the attractors having a large (reachable) strong basin will dominate an equiprobable random walk of MP dynamics. Variable-depth simulation can then give more weight to transition with low permissive depth and sooth the effect of the size of the basins while still capturing the full MP dynamics.

It should be noted that, as with usual simulations of asynchronous BNs [6, 14], the estimated attractor propensities are purely empirical and do not relate formally to the actual propensities in the modeled quantitative system. Due to its abstraction level, a BN is intrinsically non-deterministic. Assigning probabilities to transitions is a very strong assumption which, in this context, cannot be justified with any modeling or physical principle. At this level of abstraction, the behavior of the modeled system is not a Markov process and cannot be approximated by a Markov process (would the coefficient be fitted on data, for instance).

From a modeling perspective, our algorithm could be extended to have the permissive depth being different among components, enabling to fine-tune the time scale of their updates: the slower components should allow the more permissive depth. This would bring further control over the transitions added by the MP mode, while capturing trajectories precluded by the asynchronous mode.

Acknowledgments. This work was supported by the French Agence Nationale pour la Recherche (ANR), with the project "BNeDiction" (ANR-20-CE45-0001).

A   Proofs

A.1   Proof of Lemma 1

Consider any $(x, H, L) \in S$. Let $h \in \{0, 1, *\}^n$ be the sub-hypercube where dimensions in H are free, and otherwise fixed as in x: for each $i \in \{1, \ldots, n\}$, if $i \in H$, then $h_i = *$, otherwise, $h_i = x_i$. In such a configuration y, for each $i \in H$, there exists a configuration $z \in c(h)$ such that $f_i(z) = y_i$: indeed, if $y_i = x_i$, then $i \notin L$. Thus, $x \rightarrow_{\mathrm{MP}} y$ by instantiating Eq. (2) with $K = H$.

Conversely, let us assume that there exists $K \subseteq \{1, \ldots, n\}$ leading to $x \rightarrow_{\mathrm{MP}} y$, and let H^0 be the set of free components in $h^{\langle x, K \rangle}$ (note that $\Delta(x, y) \subseteq H^0$). Let (x, H^1, L^1) be the first space computed by our algorithm (i.e., from $K = \{1, \ldots, n\}$). By construction, H^1 is the set of free components in $h^{\langle x, \{1, \ldots, n\} \rangle}$, thus $H^0 \subseteq H^1$. Two cases arise: either $L^1 \subseteq \Delta(x, y)$ and then the transition to y is generated from this first space, or there exists at least one component $i \in L^1$ whereas $x_i = y_i$. Recall that a component i can be in L^1 only if from each vertex z of the sub-hypercube $f_i(z) \neq x_i$. Thus K is necessarily a strict subset of H^1, excluding at least the component i. Therefore, $H^0 \subseteq K \subseteq H^1 \setminus \{i \in L^1 \mid x_i = y_i\} = H^2$, and by construction, $(x, H^2, L^2) \in S$ with $L^2 \subseteq \Delta(x, y)$.

A.2   Proof of Lemma 2

The set of spaces computed by `reachable_spaces` can be inductively characterized by a map S^l for some $l \in \mathbb{N}$ from sets of components K to their associated sub-hypercube $h^{\langle x, K \rangle}$ characterized by (H, L), as defined below (x being fixed it is omitted):

$$S^1 = \{\{1, \ldots, n\} \mapsto (H^0, L^0)\} \tag{7}$$

$$S^{k+1} = S^k \cup \{K \setminus M \mapsto (H', L') \mid K \mapsto (H, L) \in S^i, \emptyset \subsetneq M \subseteq L\} \tag{8}$$

Recall that $\forall K \mapsto (H, L) \in S^k$, $L \subseteq H \subseteq K$, and $\mathrm{tr}(x, H, L) = \{y \in \mathbb{B}^n \mid L \subseteq \Delta(x, y) \subseteq H\}$.

We prove that for any $k \in \mathbb{N}$, $\forall K \mapsto (H, L), K' \mapsto (H', L') \in S^k$, $K \neq K'$,

$$\mathrm{tr}(x, H, L) \cap \mathrm{tr}(x, H', L') = \emptyset \ .$$

Let us first consider the cases whenever $L \not\subseteq H'$ or $L' \not\subseteq H$: by tr definition, $\mathrm{tr}(x, H, L) \cap \mathrm{tr}(x, H', L') = \emptyset$. Indeed, in the first case, remark that $\forall y \in \mathrm{tr}(x, H, L), L \setminus H' \subseteq \Delta(x, y)$ while $\forall y' \in \mathrm{tr}(x, H', L'), (L \setminus H') \cap \Delta(x, y') = \emptyset$, thus $y \neq y'$. The second case is a symmetry.

We establish the following propositions:

- (P1) Any component $i \in \{1, \ldots, n\}$ such that there exists $K \mapsto (H, L)$ with $i \in L$ verifies $f_i(x) = \neg x_i$. Thus, $\forall K' \mapsto (H', L'), i \in K' \implies i \in H'$.
- (P2) By definition of S^k, $K \mapsto (H, L) \in S^k \implies \forall i \in \{1, \ldots, n\} \setminus K$ there exists $K' \mapsto (H', L') \in S^{k-1}$ with $i \in L'$. Thus by P1, $f_i(x) = \neg x_i$.

- (P3) By sub-hypercube K-closeness definition and minimality,
  $\forall K \mapsto (H, L), K' \mapsto (H', L') \in S^k, K' \subseteq K \implies K' \cap L \subseteq L' \subseteq H'$.
- (P4) for any $K \neq \{1, \ldots, n\}$, $K \mapsto (H, L) \in S^k \setminus S^{k-1} \implies \exists K' \mapsto (H', L') \in S^{k-1}$ with $K \subsetneq K'$ and $K' \setminus K \subseteq L'$ (by P2).

Consider any $K \mapsto (H, L), K' \mapsto (H', L') \in S^k, K \neq K'$, such that both $L \subseteq H'$ and $L' \subseteq H'$. By P4, there exists $K'' \mapsto (H'', L'') \in S^{k-1}$ with $(K \cup K') \subseteq K''$ and such that there exists $i \in L''$ where $i \notin K$ (note: it always work with $K'' = \{1, \ldots, n\}$). If $i \in K'$, then $i \in L'$ (by P3), thus $L' \not\subseteq H$, a contradiction. Thus, $i \notin K'$. By induction using P2 and P4, we obtain that $K' \subseteq K'$. By symmetry (apply the same reasoning by swapping K and K'), $K = K'$.

References

1. Calzone, L., et al.: Mathematical modelling of cell-fate decision in response to death receptor engagement. PLOS Comput. Biol. **6**(3), e1000702 (2010). https://doi.org/10.1371/journal.pcbi.1000702
2. Cohen, D.P.A., Martignetti, L., Robine, S., Barillot, E., Zinovyev, A., Calzone, L.: Mathematical modelling of molecular pathways enabling tumour cell invasion and migration. PLOS Comput. Biol. **11**(11), e1004571 (2015). https://doi.org/10.1371/journal.pcbi.1004571
3. Crama, Y., Hammer, P.L.: Boolean Functions. Cambridge University Press, Cambridge (2011). https://doi.org/10.1017/cbo9780511852008
4. Ishihara, S., Fujimoto, K., Shibata, T.: Cross talking of network motifs in gene regulation that generates temporal pulses and spatial stripes. Genes Cells **10**(11), 1025–1038 (2005). https://doi.org/10.1111/j.1365-2443.2005.00897.x
5. Kauffman, S.A.: Metabolic stability and epigenesis in randomly connected nets. J. Theor. Biol. **22**, 437–467 (1969). https://doi.org/10.1016/0022-5193(69)90015-0
6. Mendes, N.D., Henriques, R., Remy, E., Carneiro, J., Monteiro, P.T., Chaouiya, C.: Estimating attractor reachability in asynchronous logical models. Front. Physiol. **9**, 1161 (2018). https://doi.org/10.3389/fphys.2018.01161
7. Montagud, A., Béal, J., et al.: Patient-specific boolean models of signalling networks guide personalised treatments. eLife **11** (2022). https://doi.org/10.7554/elife.72626
8. Paulevé, L., Kolčák, J., Chatain, T., Haar, S.: Reconciling qualitative, abstract, and scalable modeling of biological networks. Nat. Commun. **11**(1), 1–7 (2020). https://doi.org/10.1038/s41467-020-18112-5
9. Paulevé, L., Sené, S.: Non-deterministic updates of boolean networks. In: 27th IFIP WG 1.5 International Workshop on Cellular Automata and Discrete Complex Systems (AUTOMATA 2021). Open Access Series in Informatics (OASIcs), vol. 90, pp. 10:1–10:16. Schloss Dagstuhl - Leibniz-Zentrum für Informatik, Dagstuhl, Germany (2021). https://doi.org/10.4230/OASIcs.AUTOMATA.2021.10
10. Paulevé, L.: VLBNs - very large boolean networks (2020). https://doi.org/10.5281/zenodo.3714876
11. Remy, E., Rebouissou, S., Chaouiya, C., Zinovyev, A., Radvanyi, F., Calzone, L.: A modeling approach to explain mutually exclusive and co-occurring genetic alterations in bladder tumorigenesis. Cancer Res. **75**(19), 4042–4052 (2015). https://doi.org/10.1158/0008-5472.can-15-0602

12. Rodrigo, G., Elena, S.F.: Structural discrimination of robustness in transcriptional feedforward loops for pattern formation. PLOS ONE **6**(2), e16904 (2011). https://doi.org/10.1371/journal.pone.0016904

13. Schaerli, Y., Munteanu, A., Gili, M., Cotterell, J., Sharpe, J., Isalan, M.: A unified design space of synthetic stripe-forming networks. Nat. Commun. **5**(1), 1–10 (2014). https://doi.org/10.1038/ncomms5905

14. Stoll, G., Viara, E., Barillot, E., Calzone, L.: Continuous time boolean modeling for biological signaling: Application of gillespie algorithm. BMC Syst. Biol. **6**(1), 116 (2012). https://doi.org/10.1186/1752-0509-6-116

15. Thomas, R.: Boolean formalization of genetic control circuits. J. Theor. Biol. **42**(3), 563–585 (1973). https://doi.org/10.1016/0022-5193(73)90247-6

16. Zañudo, J.G.T.: Cell line-specific network models of ER+ breast cancer identify potential PI3kα inhibitor resistance mechanisms and drug combinations. Cancer Res. **81**(17), 4603–4617 (2021). https://doi.org/10.1158/0008-5472.can-21-1208

Minimal Trap Spaces of Logical Models are Maximal Siphons of Their Petri Net Encoding

Van-Giang Trinh[1] , Belaid Benhamou[1], Kunihiko Hiraishi[2] ,
and Sylvain Soliman[3(✉)]

[1] LIS, Aix-Marseille Université, Marseille, France
{trinh.van-giang,belaid.benhamou}@lis-lab.fr
[2] School of Information Science, Japan Advanced Institute of Science
and Technology, 1-1 Asahidai, Nomi, Ishikawa 923-1292, Japan
hira@jaist.ac.jp
[3] Lifeware Team, Inria Saclay Center, Palaiseau, France
Sylvain.Soliman@inria.fr

Abstract. Boolean modelling of gene regulation but also of post-transcriptomic systems has proven over the years that it can bring powerful analyses and corresponding insight to the many cases where precise biological data is not sufficiently available to build a detailed quantitative model. This is even more true for very large models where such data is frequently missing and led to a constant increase in size of logical models *à la* Thomas. Besides simulation, the analysis of such models is mostly based on attractor computation, since those correspond roughly to observable biological *phenotypes*. The recent use of trap spaces made a real breakthrough in that field allowing to consider medium-sized models that used to be out of reach. However, with the continuing increase in model-size, the state-of-the-art computation of minimal trap spaces based on *prime-implicants* shows its limits as there can be a huge number of implicants.

In this article we present an alternative method to compute minimal trap spaces, and hence complex attractors, of a Boolean model. It replaces the need for prime-implicants by a completely different technique, namely the enumeration of maximal siphons in the Petri net encoding of the original model. After some technical preliminaries, we expose the concrete need for such a method and detail its implementation using Answer Set Programming. We then demonstrate its efficiency and compare it to implicant-based methods on some large Boolean models from the literature.

Keywords: Logical models · Boolean models · Trap spaces · Attractor computation · Petri nets · Siphons

© The Author(s), under exclusive license to Springer Nature Switzerland AG 2022
I. Petre and A. Păun (Eds.): CMSB 2022, LNBI 13447, pp. 158–176, 2022.
https://doi.org/10.1007/978-3-031-15034-0_8

1   Introduction

From the observation that the transcriptional regulation behaved in a sigmoid step-like way, came the original idea to represent models of gene regulation as discrete event systems. Those Gene Regulation Networks (GRN) use thresholds or equivalently logical functions to represent the different regulations [18,46–48].

Boolean modelling has proven over the years that it can bring powerful analyses and corresponding insight to the many cases where precise biological data is not sufficiently available to build a detailed quantitative model [50], even for modelling post-transcriptional mechanisms. This is even more true for very large models where such data is frequently missing and led to a constant increase in size of logical models *à la* Thomas [1]. Besides simulation, the analysis of such models is mostly based on attractor computation, since those correspond roughly to observable biological *phenotypes*. The recent use of trap spaces [27] made a real breakthrough in that field allowing to consider medium-sized models that used to be out of reach. However, with the continuing increase in model-size, the state-of-the-art computation of minimal trap spaces based on *prime-implicants* shows its limits as there can be a huge number of implicants.

It is worth noting that the recent method presented in [11] for computing minimal trap spaces avoids the prime-implicants computation by relying on the *most-permissive* semantics of Boolean models. This method has been implemented in the tool mpbn[1] demonstrated in [41] for handling medium-sized models from the literature and very large synthetic models (up to 100,000 nodes). However, this method is only applicable for *locally-monotonic* Boolean models, whereas the prime-implicants based method [27] is applicable for *general* Boolean models (i.e., including both locally-monotonic and non-locally-monotonic ones). The study [37] highlights the need for non-locally-monotonic Boolean models in both biological and theoretical aspects. Hence, it is still necessary to develop efficient methods for computing minimal trap spaces of large-scale general Boolean models.

Petri nets were introduced in the 60 s as simple formalism for describing and analyzing information-processing systems that are characterized as being concurrent, asynchronous, non-deterministic and possibly distributed [33,42]. The use of Petri nets for representing biochemical reaction systems, by mapping molecular species to places and reactions to transitions, hinted at already in [33,42] was used more thoroughly quite late in [43], together with some Petri net concepts and tools for the analysis of metabolic networks. Siphons are such a concept, but they have not been used a lot for the study of biochemical systems [4,51] even if the practical cost of computing their minimal/maximal elements appear much more manageable than the theoretical complexity would indicate [35,38].

In this article we present an alternative method to compute minimal trap spaces, and hence complex attractors, of a general Boolean model. It replaces the need for prime-implicants by a completely different technique, namely the enumeration of maximal siphons in the Petri net encoding of the original model.

---

[1] https://github.com/bnediction/mpbn.

After some technical preliminaries, we expose the concrete need for such a method and detail its implementation using Answer Set Programming. We then demonstrate its efficiency and compare it to implicant-based methods on several large Boolean models from the literature.

All models used for evaluation and the implementation of the presented method are available at https://github.com/soli/trap-spaces-as-siphons and executable as a CoLoMoTo docker image.

2 Preliminaries

We will briefly recall here some preliminaries on Boolean models related to trap spaces and Petri nets. In the case of multi-level Logical models, an encoding into a Boolean model is always possible [15].

2.1 Traps Spaces

We recall here some definitions from [27] for the introduction of *trap spaces*. Minimal trap spaces prove to be a very good approximation of the attractors of a Boolean model under asynchronous update schemes and have become the *de facto* standard way to analyze models of a few tens of *genes* [16,28].

Given a Boolean model $\mathcal{M} = (V, F)$ with nodes $V = (v_1, \ldots, v_n)$ and Boolean functions $F = (f_1, \ldots, f_n)$, its state-space is $\mathcal{S}_{\mathcal{M}} = \mathbb{B}^n$ with $\mathbb{B} = \{0, 1\}$. A state $s \in \mathbb{B}^n$ is a mapping $s \colon V \mapsto \mathbb{B}$ that assigns either 0 (inactive) or 1 (active) to each node. At each time step t, node v_i can update its state by

$$v_i(t + 1) \in \{v_i(t), f_i(v(t))\}$$

where $v(t)$ is the state of $\mathcal{M}$ at time t, $v_i(t+1)$ is the state of node v_i at time $t+1$ and an update scheme specifies which value is chosen [47]. There are two main types of update schemes: synchronous, where for all $i, v_i(t + 1) = f_i(v(t))$, and asynchronous, where an i is chosen among the possible updates $(f_i(v(t)) \neq v_i(t))$, and all the other nodes j remain fixed at $v_j(t)$.

A non-empty set $T \subseteq \mathcal{S}_{\mathcal{M}}$ is a *trap set* of $\mathcal{M}$ if and only if for every $x \in T$ and $y \in \mathcal{S}_{\mathcal{M}}$ with y is reachable from x it holds that $y \in T$.

A *subspace* m of $\mathcal{S}_{\mathcal{M}}$ is characterized by its fixed nodes (denoted by D_m) and free nodes. This subspace can be specified by an assignment $m \colon D_m \mapsto \mathbb{B}$ where $D_m \subseteq V$ and $m(u)$ is the value of node $u \in D_m$. The remaining nodes of $\mathcal{M}$, $V \setminus D_m$, are said to be *free*, i.e., they can receive any Boolean value. We write subspaces like states but use in addition the symbol $\star$ to indicate that a node is free. A subspace m thus corresponds to the set of states $\mathcal{S}_{\mathcal{M}}[m] := \{s \in \mathcal{S}_{\mathcal{M}} \mid \forall v \in D_m : s(v) = m(v)\}$. For example, $m = \star \star 1$ means that $D_m = \{v_3\}$, $m(v_3) = 1$, and corresponds to the set of states $\{001, 011, 101, 111\}$. Let $\mathcal{S}_{\mathcal{M}}^{\star}$ denote the set of all possible subspaces of $\mathcal{M}$. Note that $|\mathcal{S}_{\mathcal{M}}^{\star}| = 3^n$ and $\mathcal{S}_{\mathcal{M}} \subset \mathcal{S}_{\mathcal{M}}^{\star}$ [27].

A *trap space* is defined as a subspace that is also a trap set. It is noted that trap spaces of a Boolean model are independent of the update scheme of this model [27]. Then, we define a partial order $<$ on $\mathcal{S}_{\mathcal{M}}^{\star}$ as: $m < m'$ if and only if $\mathcal{S}_{\mathcal{M}}[m] \subseteq \mathcal{S}_{\mathcal{M}}[m']$ and $\mathcal{S}_{\mathcal{M}}[m] \neq \mathcal{S}_{\mathcal{M}}[m']$. Consequently, a trap space m is minimal if and only if there is no trap space $m' \in \mathcal{S}_{\mathcal{M}}^{\star}$ such that $m' < m$.

For example, let us consider the Boolean model shown in Example 1. Figure 1(a) shows the dynamics of this model under the fully asynchronous update (i.e., only one node is nondeterministically selected in order to be updated at each time step). The model has all two trap spaces, $m_1 = 11$ and $m_2 = \star\star$. Since $m_1 < m_2$, m_1 is a minimal trap space of the Boolean model.

Example 1. We give a Boolean model $\mathcal{M} = (V, F)$, where $V = (x_1, x_2)$ and $F = (f_1, f_2)$ with $f_1 = (x_1 \wedge x_2) \vee (\neg x_1 \wedge \neg x_2)$, $f_2 = (x_1 \wedge x_2) \vee (\neg x_1 \wedge \neg x_2)$. Herein, $\wedge$, $\vee$, and $\neg$ denote the conjunction, disjunction, and negation logical operators, respectively.

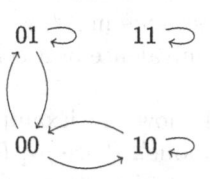

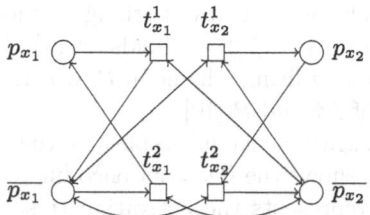

(a) State transition graph, under the fully asynchronous update.

(b) Petri net encoding of the model. Circles denote places, whereas rectangles denote transitions.

Fig. 1. Dynamics and encoding of the Boolean model of Example 1

2.2   Petri Net Encoding of Boolean Models

Definition 1. *A Petri net is a weighted bipartite directed graph (P, T, W), where P is a non-empty finite set of vertices called* places, *T is a non-empty finite set of vertices called* transitions, *$P \cap T = \emptyset$, and $W : (P \times T) \cup (T \times P) \mapsto \mathbb{N}$ is a weight function attached to the arcs.*

A *marking* for a Petri net is a mapping $m : P \mapsto \mathbb{N}$ that assigns a number of tokens to each place. A place p is marked by a marking m if and only if $m(p) > 0$. We shall write $pred(x)$ (resp. $succ(x)$) to represent the set of vertices that have a (non-zero weighted) arc leading to (resp. coming from) x.

The link between Boolean models à la Thomas and Petri nets was originally established in [8] in order to make available formal methods like model-checking for the analysis of such systems. The basic encoding into 1-safe (i.e., never more than one token in each place) nets only holds for purely Boolean models but was later extended to multivalued Logical models in two ways, either in [6] with non 1-safe Petri nets or more recently in [9] with 1-safe nets but many more places.

Since our study is focused on Boolean models, we briefly recall the original encoding here. Its basis is that every node (*gene*) v of the original model $\mathcal{M} = (V, F)$ is represented by two separate places (p_v and $\overline{p}_v$), corresponding to its two states, active, and inactive, respectively. Each conjunct of the logical function that activates the *gene* will lead to a transition t, consuming the inactive place (i.e., a directional arc from $\overline{p}_v$ to t), producing the active place (i.e., a directional arc from t to p_v), and with all other literals both consumed and produced (i.e., a bidirectional arc). And conversely for the inactivation. Let s be a state of the Boolean model and m_s be its corresponding marking in the encoded Petri net. It holds that $\forall v \in V$, $s(v) = 0$ if and only if $m_s(\overline{p}_v) = 1$ and $s(v) = 1$ if and only if $m_s(p_v) = 1$. Note also that at any marking m of the Petri net encoding a Boolean model, it always holds that $m(p_v) + m(\overline{p}_v) = 1$.

The main property of this encoding is that it is completely faithful with respect to the update scheme of the original Boolean model. For each node v of $\mathcal{M}$, only transitions corresponding to v can change the current marking of p_v or $\overline{p}_v$. In addition, at any marking at most one of such transitions is enabled because $m(p_v) + m(\overline{p}_v) = 1$ holds. Hence, for any update scheme in $\mathcal{M}$, we have a corresponding firing scheme in $\mathcal{P}$, which preserves the equivalence between the dynamics of $\mathcal{M}$ and $\mathcal{P}$ [10].

For illustration, let us reconsider the Boolean model shown in Example 1. Figure 1(b) shows the Petri net encoding of this Boolean model. Place p_{x_1} (resp. $\overline{p}_{x_1}$) in $\mathcal{P}$ represents the activation (resp. the inactivation) of node x_1 in $\mathcal{M}$. Marking $\{p_{x_1}, \overline{p}_{x_2}\}$ in $\mathcal{P}$ represents state 10 in $\mathcal{M}$. Transitions $t_{x_1}^1$ and $t_{x_1}^2$ represent the update of node x_1. Of course, in any marking $t_{x_1}^1$ and $t_{x_1}^2$ cannot be both enabled. Then, the fully asynchronous update scheme in $\mathcal{M}$ corresponds to the classical firing scheme in $\mathcal{P}$ where only one of the enabled transitions for a given marking will be fired [33].

Note that given a Boolean model in the standard SBML-Qual format [5], i.e., the package of SBML v3 [24] for such models, one can easily obtain its Petri net encoding in the Petri Net Markup Language (PNML)[2] standard using the BioLQM[3] library. This piece of software extracted from GINsim [7] and part of the CoLoMoTo[4] [36] software suite allows for easy conversion between standard formats. It also accepts many other common formats for Boolean models, notably the .bnet files of the BoolNet [28,34] tools. The conversion is executed as follows:

```
java -jar GINsim.jar -lqm <input.{sbml,bnet,zginml,...}> <output.pnml>
```

Note that transforming a Boolean model defined by its functions into its Petri net encoding roughly relies on obtaining conditions for the activation and inactivation of the states. In [8] this took the form of the whole truth table of the Boolean functions, but as shown in Appendix 1 of [9] computing Disjunctive Normal Forms (DNF) of each Boolean function is enough. Though this might appear quite computationally intensive it is important to remark first that contrary to the prime-implicants case, there is no need to find *minimal* DNFs. One

---

[2] https://www.pnml.org/.
[3] http://www.colomoto.org/biolqm/.
[4] http://colomoto.org/.

way to look at this is to consider that this amounts to a similar approach as that used in [11] but with the encoding of both activation and inhibition functions as DNFs in order to take into account possible non-local-monotonicity. This does not change the worst-case-complexity (obtaining a single DNF being exponential) but might matter a lot in practice. As such, we will explore how this transformation, here using BDDs in BioLQM, and the one based on the most-permissive semantics compare in the Sect. 6 on evaluation.

2.3 Siphons

Siphons are a static and classical property of Petri nets [42]. Note however that the use of siphons for the analysis of biological models, though it is not new, has been mostly relevant to the ODE-based continuous semantics of Chemical Reaction Networks [2,3,14].

We recall here the basic definition establishing that to produce something in a siphon you must consume something from the siphon. This corresponds to the idea that a siphon is a set of places that once unmarked remains unmarked.

Definition 2. *A* siphon *of a Petri net* (P, T, W) *is a set of places* S *such that:*

$$\forall t \in T, S \cap succ(t) \neq \emptyset \Rightarrow S \cap pred(t) \neq \emptyset.$$

Note that $\emptyset$ is trivially a siphon.

3   Minimal Trap Spaces as Maximal Conflict-Free Siphons

First, we add a definition related to any set of places of a Petri net encoding a Boolean model, and notably a siphon of such a net.

Definition 3. *A set of places of Petri net* $\mathcal{P}$ *encoding Boolean model* $\mathcal{M}$ *is* conflict-free *if it does not contain any two places corresponding to the active and inactive states of the same gene of* $\mathcal{M}$. *Then, a conflict-free siphon* S *is said to be* maximal *if and only if there is no other conflict-free siphon* S' *such that* $S \subset S'$.

Intuitively, a siphon is a set of places that once unmarked remains so. If it is conflict-free then its dual corresponds to a partial-state of the model such that whatever update, the fixed values remain so (since the unmarked places remain unmarked). This is precisely the definition of a trap space and maximality of the siphon is equivalent to as many fixed values as possible, hence minimality of the trap space. For example, the Boolean model given in Example 1 has two trap spaces, $m_1 = 11$ and $m_2 = \star\star$. The Petri net encoding of this Boolean model has five generic siphons, $S_1 = \emptyset$, $S_2 = \{p_{x_1}, \overline{p_{x_1}}\}$, $S_3 = \{p_{x_2}, \overline{p_{x_2}}\}$, $S_4 = \{\overline{p_{x_1}}, \overline{p_{x_2}}\}$, and $S_5 = \{p_{x_1}, \overline{p_{x_1}}, p_{x_2}, \overline{p_{x_2}}\}$. However, only S_1 and S_4 are conflict-free siphons and correspond to m_2 and m_1, respectively. Since $S_1 \subset S_4$, S_4 is a maximal siphon corresponding to the minimal trap space m_1. Hereafter, we formally prove that a maximal conflict-free siphon is equivalent to a minimal trap space.

Definition 4. *Let* m *be a subspace of Boolean model* $\mathcal{M} = (V, F)$. *A* mirror *of* m *is a set of places* S *in the Petri net encoding* $\mathcal{P}$ *of* $\mathcal{M}$ *such that:*

$$\forall v \in D_m, m(v) = 0 \Leftrightarrow p_v \in S, m(v) = 1 \Leftrightarrow \overline{p}_v \in S$$

and

$$\forall v \in V \setminus D_m, p_v \notin S, \overline{p}_v \notin S.$$

Theorem 1. *Let* $\mathcal{M} = (V, F)$ *be a Boolean model and* $\mathcal{P}$ *be its Petri net encoding. A subspace* m *is a trap space of* $\mathcal{M}$ *if and only if its mirror* S *is a conflict-free siphon of* $\mathcal{P}$.

Proof. First, we show that if m is a trap space of $\mathcal{M}$, then S is a conflict-free siphon of $\mathcal{P}$ (*). If $D_m = \emptyset$, then $S = \emptyset$ is trivially a conflict-free siphon of $\mathcal{P}$. Thus, we consider the case that $D_m \neq \emptyset$ (resp. $S \neq \emptyset$). Assume that S is not a siphon of $\mathcal{P}$. Then, there is a transition $t \in T$ such that $S \cap succ(t) \neq \emptyset$ but $S \cap pred(t) = \emptyset$. In other words, there is a place $p \in S$ such that $p \in succ(t)$ but $p \notin pred(t)$. Let v be the corresponding node in $\mathcal{M}$ of p. By the characterization of the encoding [8], there is a directional arc from t to p and a directional arc from the complementary place of p to t. Without loss of generality, we assume that $p = p_v$, then there is a directional arc from t to p_v and a directional arc from $\overline{p}_v$ to t. In addition, there is also no arc or a bidirectional arc between t and another place rather than p_v and $\overline{p}_v$. Thus, there is no connecting arc between t and any place in $S \setminus \{p_v\}$ because $S \cap pred(t) \neq \emptyset$. In $\mathcal{S}_{\mathcal{M}}[m]$, a node in $V \setminus D_m$ can receive any Boolean value. Hence, there is a state $s \in \mathcal{S}_{\mathcal{M}}[m]$ such that $m_s(p') = 1, \forall p' \in pred(t) \setminus \{\overline{p_v}\}$ where m_s is the corresponding marking in $\mathcal{P}$ of s. We also have $m_s(p_v) = 0$, leading to $m_s(\overline{p}_v) = 1$ by the characterization of the encoding [8]. Now, t is enabled at marking m_s. Its firing leads to a new marking m'_s such that $m'_s(p_v) = 1$ and $m'_s(\overline{p}_v) = 0$. Let s' be the corresponding state in $\mathcal{M}$ of m'_s. Since m is a trap space of $\mathcal{M}$, $s' \in \mathcal{S}_{\mathcal{M}}[m]$. Then, $s'(v) = m(v)$, leading to $m'_s(p_v) = 0$, which is a contradiction. Hence, S is a siphon of $\mathcal{P}$. By the definition of a mirror, S is also a conflict-free one.

Second, we show that if S is a conflict-free siphon of $\mathcal{P}$, then m is a trap space of $\mathcal{M}$ (**). By the definition of a mirror, m is a subspace of $\mathcal{M}$. Let s be an arbitrary state in $\mathcal{S}_{\mathcal{M}}[m]$ and m_s be its corresponding marking in $\mathcal{P}$. By the characterization of the encoding [8], $m_s(p) = 0, \forall p \in S$. In any marking m'_s reachable from m_s regardless of the firing scheme of $\mathcal{P}$, we have $m'_s(p) = 0, \forall p \in S$ by the dynamical property on markings of a siphon [31]. Equivalently, in any state s' reachable from s regardless of the update scheme of $\mathcal{M}$, we have $s'(v) = s(v) = m(v), \forall v \in D_m$. Then, $s' \in \mathcal{S}_{\mathcal{M}}[m]$. By the definition of a trap space and the arbitrariness of s, m is a trap space of $\mathcal{M}$.

From (*) and (**), we can conclude the proof. $\square$

Theorem 2. *Let* $\mathcal{M}$ *be a Boolean model and* $\mathcal{P}$ *be its Petri net encoding. A subspace* m *is a minimal trap space of* $\mathcal{M}$ *if and only if its mirror* S *is a maximal conflict-free siphon of* $\mathcal{P}$.

Proof. First, we show that if m is a minimal trap space of $\mathcal{M}$, then S is a maximal conflict-free siphon of $\mathcal{P}$ (*). Since m is a trap space of $\mathcal{M}$, S is a conflict-free siphon of $\mathcal{P}$ by Theorem 1. Assume that S is not maximal. Then, there is another conflict-free siphon S' such that $S \subset S'$. By Theorem 1, there is a trap space m' corresponding to S'. Following the definition of a mirror, $\mathcal{S}_{\mathcal{M}}[m'] \subset \mathcal{S}_{\mathcal{M}}[m]$, thus $m' < m$. This is a contradiction because m is a minimal trap space. Hence, S is a maximal conflict-free siphon of $\mathcal{P}$.

Second, we show that if S is a maximal conflict-free siphon of $\mathcal{P}$, then m is a minimal trap space of $\mathcal{M}$ (**). Since S is a conflict-free siphon of $\mathcal{P}$, m is a trap space of $\mathcal{M}$ by Theorem 1. Assume that m is not minimal. Then, there is another trap space m' such that $m' < m$. In other words, $\mathcal{S}_{\mathcal{M}}[m'] \subset \mathcal{S}_{\mathcal{M}}[m]$. Let S' be the mirror of m'. S' is a conflict-free siphon by Theorem 1. Following the definition of a mirror, $S \subset S'$, which is a contradiction because S is a maximal conflict-free siphon. Hence, m is a minimal trap space of $\mathcal{M}$.

From (*) and (**), we can conclude the proof. □

By Theorem 2, we can reduce the problem of computing all minimal trap spaces of a Boolean model to the problem of computing all maximal conflict-free siphons of its Petri net encoding. Note that in the case of stable states, this can be put in regard to the classical relationship between siphons and deadlocks in Petri nets. It might actually be possible to generalize our result to any 1-safe place-complementary Petri net to define a notion of trap space that might be useful for the analysis of Petri nets, but this is out of the scope of this article.

It is noted that there are no existing methods specifically designed for computing maximal conflict-free siphons (even maximal siphons) of a Petri net. The reason might be that researchers mainly focus on minimal siphons [31]. Hence, we here propose a new method based on Answer Set Programming (ASP) [17] for computing maximal conflict-free siphons of a Petri net. The details of the proposed method shall be given in the next section.

4   Answer Set Programming-Based Method

First, we show the characterization of all conflict-free siphons of the encoded Petri net $\mathcal{P} = (P, T, W)$. Suppose that S is a generic siphon of $\mathcal{P}$. If a place p should belong to S, then by definition all the transitions in $pred(p)$ must belong to $succ(S)$. Note that $succ(S) = \bigcup_{p \in S} succ(p)$. A transition t belongs to $succ(S)$ if and only if there is at least one place p' in S such that $p' \in pred(t)$. Hence, for each transition $t \in pred(p)$, we can state that

$$p \in S \Rightarrow \bigvee_{p' \in pred(t)} p' \in S. \tag{1}$$

The system of all the rules of the above form with respect to all pairs (p, t) where $p \in P, t \in T, t \in pred(p)$ fully characterizes all generic siphons of a Petri net and has been used with SAT solvers in [35, 38]. To make S to be a conflict-free siphon, we need to add to the system the rule

$$p_v \in S \Rightarrow \overline{p_v} \notin S \land \overline{p_v} \in S \Rightarrow p_v \notin S \tag{2}$$

for each node $v \in V$. By definition, the final system fully characterizes all conflict-free siphons of the encoded Petri net.

Then, we translate the above characterization into the ASP $\mathcal{L}$ as follows. We introduce atom p-v (resp. n-v) to denote place p_v (resp. $\overline{p_v}$), $\forall v \in V$. The set of all atoms in $\mathcal{L}$ is given as $\mathcal{A} = \bigcup_{v \in V} \{p\text{-}v, n\text{-}v\}$. For each pair (p, t) where $p \in P, t \in T, t \in pred(p)$, we translate the rule (1) into the ASP rule

$$a\_1; \ \ldots \ ; \ a\_k \ :\text{-} \ a.$$

where $a \in \mathcal{A}$ is the atom representing place p and $\{a\_1, \ldots, a\_k\} \subseteq \mathcal{A}$ is the set of atoms representing places in $pred(t)$. The rule (2) is translated into the ASP rule

$$:\text{-} \ p\text{-}v, \ n\text{-}v.$$

for each $v \in V$. This ASP rule guarantees that two places representing the same node in $\mathcal{M}$ never belong to the same siphon of $\mathcal{P}$, representing the conflict-freeness. Naturally, a Herbrand model (see, e.g., [17]) of $\mathcal{L}$ is equivalent to a conflict-free siphon of $\mathcal{P}$. To guarantee that a Herbrand model is also a stable model (an answer set), we need to add to $\mathcal{L}$ the two choice rules

$$p\text{-}v. \ n\text{-}v.$$

for each $v \in V$. Note that the number of atoms of $\mathcal{L}$ is only $2n$, whereas the ASP encoding shown in [27] has as many atoms as the number of prime-implicants of the Boolean model and that number might be exponential in n. In [11], there is an ASP characterization of trap spaces that does not rely on minimal DNFs either and thus seems very similar to our ASP encoding. Remarkably it only requires the DNF for the *activation* part, using the information that it will only be used for locally-monotonic Boolean models. We would therefore expect that, when available, it will have comparable performance on the ASP part (the ASP program would be approximately twice smaller, though redundancy is not always bad in that field), but can also avoid combinatorial explosion of the Petri net encoding for some formula where the activation DNF is simple but the inhibition is not. Since mpbn is included in our benchmark this will be evaluated in our experiments.

Now, a solution (simply an answer set) $A \subseteq \mathcal{A}$ of $\mathcal{L}$ is equivalent to a conflict-free siphon S of $\mathcal{P}$, thus a trap space m of $\mathcal{M}$. The conversion from A to m is straightforward. If p-v $\in A$ then $v \in D_m$ and $m(v) = 0$. Conversely, if n-v $\in A$ then $v \in D_m$ and $m(v) = 1$. Otherwise, $v \notin D_m$. Computing multiple answer sets is built into ASP solvers and the solving collection POTASSCO [17] also features the option to find set-inclusion maximal answer sets with respect to the set of atoms. Naturally, a set-inclusion maximal answer set of $\mathcal{L}$ is equivalent to a maximal conflict-free siphon of $\mathcal{P}$, thus a minimal trap space of $\mathcal{M}$. By using this built-in option, we can compute all the set-inclusion maximal answer sets of $\mathcal{L}$ (resp. all the minimal trap spaces of $\mathcal{M}$) in one execution.

5 Motivating Example

For a few years now we have been collaborating with biologists who build very large detailed and annotated maps and now wish to analyze the dynamics of the corresponding models. One of the main maps studied this way represents knowledge about the Rheumatoïd Arthritis [45], and was the main motivation for the development of a tool to automatically transform it into an executable Boolean model [1]. In the supplementary material of the paper, an excerpt of the map, focused around the apoptosis (cell death) module is transformed into a model of *reasonable* size, namely 180 Boolean variables (model F5_RA_apoptosis_executable_module.sbml of supplementary material S3, and model "RA-apoptosis" of Sect. 6). The study of such model, though, is a big hurdle. Indeed, as stated in the article about another model of the same size: *"The size of the CaSQ-inferred MAPK model (181 nodes) made the calculation of stable states a non-realistic endeavour."*

In practice, even if there is a huge number of attractors in such a model, obtaining a sample of those can reveal very useful to invalidate the model and lead to further refinement. In particular, it provides a feature-rich alternative to random simulations for this type of very non-deterministic model. Being able to detect that there are inconsistencies with published experimental data in some of the first 1000 attractors, for instance, can lead to a much quicker Systems Biology loop: model, invalidate, refine.

However, using a state-of-the-art tool like PyBoolNet [27] on that model actually fails at the phase of prime-implicant generation. mpbn [41] does not give any answer either because it recognizes that model as non-locally-monotonic. And hence, it is not possible to extract any (complex) attractor at all. This is also true for the Alzheimer model also mentioned in that same article and originally from [39] (**F4** file in the original supplementary material, and "Alzheimer" in Table 2), but actually not for the MAPK model for which the first trap spaces can be obtained in reasonable time. The current practice usually revolves then around fixing some inputs to plausible values and reducing the model accordingly. While this approach makes sense, it relies on potentially arbitrary decisions, and *hides away* critical modelling choices that were actually not part of the original Boolean model or even of the starting map.

Using the method presented above, it is possible to convert the model to PNML in about one second and to obtain the first 1000 minimal trap spaces (including ones that contain more than one state) in a few milliseconds. Unfortunately since this was not available at the time, the analysis of the model remained very high-level and qualitative, instead of being able to use the rich information of computed minimal trap spaces.

6 Evaluation

To assess the efficiency of the proposed method, implemented as a Python package named Trappist, we compare it with the state-of-the-art method implemented in the tool PyBoolNet [27, 28] on both its own repository of models and large models from the literature. In addition, we also include the tool mpbn [41] to the benchmarks although it only handles locally-monotonic models, whereas both Trappist and PyBoolNet can handle general models.

To solve the ASP problems, we used the same ASP solver CLINGO [17] and the same configuration as that used in PyBoolNet [27,28]. Specifically, we used the configuration -heuristic=Domain -enum-mod=domRec -dom-mod=3 (subset maximality, equivalent to the deprecated --dom-pref=32 --heuristic=domain --dom-mod=7 used by PyBoolNet). We ran all the benchmarks on an apple laptop whose environment is CPU: Intel® Core™ i7 1.20GHz x 4, 16 GB DDR4 RAM, MacOS 12.3.1. Finally, we set a time limit of two minutes for each model. Note that we did not get the opportunity to run this benchmark on a proper dedicated computing workstation, the results are therefore only indicative of global trends and should not be interpreted as precise performance. This is also why in some rare cases finding all minimal trap spaces was faster than with the limit to 1000.

For all the above tools, namely PyBoolNet, mpbn and CLINGO, we used the version available in the latest CoLoMoTo docker image tagged 2022-05-01. All the models and a CoLoMoTo notebook realizing the benchmarks can be found at https://github.com/soli/trap-spaces-as-siphons. These can be run on a Docker image in the cloud by clicking the "Binder" button.

6.1 PyBoolNet Repository

As shown in Table 1, for most of the models of the official PyBoolNet repository[5], the results are comparable with all minimal trap spaces found very fast. For 5 of the 29 models, mpbn did not give any answer because it recognized these models as not locally-monotonic. Note that on some very small models, Trappist is sometimes slower than PyBoolNet and/or mpbn, but still significantly under one second. Moreover, we believe that the result on the arellano_rootstem model is caused by the cold start of the JVM for BioLQM. On the contrary, on every model that was a bit challenging for PyBoolNet or mpbn, the new method is far more efficient with speedups between one and two orders of magnitude.

6.2 Selected Models

We used a set of real-world Boolean models lying in various scales collected from numerous bibliographic sources. These models are quite big (in size), complex (i.e., having high average in-degree, which is related to the number of prime-implicants) and most of them have never been fully analyzed. We then applied PyBoolNet, mpbn, and Trappist to computing minimal trap spaces of these real-world models. It is notable that unlike existing analysis shown in the literature, we did not fix specific values for source nodes (i.e., some node v such that $f_v = v$) in these models. Table 2 shows the experimental results on those models. Hereafter, we analyze in detail the results with respect to minimal trap space computation.

---

[5] https://github.com/hklarner/pyboolnet/tree/master/pyboolnet/repository.

Table 1. Timing comparisons between PyBoolNet, mpbn and Trappist on the PyBool-Net repository. Column n denotes the number of nodes of each model. Column $|M|$ denotes the number of minimal trap spaces and for each method is given the computation time in seconds, asking only for the first 1000 trap spaces. A number in bold indicates a ratio greater than three compared to the best result. "NM" indicates a non-locally-monotonic model.

	model	n	$|M|$	PyBoolNet	mpbn	Trappist
1	arellano_rootstem	9	4	0.05	0.01	**0.20**
2	calzone_cellfate	28	27	0.03	**NM**	0.03
3	dahlhaus_neuroplastoma	23	32	0.05	0.02	0.03
4	davidich_yeast	10	12	0.04	0.01	0.02
5	dinwoodie_life	15	7	0.03	0.01	0.01
6	dinwoodie_stomatal	13	1	0.03	0.01	0.01
7	faure_cellcycle	10	2	0.04	0.01	0.02
8	grieco_mapk	53	18	0.04	0.02	0.02
9	irons_yeast	18	1	0.05	0.01	0.02
10	jaoude_thdiff	103	>1000	**1.44**	**0.90**	0.09
11	klamt_tcr	40	8	0.04	0.01	0.03
12	krumsiek_myeloid	11	6	0.03	0.01	0.01
13	multivalued	13	4	0.03	0.01	0.01
14	n12c5	11	5	**35.16**	0.01	0.02
15	n3s1c1a	2	2	0.02	0.01	0.01
16	n3s1c1b	2	2	0.02	0.01	0.01
17	n5s3	4	3	0.03	**NM**	0.01
18	n6s1c2	5	3	0.03	0.01	0.01
19	n7s3	6	3	0.02	0.01	0.01
20	raf	3	2	0.02	0.01	0.01
21	randomnet_n15k3	15	3	0.03	**NM**	0.01
22	randomnet_n7k3	7	10	0.03	**NM**	0.01
23	remy_tumorigenesis	34	25	**2.14**	0.02	0.02
24	saadatpour_guardcell	13	1	0.03	0.01	0.01
25	selvaggio_emt	56	>1000	**1.02**	**0.52**	0.09
26	tournier_apoptosis	12	3	0.04	0.01	0.01
27	xiao_wnt5a	7	4	0.03	0.01	0.01
28	zhang_tlgl	60	156	0.22	**NM**	0.05
29	zhang_tlgl_v2	60	258	0.09	0.15	0.02

The first observation is that for 26 of the 33 models (more than 78%), mpbn did not give any answer because it recognized that these models as not locally-monotonic. For 6 of the 33 models where mpbn returned the answers, mpbn

Table 2. Timing comparisons between PyBoolNet (PBN), mpbn and Trappist on selected models from the literature. Column n (resp. s) denotes the number of nodes (resp. source nodes) of each model. Column $|M|$ denotes the number of minimal trap spaces and for each method is given the computation time in seconds. "DNF" means that the method did not finish the computation (stopping at the first 1000 minimal trap spaces, or all for the last column) within the timeout of two minutes. "NM" indicates a non-locally-monotonic model.

	model	n	s	$|M|$	PBN 1000	mpbn 1000	Trappist 1000	All
1	inflammatory-bowel [22]	47	0	1	DNF	NM	1.26	0.69
2	T-LGL-survival [22]	61	7	318	0.84	NM	0.01	0.01
3	butanol-production [22]	66	13	8192	0.71	NM	0.01	0.02
4	colon-cancer [22]	70	1	10	0.20	NM	0.02	0.01
5	mast-cell-activation [1]	73	19	>1000	0.76	NM	0.01	DNF
6	IL-6-signalling [22]	86	15	32768	1.29	NM	0.01	0.01
7	Corral-ThIL-17-diff [13]	92	16	>1000	DNF	NM	0.03	DNF
8	Korkut-2015 [30]	99	12	18556	DNF	1.04	0.06	0.08
9	adhesion-cip-migration [19]	121	4	78	36.80	0.35	0.27	0.09
10	interferon-1 [40]	121	55	>1000	10.11	NM	0.02	DNF
11	TCR-TLR5-signaling [44]	130	5	48	2.06	NM	0.03	0.02
12	influenza-replication [22]	131	11	10128	46.67	NM	0.02	0.02
13	prostate-cancer [32]	133	11	2760	DNF	NM	0.16	0.04
14	HIV-1 [22]	138	14	39424	DNF	NM	0.06	0.06
15	fibroblasts [21]	139	9	>1000	DNF	NM	0.08	DNF
16	HMOX-1-pathway [40]	145	56	>1000	6.52	NM	0.02	DNF
17	kynurenine-pathway [40]	150	72	>1000	DNF	NM	0.19	DNF
18	virus-replication-cycle [40]	154	25	>1000	DNF	NM	0.03	DNF
19	immune-system [22]	164	13	>1000	DNF	NM	0.08	DNF
20	RA-apoptosis [1]	180	59	>1000	DNF	NM	0.02	DNF
21	MAPK [1]	181	37	>1000	87.13	NM	0.03	DNF
22	er-stress [40]	182	75	>1000	17.88	NM	0.03	DNF
23	cascade-3 [49]	183	0	1	101.18	NM	0.40	0.07
24	CHO-2016 [30]	200	13	13312	DNF	2.45	0.07	0.08
25	T-cell-check-point [23]	218	14	>1000	69.94	NM	0.05	DNF
26	ErbB-receptor-signaling [20]	247	22	>1000	DNF	NM	0.24	DNF
27	macrophage-activation [22]	321	19	>1000	16.43	NM	0.05	DNF
28	cholocystokinin [1]	383	74	>1000	1.42	NM	0.08	DNF
29	Alzheimer [1]	762	237	>1000	DNF	NM	0.29	DNF
30	KEGG-network [29]	1659	521	>1000	DNF	20.76	3.40	DNF
31	human-network [25]	1953	669	>1000	DNF	23.62	5.93	DNF
32	SN-5 [26]	2746	829	>1000	DNF	28.94	DNF	DNF
33	turei-2016 [30]	4691	1257	???	DNF	DNF	DNF	DNF

and Trappist are comparable in computation time, though surprisingly mpbn appears a bit slower on average. Note however that mpbn was the only tool to provide a solution for the SN-5 model, thus confirming that if the activation function is in the right form, not having to compute the inactivation function's disjunctive normal form can render a difficult problem tractable. However, since mbpn can handle only locally-monotonic models and Trappist can handle general models, it is difficult to further compare between them. Hence, we focus on only comparisons between PyBoolNet and Trappist in the following observations.

The second observation is that the proposed method vastly outperforms PyBoolNet in computational time, on each and every model, and sometimes with orders of magnitude of difference (e.g., for most models in the 100–1000 nodes size range). Note that for all the cases where PyBoolNet did not manage to finish before the timeout, as marked by "DNF" in Table 2, the timeout occurred during the computation of the prime-implicants. Hence, not even a single minimal trap space was output by that method. The computational advantage is therefore immediately a practical advantage since on the one hand the state-of-the-art method did not allow any analysis whatsoever of the models, and on the other hand the proposed method could provide, very often under one second, the first thousand minimal trap spaces. For modellers having a critical look at a model and in a *model, invalidate, refine* loop this means a huge difference in the models that are amenable to study.

Note that even with a very restricted time-limit of two minutes, it was possible with the proposed technique to find *all* minimal trap spaces of small models (roughly under 130 nodes, i.e., considered as quite big up to now). Though it might seem impractical to handle tens of thousands of such possible complex attractors in a manual way, i.e., to compare them to specific experimental conditions and corresponding data, we hope that an automatic analysis of such attractors might become possible with systematic verification methods, not unlike that described in [23]. Since the ASP code is declarative by nature, it is also possible to add to it supplementary constraints coming from the modeler in case one is looking for specific attractors. Finally, sampling from the ASP-generated solutions as is done in [12] would allow for a different type of exploration.

The third observation is that for all the models where PyBoolNet finished before the timeout, once PyBoolNet went through the prime-implicant phase, its ASP solving phase quickly returned the first 1000 minimal trap spaces, all under one second. For these models, the ASP solving phase of the proposed method also took very short time, all under one second. Hence, with the experimental results shown in this paper, the practical differences between our ASP encoding and that of PyBoolNet are not distinctly exposed. The fact that our new ASP encoding is guaranteed to be linear in the number of nodes of the original model does not seem to be crucial here, however a much deeper analysis of those cases remains to be done.

The last observation is that for very large models (i.e., more than two thousand nodes) the proposed method did not manage to finish the Petri net conversion before the timeout, as marked by "???" in Table 2. This points to the fact that our current choice of using a BDD-based translation to obtain that Petri net encoding, though it provides a small/efficient ASP might be too costly to handle the largest models. In such a case, a more *naive* encoding might provide a much larger ASP program, with many redundant rules, but easier/faster to obtain. The evaluation of the feasibility of such strategy, and of its impact on smaller instances, remains to be done and is out of the scope of this first article. Recognizing that a model is locally-monotonic and applying in that specific case dedicated strategies as those of mpbn might also be a partial solution as shown in the SN-5 case, but not for the turei-2016 model.

Note that though enumerating the extremal siphons of a Petri net is exponential (see [35] for instance) this is apparently not the bottleneck of the proposed method, showing once again that networks obtained from biochemical models do have a specific structure.

7 Conclusion

In this article we proposed a new method for the computation of minimal trap spaces of Boolean models, based on a new concept called maximal conflict-free siphons. This method is evaluated on large models from the literature and shows that it can scale up much better than the state-of-the-art prime-implicants based techniques. We believe that this opens up the way to a much better analysis of large Boolean models, which is needed with the advent of automatic model-generation pipelines [40].

Though we benchmarked this new approach against state-of-the-art tools, there are many more evaluations that we plan to do in the future. First, the BioLQM platform that we are using is providing another implicant-based method using BDDs in http://colomoto.org/biolqm/doc/tools-trapspace.html and though we expect it to behave mostly like PyBoolNet, that remains to be checked. Note that this also raises the question of replacing altogether the BioLQM preprocessing step we use to obtain the Petri net encoding, since its use of BDDs might not be optimal for that step. The trade-off between a small Petri net, and hence small ASP, and the time it takes to compute it (in other words, the trade-off of allowing redundant constraints) has to be evaluated in depth.

Second, the experimental results shown in this paper mostly expose the differences caused by the prime-implicant phase of PyBoolNet. There is much more information required to distinctly study the practical differences between our ASP encoding and that of PyBoolNet, and notably their size/efficiency ratio. Hence, we plan to conduct experiments on more real-world models or maybe random models that can be randomly generated by using BoolNet [34]. Such experiments should include a time-course of the number of ASP solutions found for proper comparison of the ASP encodings.

In addition, there are possibly other methods for computing maximal conflict-free siphons in Petri nets, like SAT/MaxSAT approaches [35]. Although these approaches do not directly support the maximal conflict-free siphon computation now, we plan to investigate them in the future. They could replace our ASP program if they outperform it. However, the current method appears to already perform very well even on the biggest models we have considered.

Finally, we think that the links between Petri nets and Boolean models that we stumbled upon in this method might have deeper roots. Exploring those connections might lead both to interesting topics of research for Petri nets, like a notion of trap-spaces, and for Boolean models.

References

1. Aghamiri, S.S., et al.: Automated inference of Boolean models from molecular interaction maps using CaSQ. Bioinformatics **36**(16), 4473–4482 (2020). https://doi.org/10.1093/bioinformatics/btaa484
2. Angeli, D., Leenheer, P.D., Sontag, E.: A Petri net approach to persistence analysis in chemical reaction networks. In: Queinnec, I., Tarbouriech, S., Garcia, G., Niculescu, SI. (eds.) Biology and Control Theory: Current Challenges, pp. 181–216. Springer (2007). https://doi.org/10.1007/978-3-540-71988-5_9
3. Angeli, D., Leenheer, P.D., Sontag, E.D.: Persistence results for chemical reaction networks with time-dependent kinetics and no global conservation laws. SIAM J. Appl. Math. **71**(1), 128–146 (2011). https://doi.org/10.1137/090779401
4. Blätke, M.A., Heiner, M., Marwan, W.: Biomodel engineering with Petri nets. In: Algebraic and Discrete Mathematical Methods for Modern Biology, pp. 141–192. Elsevier (2015). https://doi.org/10.1016/B978-0-12-801213-0.00007-1
5. Chaouiya, C., Bérenguier, D., Keating, S.M., Naldi, A., et al.: SBML qualitative models: a model representation format and infrastructure to foster interactions between qualitative modelling formalisms and tools. BMC Syst. Biol. **7**, 135 (2013). https://doi.org/10.1186/1752-0509-7-135
6. Chaouiya, C., Naldi, A., Remy, E., Thieffry, D.: Petri net representation of multi-valued logical regulatory graphs. Nat. Comput. **10**(2), 727–750 (2011). https://doi.org/10.1007/s11047-010-9178-0
7. Chaouiya, C., Naldi, A., Thieffry, D.: Logical modelling of gene regulatory networks with GINsim. In: van Helden, J., Toussaint, A., Thieffry, D. (eds.) Bacterial Molecular Networks, pp. 463–479. Springer (2012). https://doi.org/10.1007/978-1-61779-361-5_23
8. Chaouiya, C., Remy, E., Ruet, P., Thieffry, D.: Qualitative modelling of genetic networks: from logical regulatory graphs to standard petri nets. In: Cortadella, J., Reisig, W. (eds.) ICATPN 2004. LNCS, vol. 3099, pp. 137–156. Springer, Heidelberg (2004). https://doi.org/10.1007/978-3-540-27793-4_9
9. Chatain, T., Haar, S., Jezequel, L., Paulevé, L., Schwoon, S.: Characterization of reachable attractors using petri net unfoldings. In: Mendes, P., Dada, J.O., Smallbone, K. (eds.) CMSB 2014. LNCS, vol. 8859, pp. 129–142. Springer, Cham (2014). https://doi.org/10.1007/978-3-319-12982-2_10
10. Chatain, T., Haar, S., Kolčák, J., Paulevé, L., Thakkar, A.: Concurrency in Boolean networks. Nat. Comput. **19**(1), 91–109 (2019). https://doi.org/10.1007/s11047-019-09748-4
11. Chevalier, S., Froidevaux, C., Paulevé, L., Zinovyev, A.Y.: Synthesis of Boolean networks from biological dynamical constraints using answer-set programming. In: 31st IEEE International Conference on Tools with Artificial Intelligence, ICTAI 2019, Portland, OR, USA, 4–6 November 2019, pp. 34–41. IEEE (2019). https://doi.org/10.1109/ICTAI.2019.00014

12. Chevalier, S., Noël, V., Calzone, L., Zinovyev, A., Paulevé, L.: Synthesis and simulation of ensembles of boolean networks for cell fate decision. In: Abate, A., Petrov, T., Wolf, V. (eds.) CMSB 2020. LNCS, vol. 12314, pp. 193–209. Springer, Cham (2020). https://doi.org/10.1007/978-3-030-60327-4_11
13. Corral-Jara, K.F., et al.: Interplay between SMAD2 and STAT5A is a critical determinant of IL-17A/IL-17F differential expression. Mol. Biomed. **2**(1), 1–16 (2021). https://doi.org/10.1186/s43556-021-00034-3
14. Degrand, E., Fages, F., Soliman, S.: Graphical conditions for rate independence in chemical reaction networks. In: Abate, A., Petrov, T., Wolf, V. (eds.) CMSB 2020. LNCS, vol. 12314, pp. 61–78. Springer, Cham (2020). https://doi.org/10.1007/978-3-030-60327-4_4
15. Didier, G., Remy, E., Chaouiya, C.: Mapping multivalued onto Boolean dynamics. J. Theor. Biol. **270**(1), 177–184 (2011). https://doi.org/10.1016/j.jtbi.2010.09.017
16. Cifuentes Fontanals, L., Tonello, E., Siebert, H.: Control strategy identification via trap spaces in Boolean networks. In: Abate, A., Petrov, T., Wolf, V. (eds.) CMSB 2020. LNCS, vol. 12314, pp. 159–175. Springer, Cham (2020). https://doi.org/10.1007/978-3-030-60327-4_9
17. Gebser, M., Kaufmann, B., Kaminski, R., Ostrowski, M., Schaub, T., Schneider, M.: Potassco: the Potsdam answer set solving collection. AI Commun. **24**(2), 107–124 (2011). https://doi.org/10.3233/AIC-2011-0491
18. Glass, L., Kauffman, S.A.: The logical analysis of continuous, non-linear biochemical control networks. J. Theor. Biol. **39**(1), 103–129 (1973). https://doi.org/10.1016/0022-5193(73)90208-7
19. Guberman, E., Sherief, H., Regan, E.R.: Boolean model of anchorage dependence and contact inhibition points to coordinated inhibition but semi-independent induction of proliferation and migration. Comput. Struct. Biotechnol. J. **18**, 2145–2165 (2020). https://doi.org/10.1016/j.csbj.2020.07.016
20. Helikar, T., et al.: A comprehensive, multi-scale dynamical model of ErbB receptor signal transduction in human mammary epithelial cells. PloS One **8**(4), e61757 (2013). https://doi.org/10.1371/journal.pone.0061757
21. Helikar, T., Konvalina, J., Heidel, J., Rogers, J.A.: Emergent decision-making in biological signal transduction networks. Proc. National Acad. Sci. **105**(6), 1913–1918 (2008). https://doi.org/10.1073/pnas.0705088105
22. Helikar, T., Kowal, B.M., McClenathan, S., Bruckner, M., et al.: The Cell Collective: toward an open and collaborative approach to systems biology. BMC Syst. Biol. **6**, 96 (2012). https://doi.org/10.1186/1752-0509-6-96
23. Hernandez, C., Thomas-Chollier, M., Naldi, A., Thieffry, D.: Computational verification of large logical models-application to the prediction of T cell response to checkpoint inhibitors. Front. Physiol. 1154 (2020). https://doi.org/10.3389/fphys.2020.558606
24. Keating, S.M., Waltemath, D., König, M., Zhang, F., et al.: SBML Level 3: an extensible format for the exchange and reuse of biological models. Mol. Syst. Biol. **16**(8), e9110 (2020). https://doi.org/10.15252/msb.20199110
25. Kim, J.R., Kim, J., Kwon, Y.K., Lee, H.Y., Heslop-Harrison, P., Cho, K.H.: Reduction of complex signaling networks to a representative kernel. Sci. Signal. **4**(175), ra35 (2011). https://doi.org/10.1126/scisignal.2001390
26. Kim, J., Yi, G.S.: RMOD: a tool for regulatory motif detection in signaling network. PloS One **8**(7), e68407 (2013). https://doi.org/10.1371/journal.pone.0068407
27. Klarner, H., Bockmayr, A., Siebert, H.: Computing maximal and minimal trap spaces of Boolean networks. Nat. Comput. **14**(4), 535–544 (2015). https://doi.org/10.1007/s11047-015-9520-7

28. Klarner, H., Streck, A., Siebert, H.: PyBoolNet: a python package for the generation, analysis and visualization of Boolean networks. Bioinformatics **33**(5), 770–772 (2017). https://doi.org/10.1093/bioinformatics/btw682

29. Kwon, Y.: Properties of Boolean dynamics by node classification using feedback loops in a network. BMC Syst. Biol. **10**, 83 (2016). https://doi.org/10.1186/s12918-016-0322-z

30. Lee, D., Cho, K.H.: Signal flow control of complex signaling networks. Sci. Rep. **9**(1), 1–18 (2019). https://doi.org/10.1038/s41598-019-50790-0

31. Liu, G., Barkaoui, K.: A survey of siphons in Petri nets. Inf. Sci. **363**, 198–220 (2016). https://doi.org/10.1016/j.ins.2015.08.037

32. Montagud, A., et al.: Patient-specific Boolean models of signaling networks guide personalized treatments. BioRxiv (2021). https://doi.org/10.1101/2021.07.28.454126

33. Murata, T.: Petri nets: properties, analysis and applications. Proc. IEEE **77**(4), 541–580 (1989). https://doi.org/10.1109/5.24143

34. Müssel, C., Hopfensitz, M., Kestler, H.A.: BoolNet - an R package for generation, reconstruction and analysis of Boolean networks. Bioinformatics **26**(10), 1378–1380 (2010). https://doi.org/10.1093/bioinformatics/btq124

35. Nabli, F., Martinez, T., Fages, F., Soliman, S.: On enumerating minimal siphons in Petri nets using CLP and SAT solvers: theoretical and practical complexity. Constraints **21**(2), 251–276 (2015). https://doi.org/10.1007/s10601-015-9190-1

36. Naldi, A., et al.: Cooperative development of logical modelling standards and tools with CoLoMoTo. Bioinformatics **31**(7), 1154–1159 (2015). https://doi.org/10.1093/bioinformatics/btv013

37. Noual, M., Regnault, D., Sené, S.: About non-monotony in Boolean automata networks. Theor. Comput. Sci. **504**, 12–25 (2013). https://doi.org/10.1016/j.tcs.2012.05.034

38. Oanea, O., Wimmel, H., Wolf, K.: New algorithms for deciding the siphon-trap property. In: Lilius, J., Penczek, W. (eds.) PETRI NETS 2010. LNCS, vol. 6128, pp. 267–286. Springer, Heidelberg (2010). https://doi.org/10.1007/978-3-642-13675-7_16

39. Ogishima, S., et al.: AlzPathway, an updated map of curated signaling pathways: towards deciphering Alzheimer's disease pathogenesis. In: Castrillo, J.I., Oliver, S.G. (eds.) Systems Biology of Alzheimer's Disease. MMB, vol. 1303, pp. 423–432. Springer, New York (2016). https://doi.org/10.1007/978-1-4939-2627-5_25

40. Ostaszewski, M., Niarakis, A., Mazein, A., Kuperstein, I., Phair, R., Orta-Resendiz, A., Singh, V., Aghamiri, S.S., Acencio, M.L., Glaab, E., et al.: COVID19 disease map, a computational knowledge repository of virus-host interaction mechanisms. Mol. Syst. Biol. **17**(10), e10387 (2021). https://doi.org/10.15252/msb.202110387

41. Paulevé, L., Kolčák, J., Chatain, T., Haar, S.: Reconciling qualitative, abstract, and scalable modeling of biological networks. Nat. Commun. **11**(1), 1–7 (2020). https://doi.org/10.1038/s41467-020-18112-5

42. Peterson, J.L.: Petri Net Theory and the Modeling of Systems. Prentice Hall PTR, Hoboken (1981)

43. Reddy, V.N., Mavrovouniotis, M.L., Liebman, M.N.: Petri net representations in metabolic pathways. In: Hunter, L., Searls, D.B., Shavlik, J.W. (eds.) Proceedings of the 1st International Conference on Intelligent Systems for Molecular Biology, Bethesda, MD, USA, July 1993, pp. 328–336. AAAI (1993). http://www.aaai.org/Library/ISMB/1993/ismb93-038.php

44. Rodríguez-Jorge, O., et al.: Cooperation between T cell receptor and Toll-like receptor 5 signaling for CD4+ T cell activation. Sci. Signal. **12**(577), eaar3641 (2019). https://doi.org/10.1126/scisignal.aar3641

45. Singh, V., et al.: Computational systems biology approach for the study of rheumatoid arthritis: from a molecular map to a dynamical model. Genom. Comput. Biol. **4**(1), e100050 (2018). https://doi.org/10.18547/gcb.2018.vol4.iss1.e100050

46. Thomas, R.: Boolean formalisation of genetic control circuits. J. Theor. Biol. **42**, 565–583 (1973). https://doi.org/10.1016/0022-5193(73)90247-6

47. Thomas, R.: Regulatory networks seen as asynchronous automata: a logical description. J. Theor. Biol. **153**(1), 1–23 (1991). https://doi.org/10.1016/S0022-5193(05)80350-9

48. Thomas, R., d'Ari, R.: Biological Feedback. CRC Press, Boca Raton (1990)

49. Tsirvouli, E., Touré, V., Niederdorfer, B., Vázquez, M., Flobak, Å., Kuiper, M.: A middle-out modeling strategy to extend a colon cancer logical model improves drug synergy predictions in epithelial-derived cancer cell lines. Front. Mol. Biosci. **7**, 502573 (2020). https://doi.org/10.3389/fmolb.2020.502573

50. Wang, R.S., Saadatpour, A., Albert, R.: Boolean modeling in systems biology: an overview of methodology and applications. Phys. Biol. **9**(5), 055001 (2012). https://doi.org/10.1088/1478-3975/9/5/055001

51. Zevedei-Oancea, I., Schuster, S.: Topological analysis of metabolic networks based on Petri net theory. Silico Biol. **3**(3), 323–345 (2003). http://content.iospress.com/articles/in-silico-biology/isb00100

Continuous and Hybrid Models

Stability Versus Meta-stability in a Skin Microbiome Model

Eléa Thibault Greugny[1,2], Georgios N. Stamatas[1], and François Fages[2(✉)]

[1] Johnson & Johnson Santé Beauté France, Issy-les-Moulineaux, France
[2] Inria Saclay, Lifeware Team, Palaiseau, France
`Francois.Fages@inria.fr`

Abstract. The skin microbiome plays an important role in the maintenance of a healthy skin. It is an ecosystem, composed of several species, competing for resources and interacting with the skin cells. Imbalance in the cutaneous microbiome, also called dysbiosis, has been correlated with several skin conditions, including acne and atopic dermatitis. Generally, dysbiosis is linked to colonization of the skin by a population of opportunistic pathogenic bacteria (for example *C. acnes* in acne or *S. aureus* in atopic dermatitis). Treatments consisting in non-specific elimination of cutaneous microflora have shown conflicting results. It is therefore necessary to understand the factors influencing shifts of the skin microbiome composition. In this work, we introduce a mathematical model based on ordinary differential equations, with 2 types of bacteria populations (skin commensals and opportunistic pathogens) to study the mechanisms driving the dominance of one population over the other. By using published experimental data, assumed to correspond to the observation of stable states in our model, we derive constraints that allow us to reduce the number of parameters of the model from 13 to 5. Interestingly, a meta-stable state settled at around 2 d following the introduction of bacteria in the model, is followed by a reversed stable state after 300 h. On the time scale of the experiments, we show that certain changes of the environment, like the elevation of skin surface pH, create favorable conditions for the emergence and colonization of the skin by the opportunistic pathogen population. Such predictions help identifying potential therapeutic targets for the treatment of skin conditions involving dysbiosis of the microbiome, and question the importance of meta-stable states in mathematical models of biological processes.

Keywords: skin microbiome · atopic dermatitis · ODE model · steady-state reasoning · parameter relations · quasi-stability · meta-stability

1 Introduction

Located at the interface between the organism and the surrounding environment, the skin constitutes the first line of defense against external threats, including

I. Petre and A. Păun (Eds.): CMSB 2022, LNBI 13447, pp. 179–197, 2022.
https://doi.org/10.1007/978-3-031-15034-0_9

irritants and pathogens. In order to control potential colonization of the skin surface by pathogens, the epidermal cells, called keratinocytes, produce antimicrobial peptides (AMPs) [26]. The physiologically acidic skin surface pH also contributes to control the growth of bacterial populations [17,27]. Another contributor to the defense against pathogen colonization are commensal bacteria in the community of microorganisms living on the skin, commonly referred to as the skin microbiome. Over the past decade, several studies have highlighted the key role played by such commensal bacterial species defending against invading pathogens, as well as their contribution to the regulation of the immune system [1,2,5,15,18,19].

Alterations in the composition of the skin microbiome resulting in a dominance by a pathogenic species, also called dysbiosis, have been associated with skin conditions such as acne or atopic dermatitis (AD) [16,20]. In the case of AD, the patient skin is often colonized by *Staphylococcus aureus* (*S. aureus*), especially on the lesions [16]. Treatment strategies targeting non-specific elimination of cutaneous microflora, such as bleach baths, have shown conflicting results regarding their capacity to reduce the disease severity [4]. On the other hand, treatments involving introduction of commensal species, like *Staphylococcus hominis* [24] on the skin surface appear promising. Accordingly, the interactions between the commensal populations, pathogens and skin cells seem at the heart of maintaining microbiome balance. There is therefore a necessity to investigate further those interactions and the drivers of dominance of one population over others. Unfortunately, it is challenging to perform *in vitro* experiments involving more than one or two different species, even more so on skin explants or skin equivalents.

Mathematical models of population dynamics have been developed and used for more that 200 years [21]. Here, we introduce a model based on ordinary differential equations (ODEs), describing the interactions of a population of commensal species with one of opportunistic pathogens and the skin cells. We study the factors influencing the dominance of one population over the other on a microbiologically relevant timescale of a couple of days corresponding to biological experimental data. More specifically, we identify constraining relationships on the parameter values, based on published experimental data [14,23], corresponding to special cases of our model, allowing us to reduce the parametric dimension of our model from 13 to 5 parameters. Interestingly, we observe in the reduced model a phenomenon of meta-stability [28,30], also called quasi-stability, in which the seemingly stable state reached after 30 h following the initiation of the experiment, is followed after 300 h by a reversed stable state. On the time scale of the experiments, we show that certain changes in the environment, like an elevation of skin surface pH, create favorable conditions for the emergence and colonization of the skin by the opportunistic pathogen population. Such predictions can help identify potential therapeutic strategies for the treatment of skin conditions involving microbiome dysbiosis, and underscore the importance of meta-stable states in the real biological processes at their different time scales.

2 Initial ODE Model with 13 Parameters

The model built in this paper considers two types of bacterial populations. The first population, S_c, regroups commensal bacteria species having an overall beneficial effect for the skin, and the second population, S_p, represents opportunistic pathogens. The differential equations for both bacterial populations are based on the common logistic growth model [32], considering non-explicitly the limitations in food and space. The limited resources are included in the parameters K_{sc} and K_{sp}, representing the optimum concentration of the populations in a given environment, considering the available resources.

The bactericidal effect of antimicrobial peptides (AMPs) produced by skin cells, Amp_h, on S_p is included with a Hill function. This type of highly non-linear functions have been used previously to model the effect of antibiotics on bacterial populations [22]. For the sake of simplicity, the AMPs produced by skin cells is introduced as a constant parameter, $[Amp_h]$, in the model. It represents the average concentration of these AMPs among surface cells, under given human genetic background and environmental conditions.

Several studies revealed that commensal bacterial populations, like *S. epidermidis* or *S. hominis*, are also able to produce AMPs targeted against opportunistic pathogens, such as *S. aureus* [6,23]. For these reasons, we introduce in the model AMPs of bacterial origin, Amp_b, acting similarly to Amp_h on the pathogenic population S_p. Amp_b is produced at rate k_c by S_c, and degraded at rate d_a. Furthermore, we include a defense mechanism of S_p against S_c with a direct killing effect.

Altogether, this gives us the following ODE system with 3 variables and 13 parameters, all taking non-negative values:

$$
\begin{cases}
\dfrac{d[S_c]}{dt} = \left(r_{sc}\left(1 - \dfrac{[S_c]}{K_{sc}}\right) - \dfrac{d_{sc}[S_p]}{C_1 + [S_p]} \right)[S_c] \\[3mm]
\dfrac{d[S_p]}{dt} = \left(r_{sp}\left(1 - \dfrac{[S_p]}{K_{sp}}\right) - \dfrac{d_{spb}[Amp_b]}{C_{ab} + [Amp_b]} - \dfrac{d_{sph}[Amp_h]}{C_{ah} + [Amp_h]} \right)[S_p] \\[3mm]
\dfrac{d[Amp_b]}{dt} = k_c[S_c] - d_a\,[Amp_b]
\end{cases}
\tag{1}
$$

The model is illustrated on Fig. 1 and Table 1 recapitulates the variables and the parameters with their unit.

Such a model cannot be solved analytically. Furthermore, the use of optimization algorithms to infer the 13 parameter values from data resulted in many valid sets of parameter values. Therefore, it is clearly necessary to restrict the number of parameters by identifying some of them, to be able to analyze the model.

Table 1. List of the parameters and variables of our mathematical model with their units. CFU = Colony forming unit, AU = Arbitrary Unit, ASU = Arbitrary Surface Unit

Variable	Interpretation (unit)
$[S_c]$	Surface apparent concentration of S_c $(CFU.ASU^{-1})$
$[S_p]$	Surface apparent concentration of S_p $(CFU.ASU^{-1})$
$[Amp_b]$	Concentration of Amp_b $(AU.ASU^{-1})$
Parameter	**Interpretation (unit)**
r_{sc}	Growth rate of S_c (h^{-1})
r_{sp}	Growth rate of S_p, (h^{-1})
K_{sc}	Optimum concentration of S_c $(CFU.ASU^{-1})$
K_{sp}	Optimum concentration of S_p $(CFU.ASU^{-1})$
d_{sc}	Maximal killing rate of S_c by S_p (h^{-1})
C_1	Concentration of S_p inducing half the maximum killing rate d_{sc} $(CFU.ASU^{-1})$
d_{spb}	Maximal killing rate of S_p by Amp_b, (h^{-1})
C_{ab}	Concentration of Amp_b inducing half the maximum killing rate d_{spb} $(AU.ASU^{-1})$
d_{sph}	Maximal killing rate of S_p by Amp_h, (h^{-1})
C_{ah}	Concentration of Amp_h inducing half the maximum killing rate d_{sph} $(AU.ASU^{-1})$
$[Amp_h]$	Concentration of AMPs produced by the skin cells $(AU.ASU^{-1})$
k_c	Production rate of Amp_b by S_c $(AU.h^{-1}.CFU^{-1})$
d_a	Degradation rate of Amp_b $(AU.h^{-1})$

3 Using Published Experimental Data to Define Relations Between Model Parameters by Steady-State Reasoning

The amount of quantitative experimental data available for the model calibration is very limited due to the difficulty of carrying out experiments involving co-cultures of different bacterial species. Most of the published work focuses on single species or on measuring the relative abundances of species living on the skin, which is highly variable between individuals and skin sites [10]. In the case of AD specifically, *S. aureus* is considered pathogenic and *S. epidermidis* commensal. Published data exist however for those species which we can use to constrain the parameter values of the model.

Two series of *in vitro* experiments are considered [14,23]. While *in vitro* cultures, even on epidermal equivalent, do not entirely capture the native growth

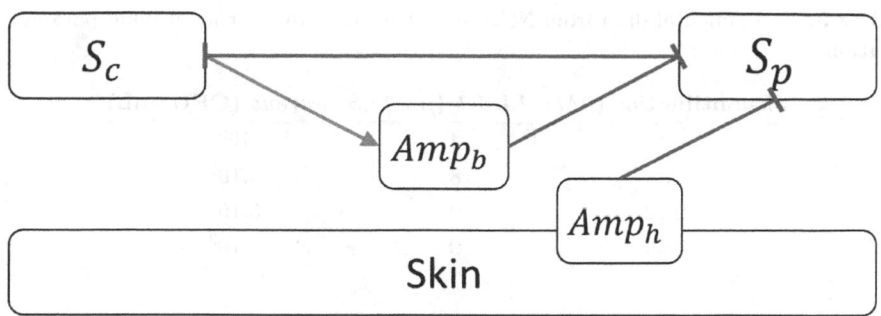

Fig. 1. Model overview, green arrow representing production and red T-lines representing killing effect.

of bacteria on human skin, they provide useful quantitative data that would be very difficult to measure *in vivo*.

In the first experiment [14], mono-cultures and co-cultures of *S. epidermidis* and *S. aureus* were allowed to develop on a 3D epidermal equivalent. Table 2 recapitulates the population sizes of the two species measured after 48 h of incubation. Kohda *et al.* also performed another co-culture experiment where *S.epidermidis* was inoculated 4 h prior to *S.aureus* in the media. This data is not used here as it requires additional manipulation to match the situation represented by the model. However, it would be interesting to use it in the future for model validation.

In the second experiment [23] the impact of human (LL-37) and bacterial (*Sh*-lantibiotics) AMPs on *S. aureus* survival was studied. The experiments were performed *in vitro*, and the *S. aureus* population size was measured after 24 h of incubation. Table 3 summarizes their observations.

Table 2. Experimental data from Kohda et al. [14] used for identifying parameter values.

	S. epidermidis (CFU/well)	*S. aureus* (CFU/well)
Mono-cultures	4.10^8	3.10^9
Co-cultures	1.10^8	1.10^9

3.1 Parameter Values Inferred from Mono-culture Experiment Data

We consider first the monocultures experiments from Kohda *et al.* [14], representing the simplest experimental conditions. *S. epidermis* is a representative of

Table 3. Experimental data from Nakatsuji et al. [23] used for identifying parameter relations.

Sh-lantibiotics (μM)	LL-37 (μM)	S. aureus (CFU/mL)
0	4	10^9
0	8	6.10^5
0.32	0	5.10^8
0.64	0	3.10^3

the commensal population S_c, and S. aureus of the pathogenic one, S_p. Since the two species are not interacting, the set of equations simplifies to:

$$\begin{cases} \dfrac{d[S_c]}{dt} = \left(r_{sc} \left(1 - \dfrac{[S_c]}{K_{sc}} \right) \right) [S_c] \\[3mm] \dfrac{d[S_p]}{dt} = \left(r_{sp} \left(1 - \dfrac{[S_p]}{K_{sp}} \right) \right) [S_p] \end{cases} \tag{2}$$

At steady-state, the population concentrations are either zero, or equal to their optimum capacities (K_{sc} or K_{sp}) when the initial population concentration is non-zero. Given the rapid growth of bacterial population, the experimental measurements done after 48 h of incubation can be considered as corresponding to a steady-state, which gives:

$$K_{sc} = 4.10^8 \ CFU.ASU^{-1} \tag{3}$$

$$K_{sp} = 3.10^9 \ CFU.ASU^{-1} \tag{4}$$

3.2   Parameter Relations Inferred from Experimental Data on AMP

The experimental conditions of Nakatsuji *et al.* [23] correspond to the special case where there is no commensal bacteria alive in the environment, only the bacterial AMPs, in addition to those produced by the skin cells. Our system of equations then reduces to:

$$\dfrac{d[S_p]}{dt} = \left(r_{sp} \left(1 - \dfrac{[S_p]}{K_{sp}} \right) - \dfrac{d_{spb}[Amp_b]}{C_{ab} + [Amp_b]} - \dfrac{d_{sph}[Amp_h]}{C_{ah} + [Amp_h]} \right) [S_p] \tag{5}$$

The concentrations in LL-37 and Sh-lantibiotics, translated in our model into $[Amp_h]$ and $[Amp_b]$ respectively, are part of the experimental settings. Therefore, we consider them as constants over time. At steady state, we get:

$$[S_p]^* = 0 \quad \text{or} \quad [S_p]^* = K_{sp}\left(1 - \frac{d_{spb}[Amp_b]}{r_{sp}(C_{ab} + [Amp_b])} - \frac{d_{sph}[Amp_h]}{r_{sp}(C_{ah} + [Amp_h])}\right)$$
(6)

Let us first focus on the special case where no Sh-lantibiotics were introduced in the media, translating into $[Amp_b] = 0$ in our model. We consider again that the biological observations after 24 h of incubation correspond to steady-state and substitute the experimental values measured $[Amp_h] = 4\,\mu M$; $[S_p]^* = 10^9$ CFU, and $[Amp_h] = 8\,\mu M$; $[S_p]^* = 6.10^5$ CFU, together with the values of K_{sc} and K_{sp} (from (3) and (4)) in (6), to obtain the following equations:

$$\begin{cases} \dfrac{d_{sph}}{r_{sp}} = \dfrac{4 + C_{ah}}{6} \\[4mm] \dfrac{d_{sph}}{r_{sp}} = \dfrac{(10^4 - 2)(C_{ah} + 8)}{8.10^4} \end{cases}$$
(7)

which reduce to $C_{ah} = 8$ and $\frac{d_{sph}}{r_{sp}} = 2$.

Following the same method with the experimental conditions without any LL-37 (i.e. $[Amp_h] = 0$) and using two data points ($[Amp_b] = 0.32\,\mu M$; $[S_p]^* = 5.10^8$ CFU) and ($[Amp_b] = 0.64\,\mu M$; $[S_p]^* = 3.10^3$ CFU), we get $C_{ab} = 0.16$ and $\frac{d_{spb}}{r_{sp}} = \frac{5}{4}$.

It is notable that the maximum killing rates of S_p by Amp_b and Amp_h are both proportional to S_p growth rate. Interestingly, such proportional relation has been observed experimentally between the killing rate of *Escherichia coli* by an antibiotic and the bacterial growth rate [31].

To be consistent with the ranges of Sh-lantibiotics concentrations described in Nakatsuji et al. [23], $[Amp_b]$ should take positive values below 10. Given that $[Amp_b]^* = \frac{k_c[S_c]^*}{d_a}$ at steady-state, and that $K_{sc} = 4.10^8$ CFU is the upper bound for $[S_c]^*$, we obtain the following constraint:

$$\frac{k_c}{d_a} \leq \frac{1}{4.10^7}$$
(8)

3.3   Parameter Relations Inferred from Co-culture Data

The initial model described earlier is representative of the experimental settings of the co-culture conditions described in Kohda et al. [14]. At steady-state, the system (1) gives:

$$[S_c]^* = 0 \quad \text{or} \quad [S_c]^* = K_{sc}\left(1 - \frac{d_{sc}[S_p]^*}{r_{sc}(C_1 + [S_p]^*)}\right)$$
(9)

$$[S_p]^* = 0 \quad \text{or} \quad [S_p]^* = K_{sp}\left(1 - \frac{d_{spb}[Amp_b]}{r_{sp}(C_{ab} + [Amp_b])} - \frac{d_{sph}[Amp_h]}{r_{sp}(C_{ah} + [Amp_h])}\right)$$
(10)

$$[Amp_b]^* = \frac{k_c[S_c]^*}{d_a} \qquad (11)$$

Considering that what is observed experimentally after 48 h of incubation is at steady-state, one can replace $[S_c]^*$ and $[S_p]^*$ with the experimental data point ($S.\ epidermidis = 10^8$ CFU; $S.\ aureus = 10^9$ CFU) in (9) and (10) to get the following parameter relation:

$$\frac{d_{sc}}{r_{sc}} = \frac{3}{4.10^9}C_1 + \frac{3}{4} \qquad (12)$$

$$\frac{2}{3}r_{sp} = \frac{d_{sph}[Amp_h]}{C_{ah} + [Amp_h]} + \frac{10^8 d_{spb}k_c}{d_a C_{ab} + 10^8 k_c} \qquad (13)$$

By integrating the values found for C_{ah} and C_{ab}, and the relations involving d_{sph} and d_{spb} into (13), we end up with:

$$d_a = 10^8 k_c \frac{56 + 31[Amp_h]}{2.56\,(4 - [Amp_h])} \quad \text{with } [Amp_h] < 4 \qquad (14)$$

4   Reduced Model with 5 Parameters

Using the previously mentioned experimental data, and assuming they represent steady state conditions of the initial model (1), we have reduced the parametric dimension of the model from 13 to 5. Specifically, out of the original 13 parameters, we could define the values of 4 of them, and derive 4 functional dependencies from the values of the remaining parameters, as summarized in Table 4).

In our skin microbiome model (1), the parameters that remain unknown are thus:

- r_{sc}, the growth rate of S_c which can reasonably take values between 0 and 2 h^{-1} following [3, 7];
- r_{sp}, the growth rate of S_p, taking similar values in the interval between 0 and 2 h^{-1};
- C_1, the concentration of S_p that induces half the maximum killing rate d_{sc} (in $CFU.ASU^{-1}$) and is thus bounded by the optimum concentration of S_p, i.e. $K_{sp} = 3.10^9\ CFU.ASU^{-1}$, as calculated in Sect. 3.1 from [14];
- k_c, the production rate of $[Amp_b]$ chosen to take values between 0 and 0.1 $AU.h^{-1}.CFU^{-1}$, and shown to have a limited impact on the steady-state values in Sect. 4.2;
- $[Amp_h]$, the concentration in $AU.ASU^{-1}$ of AMPs produced by skin cells between 0 and 4 (equation (14)).

Table 4. Summary of the parameter relations embedded in the reduced model.

Parameter	Value or relation to other parameters
K_{sc}	4.10^8
K_{sp}	3.10^9
C_{ah}	8
C_{ab}	0.16
d_{sph}	$2\,r_{sp}$
d_{spb}	$\frac{5}{4}\,r_{sp}$
d_{sc}	$r_{sc}\left(\dfrac{3}{4.10^9}\,C_1 + \dfrac{3}{4}\right)$
d_a	$10^8 k_c\,\dfrac{56 + 31[Amp_h]}{2.56\,(4 - [Amp_h])}$ with $[Amp_h] < 4$

4.1 Simulations at the Time Scale of the Experiments

In order to reproduce what was observed by Kohda et al. [14], that is a dominant pathogenic population after 50 h which can thus be considered as dysbiosis in our skin microbiome model, it is sufficient to fix a relatively low concentration of Amp produced by the skin cells, i.e. $Amp_h = 1.5$, and some fixed values for the four other parameters chosen in their intervals described above. Among a continuum of possible solutions, we chose $r_{sc} = 0.5$, $r_{sp} = 1$, $C_1 = 5.10^6$, $k_c = 0.01$.

The doses of *S. epidermidis* and *S. aureus* applied at the surface of the 3D epidermal equivalent at the beginning of the experiment ($10^5 CFU/mL$ and $10^3 CFU/mL$ respectively) are used as the initial concentrations for $[S_c]$ and $[S_p]$ respectively. Figure 2 shows the result of a numerical simulation[1] of our model with those parameters which are in accordance to the co-culture experiments of Kohda et al. and reproduce a consistent qualitative behavior [14].

Our model can also be used to reproduce what is considered a balanced microbiome, corresponding to the commensal population being significantly more abundant than the pathogenic one. This requires modifying some parameter values to represent a less virulent pathogenic population, closer to the physiological context, given that the experiments from Kohda et al. [14] were performed using a virulent methicillin-resistant *S. aureus* strain.

We chose $r_{sp} = 0.5$, $C_1 = 2.10^8$ and a higher production of AMPs by the skin cells, $[Amp_h] = 3$, to compensate for feedback loops or stimuli that might be missing in the 3D epidermal equivalent used. Figure 3 shows a simulation trace obtained under those conditions which clearly indicates the dominance of the non-pathogenic population under those conditions.

---

[1] All computation results presented in this paper have been done using the BIOCHAM software with a notebook runnable online and available at https://lifeware.inria.fr/wiki/Main/Software#CMSB22b.

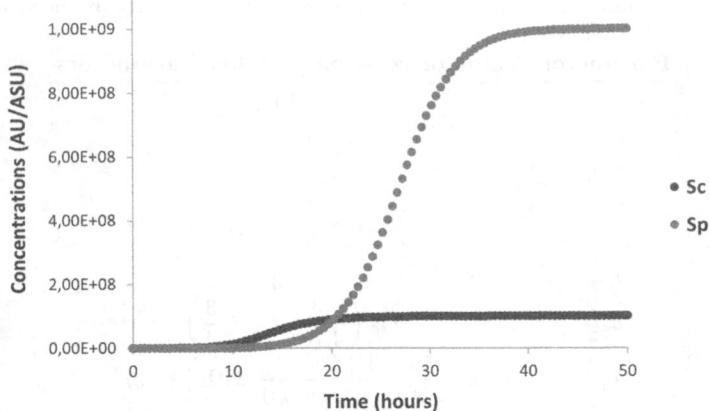

Fig. 2. Numerical simulation of the reduced ODE model over 50 h, with initial conditions $[S_c] = 10^5$, $[S_p] = 10^3$, $[Amp_b] = 0$ and parameter values $[Amp_h] = 1.5$, $r_{sc} = 0.5$, $r_{sp} = 1$, $C_1 = 5.10^6$, $k_c = 0.01$, to fit Kohda et al. co-culture data [14] (Table 2).

4.2 Parameter Sensitivity and Robustness Analyses

Since the previous simulations rely on some choices of values for the unknown parameters, it is important to evaluate the robustness of the predictions of our model by performing an analysis of sensitivity to the parameter values. This is possible in Biocham by specifying the property of interest in quantitative temporal logic [29]. The interesting property here is the stabilization at the time scale of the experiments around 48 h of the bacterial population sizes to the values given by simulation (Fig. 3), Here we use the temporal logic formula: $F(Time == 40 \wedge NSc = x1 \wedge NSp = y1 \wedge F(G(NSc = x2 \wedge NSp = y2)))$ and objective values equal to 1 for the free variables $x1, x2, y1, y2$, to express that the normalized variables NSc and NSp, i.e. current values of Sc and Sp divided by their expected value at steady state, respectively 10^8 and 10^9 in the pathogenic case of Kohda et al. experiments, is reached (F, finally) at time around 40 and finally at the end of the time horizon (FG) of 50 h. On a given simulation trace, the free variables of the formula have a validity domain (here fixed values) which is used to define a continuous degree of satisfaction of the property as a distance to the objective values, and a robustness degree by sampling parameter values around their nominal values [29].

The sensitivity analysis (Table 5) reveals that the dominance of the commensal population is highly sensitive to variations of the initial concentration of the pathogen. To a lesser extend, the dominant population is also sensitive to the

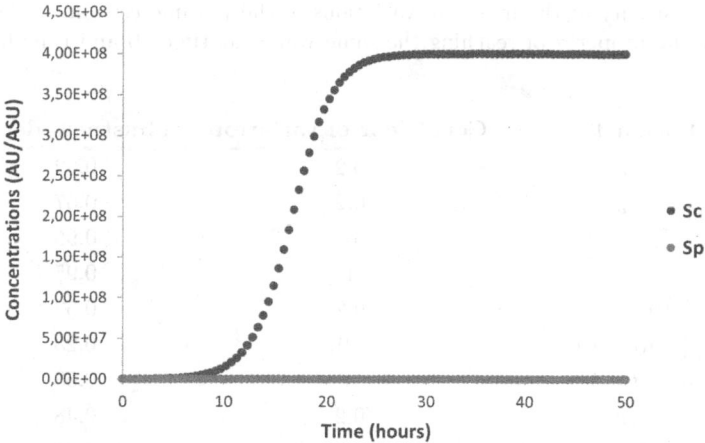

Fig. 3. Numerical simulation of the reduced ODE model over 50 h, with initial conditions $[S_c] = 10^5$, $[S_p] = 10^3$, $[Amp_b] = 0$ and parameter values $r_{sc} = r_{sp} = 0.5$, $C_1 = 2.10^8$, $k_c = 0.01$, $[Amp_h] = 3$ corresponding to Kohda et al. experiments [14].

growth rates (r_{sc} and r_{sp}) and the concentration of human AMPs ($[Amp_h]$). On the other hand, C_1 and k_c do not seem to affect the relative proportions of the bacterial populations.

4.3 Meta-stability Revealed by Simulation on a Long Time Scale

Interestingly, by extending the simulation time horizon to a longer time scale of 500 h, one can observe a meta-stability phenomenon, shown in Fig. 4. The seemingly stable state observed in Fig. 3 at the relevant time scale of 50 h of the experiments, is thus not a mathematical steady state, but a meta-stable state, also called quasi-stable state, that slowly evolves, with $\frac{d[S_c]}{dt} \neq 0$ and $\frac{d[S_p]}{dt} \neq 0$, towards a true stable state of the model reached around 300 h in which the population density are reversed.

The S_c population almost reaches its optimum capacity K_{sc} after approximately 30 h and stays relatively stable for around 100 h more, that is over 4 d, which can reasonably be considered stable on the microbiological time scale. Meanwhile, the S_p population is kept at a low concentration compared to S_c, even though it is continuously increasing and eventually leading to its overtake of S_c.

Table 5. Sensitivity of the model to variations of the parameters and initial concentrations for the property of reaching the same values at time 40 and time horizon 50 as in Fig. 3.

Parameter	Coefficient of variation	Robustness degree
r_{sc}	0.2	0.62
r_{sp}	0.2	0.57
C_1	10	0.95
k_c	1	0.95
$[Amp_h]$	0.2	0.53
$[S_p](t = 0)$	10	0.23
$[S_c](t = 0)$	10	0.58
(r_{sc}, r_{sp})	0.2	0.48
$([S_c](t = 0), [S_p](t = 0))$	10	0.31

By varying the parameters values, it appears that this meta-stability phenomenon emerges above a threshold value of 2.5 for $[Amp_h]$[2], that is for almost half of its possible values (see Sect. 4).

That phenomenon of meta-stability, also called quasi-stability, is a classical notion of dynamical systems theory, particularly well-studied in the case of oscillatory systems for which analytical solutions exist, and as models of brain activity [30]. It is worth noting that it has also been considered in the computational systems biology community with respect to model reduction methods based on the identification of different regimes corresponding to different preponderant terms of the ODEs, for which simplified dynamics can be defined, and chained within a hybrid automaton [28].

More generally, this raises the question of the existence and importance of meta-stability in real biological processes, as well as the validity of the steady state assumptions made in mathematical modeling methods to fit the models to the observed experimental data.

5   Conditions Favoring the Pathogenic Population

Whether the dysbiosis observed in AD is the cause or the result of the disease is unclear [12,13]. Infants developing AD do not necessarily have more *S. aureus* present on their skin prior to the onset of the disease compared to the healthy group [11]. This suggests that atopic skin has some characteristics enabling the dominance of *S. aureus* over the other species of the microbiome. To test this hypothesis, we investigate two changes of the skin properties observed in AD patients (skin surface pH elevation [9] and reduced production of AMPs [25])

---

[2] All computation results presented in this paper have been done using the BIOCHAM software with a notebook runnable online and available at https://lifeware.inria.fr/wiki/Main/Software#CMSB22b.

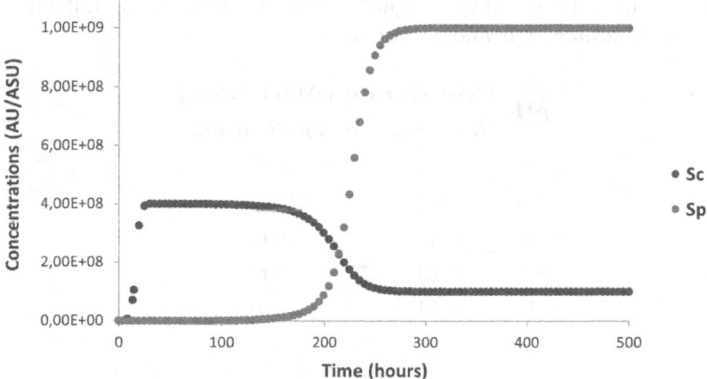

Fig. 4. Numerical simulation of the reduced ODE model on a longer time scale of 500 h, with the same initial concentrations and parameter values as in Fig. 3, showing an inversion of the dominant bacterial population after 220 h.

and their impact on the dominant species at steady-state. More specifically, we study the behavior of the system following the introduction of a pathogen and whether the pathogen will colonize the media depending on the initial concentrations of the bacterial populations and the particular skin properties mentioned before.

5.1   Skin Surface pH Elevation

According to Proksch [27], the physiological range for skin surface pH is 4.1-5.8. However, in certain skin conditions, like AD, an elevation of this pH has been observed. Dasgupta *et al.* studied *in vitro* the influence of pH on the growth rates of *S. aureus* and *S. epidermidis*[8]. Their experimental results show that, when the pH is increased from 5 to 6.5, the growth rate of *S. epidermidis* is multiplied by 1.8, whereas the one of *S. aureus* is multiplied by more than 4 (Table 6).

Their data can be used to select values for the growth rates r_{sc} and r_{sp} in our model, corresponding to healthy skin with a skin surface pH of 5 and compromised skin with a pH of 6.5. Because the experiments from Dasgupta *el al.* were performed *in vitro* and the bacterial population sizes measured with optical density (OD) instead of CFU, the growth rates cannot be directly translated into r_{sc} and r_{sp}. We use $r_{sc} = 0.5$ as the reference value for the commensal growth rate at pH 5, following on from previous simulation (Fig. 3). Maintaining the ratio between the two population growth rates at pH 5 and the multiplying factors following the pH elevation from Dasgupta *et al.* experimental data, we can define two sets of values for r_{sc} and r_{sp}:

Table 6. Experimental data from Dasgupta et al. [8] showing the influence of pH on growth rates of *S. epidermidis* and *S. aureus*

pH	Growth rate (ΔOD/hour)	
	S. aureus	*S. epidermidis*
5	0.03	0.05
5.5	0.04	0.07
6	0.09	0.08
6.5	0.13	0.09
7	0.14	0.10

$$\text{skin surface pH of 5} \quad \Rightarrow r_{sc} = 0.5, \ r_{sp} = 0.3$$

$$\text{skin surface pH of 6.5} \quad \Rightarrow r_{sc} = 0.9, \ r_{sp} = 1.3$$

Considering the healthy skin scenario with a skin surface pH of 5, the influence of the bacterial populations initial concentrations on the dominant species after 50 h is evaluate using the temporal logic formula:

$$F(\text{Time} == 40 \land ([S_c] > u1\,[S_p]) \land F(G([S_c] > u2\,[S_p])))$$

where $u1$ and $u2$ are free variables representing the abundance factors between both populations, evaluated at Time= 40 and at the last time point of the trace respectively (F stands for finally and G for globally at all future time points), i.e. at the time horizon of the experiments of 50 h.

When given with an objective value, e.g. $u1 = 10$, the distance between that value and the validity domain of the formula, i.e. the set of values for $u1$ that satisfy the formula, provides a violation degree which is used to evaluate the satisfaction degree of the property.

Here, we evaluate how much the temporal formula $F(\text{Time} == 40 \land ([S_c] > u1\,[S_p]) \land F(G([S_c] > u2\,[S_p])))$, $u1 \to 10$, $u2 \to 10$, is satisfied given variations of the initial concentrations of two populations (Fig. 5). The model predicts that, under the healthy skin condition, the commensal population will always dominate after 50 h, except when introduced at a relatively low concentration ($< 2.10^4$) while the initial concentration of the pathogenic population is high ($> 5.10^5$).

The model predicts a higher vulnerability of the skin regarding invading pathogens with an elevated skin surface pH. When evaluating the same temporal formula with growth rates values corresponding to a skin surface pH of 6.5, we observe that even when the initial concentration of commensal is high ($> 10^7$), the pathogenic population is able to colonize the skin when introduced at a concentration as low as 3.10^4 (Fig. 6).

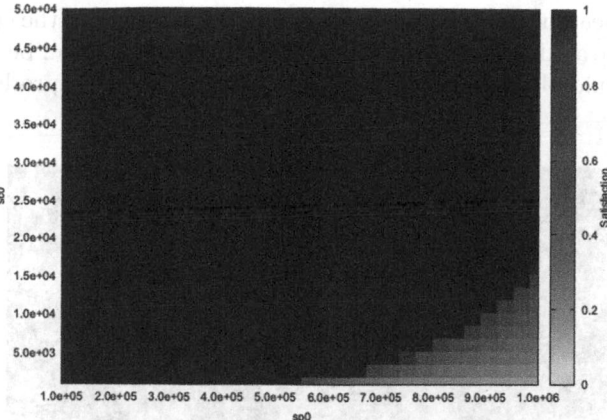

Fig. 5. Landscape of satisfaction degree of the temporal formula corresponding to healthy skin with a skin surface pH of 5 ($r_{sc} = 0.5$ and $r_{sp} = 0.3$). The x and y axis represent variations of the initial quantities of $[S_p]$ and $[S_c]$ respectively. The color coding corresponds to the satisfaction degree of the temporal logic formula. Values used for the other parameters: $C_1 = 2.10^8$, $k_c = 0.01$, $[Amp_h] = 3$.

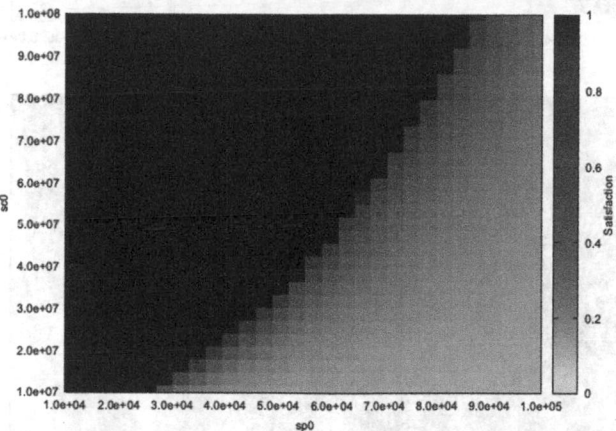

Fig. 6. Landscape of satisfaction degree of the temporal formula corresponding to compromised skin with a skin surface pH of 6.5 ($r_{sc} = 0.9$ and $r_{sp} = 1.3$). The x and y axis represent variations of the initial quantities of $[S_p]$ and $[S_c]$ respectively. The color coding corresponds to the satisfaction degree of the temporal logic formula. Values used for the other parameters: $C_1 = 2.10^8$, $k_c = 0.01$, $[Amp_h] = 3$.

Such predictions highlight the protective effect of the skin surface acidic pH against the invasion of pathogenic bacteria.

5.2 Reduced Production of Skin AMPs

As mentioned before, human keratinocytes constitutively produce AMPs as a defense against pathogens. In atopic dermatitis, the expression of AMPs is dysregulated, leading to lower concentration levels of AMPs in the epidermis [23]. Similarly to the analysis done for skin surface pH, our model can be used to study how the skin microbiome reacts to modulation of the AMPs production by the

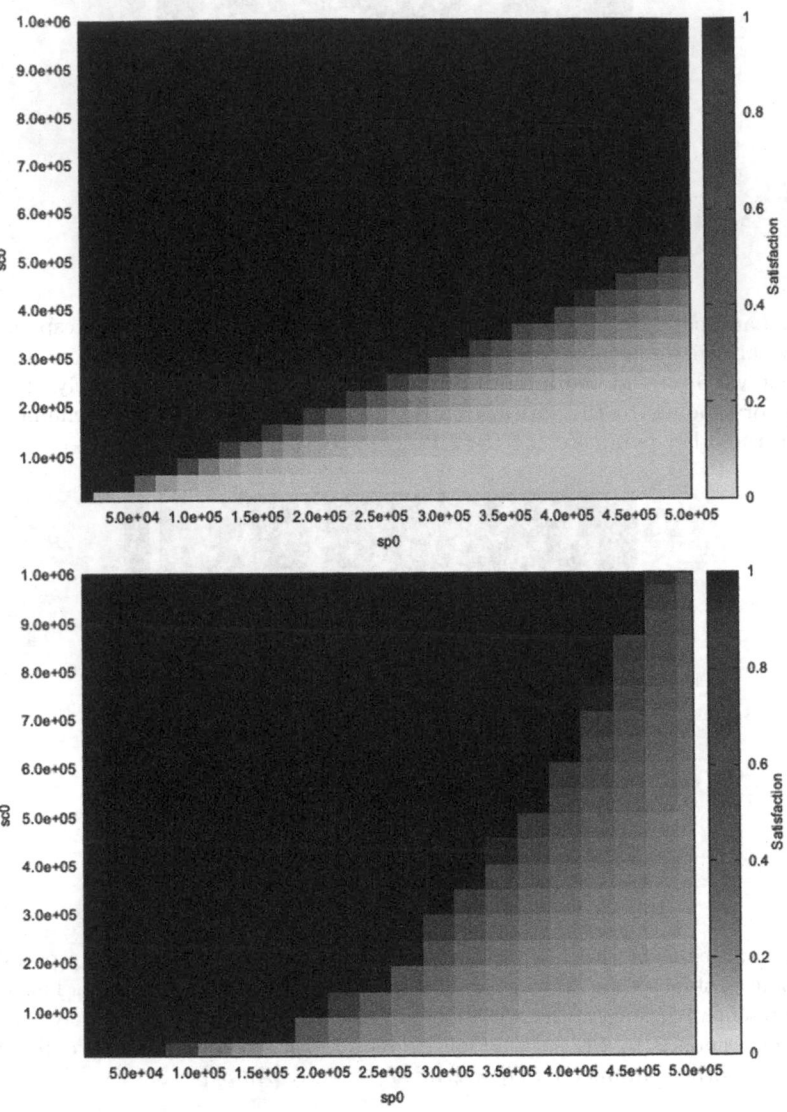

Fig. 7. Landscape of satisfaction degree of the healthy condition formula with a low concentration of human AMPs on the upper graph ($[Amp_h] = 0.5$) and a high concentration at the bottom ($[Amp_h] = 3$). The x and y axis represent variations of the initial quantities of $[S_p]$ and $[S_c]$ respectively. The color coding corresponds to the satisfaction degree of the temporal logic formula. Values used for the other parameters: $r_{sc} = r_{sp} = 0.5$, $C_1 = 2.10^8$, $k_c = 0.01$.

skin cells. Two situations are considered: an impaired production of AMPs by the skin cells ($[Amp_h] = 0.5$) and a higher concentration with $[Amp_h] = 3$. Using the same methodology as in the case of skin surface pH, the temporal logic formula $F(\text{Time} == 40 \wedge ([S_c] > u1\,[S_p]) \wedge F(G([S_c] > u2\,[S_p])))$, $u1 \to 10$, $u2 \to 10$, is evaluated for variations of the initial concentrations of both populations for $[Amp_h] = 0.5$ and $[Amp_h] = 3$ (Fig. 7).

The model predicts a slightly protective effect of Amp_h regarding the colonization of the skin by a pathogenic population, for low initial concentrations. However when both populations are introduced in high concentrations, the increase of $[Amp_h]$ appears to have the opposite effect of facilitating the colonization by the pathogenic population.

This mitigated effect might be due to the presence of $[Amp_h]$ in the constraint related to the degradation rate of $[Amp_b]$ (equation (14)) and deserves further investigation.

6   Conclusion

The objective of this research is the identification of conditions which might favor or inhibit the emergence of pathogenic populations in the skin microbiome. Such analyses can lead to insights about potential treatment strategies aiming at restoring a dysbiotic condition.

We have developed a simple ODE model of skin microbiome with 3 variables and 13 parameters which could be reduced to 5 parameters by using published data from the literature and steady state reasoning on the observations made in the biological experiments. Our bacterial population model is generic in the sense that we did not take into account the peculiarities of some specific bacterial populations, but on some general formulas of adversary population dynamics and influence factors. We showed through sensitivity analyses that our model predictions are particularly robust with respect to parameter variations.

Perhaps surprisingly, we also showed that this simple model exhibits over a large range of biologically relevant parameter values, a meta-stability phenomenon, revealed by allowing the simulation to continue for times one order of magnitude longer than the reported experimental times. This observation questions the existence and importance of meta-stability phenomena in real biological processes, whereas a natural assumption made in mathematical modeling, and model fitting to data, is that the experimental data are observed in states corresponding to real stable states of the mathematical model.

Acknowledgment. We are grateful to Mathieu Hemery, Aurélien Naldi and Sylvain Soliman for interesting discussions on this work.

References

1. Belkaid, Y., Segre, J.A.: Dialogue between skin microbiota and immunity. Science **346**(6212), 954–959 (2014)
2. Byrd, A.L., Belkaid, Y., Segre, J.A.: The human skin microbiome. Nat. Rev. Microbiol. **16**(3), 143–155 (2018). https://doi.org/10.1038/nrmicro.2017.157
3. Campion, J.J., McNamara, P.J., Evans, M.E.: Pharmacodynamic modeling of ciprofloxacin resistance in staphylococcus aureus. Antimicrob. Agents Chemother. **49**(1), 209–219 (2005)
4. Chopra, R., Vakharia, P.P., Sacotte, R., Silverberg, J.I.: Efficacy of bleach baths in reducing severity of atopic dermatitis: a systematic review and meta-analysis. Ann. Allergy Asthma Immunol. **119**(5), 435–440 (2017)
5. Cogen, A.L., et al.: Staphylococcus epidermidis antimicrobia δ-toxin (phenol-soluble modulin-γ) cooperates with host antimicrobial peptides to kill group a streptococcus. PLOS ONE **5**(1), e8557 (2010)
6. Cogen, A.L., et al.: Selective antimicrobial action is provide by phenol-soluble modulins derived from staphylococcus epidermidis, a normal resident of the skin. J. Invest. Dermatol. **130**(1), 192–200 (2010)
7. Czock, D., Keller, F.: Mechanism-based pharmacokinetic-pharmacodynamic modeling of antimicrobial drug effects. J. Pharmacokinet. Pharmacodyn. **34**(6), 727–751 (2007). https://doi.org/10.1007/s10928-007-9069-x
8. Dasgupta, A., Iyer, V., Raut, J., Qualls, A.: 16502 effect of ph on growth of skin commensals and pathogens. J. Am. Acad. Dermatol. **83**(6), AB180 (2020). https://doi.org/10.1016/j.jaad.2020.06.808, https://www.jaad.org/article/S0190-9622(20)31908-3/abstract
9. Eberlein-König, B., et al.: Skin surface pH, stratum corneum hydration, transepidermal water loss and skin roughness related to atopic eczema and skin dryness in a population of primary school children. Acta Derm. Venereol. **80**(3), 188–191 (2000). https://doi.org/10.1080/000155500750042943
10. Grice, E.A., et al.: Topographical and temporal diversity of the human skin microbiome. Science **24**(5931), 1190–1192 (2009). https://doi.org/10.1126/science.1171700, https://www.ncbi.nlm.nih.gov/pmc/articles/PMC2805064/
11. Kennedy, E.A., et al.: Skin microbiome before development of atopic dermatitis: early colonization with commensal staphylococci at 2 months is associated with a lower risk of atopic dermatitis at 1 year. J. Allergy Clin. Immunol. **139**(1), 166–172 (2017)
12. Kobayashi, T., et al.: Dysbiosis and staphylococcus aureus colonization drives inflammation in atopic dermatitis. Immunity **42**(4), 756–766 (2015)
13. Koh, L.F., Ong, R.Y., Common, J.E.: Skin microbiome of atopic dermatitis. Allergol. Int. (2021). https://doi.org/10.1016/j.alit.2021.11.001, https://linkinghub.elsevier.com/retrieve/pii/S1323893021001404
14. Kohda, K.: An in vitro mixed infection model with commensal and pathogenic staphylococci for the exploration of interspecific interactions and their impacts on skin physiology. Front. Cell. Infect. Microbiol. **11**, 712360 (2021)
15. Kong, H.H.: Skin microbiome: genomics-based insights into the diversity and role of skin microbes. Trends Mol. Med. **17**(6), 320–328 (2011)
16. Kong, H.H., et al.: Temporal shifts in the skin microbiome associated with disease flares and treatment in children with atopic dermatitis. Genome Res. **22**(5), 850–859 (2012). https://doi.org/10.1101/gr.131029.111, https://genome.cshlp.org/lookup/doi/10.1101/gr.131029.111

17. Korting, H.C., Hübner, K., Greiner, K., Hamm, G., Braun-Falco, O.: Differences in the skin surface pH and bacterial microflora due to the long-term application of synthetic detergent preparations of pH 5.5 and pH 7.0. results of a crossover trial in healthy volunteers. Acta Derm. Venereol. **70**(5), 429–431 (1990)
18. Lai, Y., et al.: Activation of TLR2 by a small molecule produced by staphylococcus epidermidis increases antimicrobial defense against bacterial skin infections. J. Invest. Dermatol. **130**(9), 2211–2221 (2010)
19. Lai, Y., et al.: Commensal bacteria regulate toll-like receptor 3-dependent inflammation after skin injury. Nat. Med. **15**(12), 1377–1382 (2009). https://doi.org/10. 1038/nm.2062
20. Leyden, J.J., McGinley, K.J., Mills, O.H., Kligman, A.M.: Propionibacterium levels in patients with and without acne vulgaris. J. Invest. Dermatol. **65**(4), 382–384 (1975)
21. Malthus, T.R.: An essay on the principle of population, as it affects the future improvement of society. With remarks on the speculations of Mr. Godwin, M. Condorcet and other writers. J. Johnson,London (1798). https://archive.org/details/ essayonprincipl00malt
22. Meredith, H.R., Lopatkin, A.J., Anderson, D.J., You, L.: Bacterial temporal dynamics enable optimal design of antibiotic treatment. PLOS Comput. Biol. **11**(4), e1004201 (2015)
23. Nakatsuji, T., et al.: Antimicrobials from human skin commensal bacteria protect against staphylococcus aureus and are deficient in atopic dermatitis. Sci. Trans. Med. **9**(378), eaah4680 (2017). https://doi.org/10.1126/scitranslmed. aah4680, https://www.ncbi.nlm.nih.gov/pmc/articles/PMC5600545/
24. Nakatsuji, T., et al.: Development of a human skin commensal microbe for bacteriotherapy of atopic dermatitis and use in a phase 1 randomized clinical trial. Nat. Med. **27**(4), 700–709 (2021). https://doi.org/10.1038/s41591-021-01256-2, https://www.nature.com/articles/s41591-021-01256-2
25. Ong, P.Y., et al.: Endogenous antimicrobial peptides and skin infections in atopic dermatitis. N. Engl. J. Med. **347**(15), 1151–1160 (2002). https://doi.org/10.1056/ NEJMoa021481
26. Pazgier, M., Hoover, D.M., Yang, D., Lu, W., Lubkowski, J.: Human β-defensins. Cell. Mol. Life Sci. CMLS **63**(11), 1294–1313 (2006)
27. Proksch, E.: pH in nature, humans and skin. J. Dermatol. **45**(9), 1044–1052 (2018)
28. Radulescu, O., Swarup Samal, S., Naldi, A., Grigoriev, D., Weber, A.: Symbolic dynamics of biochemical pathways as finite states machines. In: Roux, O., Bourdon, J. (eds.) CMSB 2015. LNCS, vol. 9308, pp. 104–120. Springer, Cham (2015). https://doi.org/10.1007/978-3-319-23401-4_10
29. Rizk, A., Batt, G., Fages, F., Soliman, S.: Continuous valuations of temporal logic specifications with applications to parameter optimization and robustness measures. Theor. Comput. Sci. **412**(26), 2827–2839 (2011). https://doi.org/10.1016/j. tcs.2010.05.008
30. Tognoli, E., Kelso, J.A.S.: The metastable brain. Neuron **81**(1), 35–48 (2014). https://doi.org/10.1016/j.neuron.2013.12.022, https://pubmed.ncbi.nlm.nih.gov/ 24411730
31. Tuomanen, E., Cozens, R., Tosch, W., Zak, O., Tomasz, A.: The rate of killing of Escherichia coli by beta-lactam antibiotics is strictly proportional to the rate of bacterial growth. J. Gen. Microbiol. **132**(5), 1297–1304 (1986). https://doi.org/10. 1099/00221287-132-5-1297
32. Zwietering, M.H., Jongenburger, I., Rombouts, F.M., van't Riet, K.: Modeling of the bacterial growth curve. Appl. Environ. Microbiol. **56**(6), 1875–1881 (1990). https://www.ncbi.nlm.nih.gov/pmc/articles/PMC184525/

Exact Linear Reduction for Rational Dynamical Systems

Antonio Jiménez-Pastor[1]([✉]), Joshua Paul Jacob[2], and Gleb Pogudin[1]

[1] LIX, CNRS, École Polytechnique, Institute Polytechnique de Paris,
Palaiseau, France
{jimenezpastor,gleb.pogudin}@lix.polytechnique.fr
[2] Univeristy of California, Berkeley, USA
joshuapjacob@berkeley.edu

Abstract. Detailed dynamical systems models used in life sciences may include dozens or even hundreds of state variables. Models of large dimension are not only harder from the numerical perspective (e.g., for parameter estimation or simulation), but it is also becoming challenging to derive mechanistic insights from such models. Exact model reduction is a way to address this issue by finding a self-consistent lower-dimensional projection of the corresponding dynamical system. A recent algorithm CLUE allows one to construct an exact linear reduction of the smallest possible dimension such that the fixed variables of interest are preserved. However, CLUE is restricted to systems with polynomial dynamics. Since rational dynamics occurs frequently in the life sciences (e.g., Michaelis-Menten or Hill kinetics), it is desirable to extend CLUE to the models with rational dynamics. In this paper, we present an extension of CLUE to the case of rational dynamics and demonstrate its applicability on examples from literature. Our implementation is available in version 1.5 of CLUE (https://github.com/pogudingleb/CLUE).

Keywords: exact reduction · dynamical systems · constrained lumping

1 Introduction

Dynamical systems modeling is one of the key mathematical tools for describing phenomena in life sciences. Making such models realistic often requires taking into account a wide range of factors yielding models of high dimension (dozens or hundreds of state variables). Because of their size, these models are challenging from the numerical standpoint (e.g., parameter estimation) and it may be hard to derive mechanistic insight from them. One possible workaround is to use *model reduction* techniques that replace a model with a simpler one while preserving,

This work was supported by the Paris Ile-de-France region (via project "XOR"). GP was partially supported by NSF grants DMS-1853482, DMS-1760448, and DMS-1853650; and AJP was partially supported by Poul Due Jensen Grant 883901.

I. Petre and A. Păun (Eds.): CMSB 2022, LNBI 13447, pp. 198–216, 2022.
https://doi.org/10.1007/978-3-031-15034-0_10

at least approximately, some of the important features of the original model. For approximate model reduction, many powerful techniques have been developed including singular value decomposition [1] and time-scale separation [24].

A complementary approach is to perform *exact model reduction*, that is, lower the dimension of the model without introducing approximation errors. For example, exact linear lumping aims at writing a self-consistent system of differential equations for a set of *macro-variables* in which each macro-variable is a linear combination of the original variables. The case of the macro-variables being sums of the original variables has been studied for some important classes of biochemical models (see, e.g., [5,13,17]) and for general rational dynamical systems in [10,11]. The latter line of research has culminated in the powerful ERODE software [10] which finds the optimal partition of the original variables into macro-variables.

A recent algorithm from [25] (and implemented in the software CLUE) allows, for a given set of linear forms in the state variables (the *observables*), constructs a linear lumping of the smallest possible dimension such that the observables can be written as combinations of the macro-variables (i.e., the observables are preserved). Unlike the prior approaches, the macro-variables produced by CLUE are allowed to involve any coefficients (not just zeroes and ones as before). Thanks to this, one may obtain reductions of significantly lower dimension than it was possible before, see [25, Table 1].

However, unlike ERODE, the algorithm used in CLUE was restricted to the models with polynomial dynamics. Since rational dynamics appears frequently in life sciences (e.g., via Michaelis-Menten or Hill kinetics), it is desirable to extend CLUE to such models. The goal of the present paper is to fill this gap. We design and implement an algorithm computing an optimal linear lumping of a rational dynamical system given a set of observables. We demonstrate the efficiency and applicability of this algorithm on several models from literature. Our new algorithm is available in CLUE starting from version 1.5 at https:// github.com/pogudingleb/CLUE.

In fact, we present two algorithms: one is a relatively direct generalization of the original algorithm from [25] which works well if there are only few different denominators in the model, and another is a randomized algorithm based on evaluation-interpolation techniques which can efficiently handle models with multiple denominators (see Table 4 for comparison). Both algorithms are based on the same key mathematical fact, going back to [20], that linear lumpings are in a bijective correspondence with the invariant subspaces of the Jacobian $J(\mathbf{x})$ of the system [25, Proposition II.1]. However, the construction of such invariant subspaces as common invariant subspaces of the coefficients of $J(\mathbf{x})$ used in [25] relies on the polynomiality of the ODE model. The workaround used in the first algorithm is to make $J(\mathbf{x})$ a polynomial matrix by multiplying the common denominator if the entries. However, if there is more than a couple of denominators, the size of the intermediate expressions becomes prohibitive. Therefore, in the second algorithm, we used a different approach to replace a non-constant matrix $J(\mathbf{x})$ with a collection of constant matrices: we replace $J(\mathbf{x})$

by its evaluations at several points which can be exactly computed by means of automatic differentiation without writing down $J(\mathbf{x})$ itself. Since the algorithm is sampling-based, it becomes probabilistic, and we guarantee that the result will be correct with a user-specified probability and also provide a possibility to perform a post-verification of the result.

The rest of the paper is organized as follows. Section 2 contains the preliminaries on lumping and summarizes the main steps of the algorithms from [25] for polynomial dynamics. In Sect. 3, we describe our two algorithms mentioned above. Section 4 described our implementation of the algorithms and reports its runtimes. In Sect. 5, we illustrate how our algorithms can be applied to models from the literature. We conclude in Sect. 6. Some proofs omitted in the main text are collected in the Appendix.

2   Preliminaries and Prior Results

2.1   Preliminaries on Lumping

In this paper we study models defined by ODE systems of the form

$$\dot{\mathbf{x}} = \mathbf{f}(\mathbf{x}), \tag{1}$$

where $\mathbf{x} = (x_1, \ldots, x_n)^T$ are the state variables and $\mathbf{f} = (f_1, \ldots, f_n)^T$ with $f_1, \ldots, f_n \in \mathbb{R}(\mathbf{x})$ (where $\mathbb{R}(\mathbf{x})$ denotes the set of rational function in $\mathbf{x}$ with real coefficients). We will describe lumping and the related notions following [25, Sect. 2] but extending to the case of rational dynamics.

Definition 1 (Lumping). *Consider a system of the form* (1). *A linear transformation* $\mathbf{y} = L\mathbf{x}$, *where* $\mathbf{y} = (y_1, \ldots, y_m)^T$ *and* $L \in \mathbb{R}^{m \times n}$, *is called* lumping *if* $\mathrm{rank}(L) = m$ *and there exist rational functions* $g_1, \ldots, g_m \in \mathbb{R}(\mathbf{y})$ *such that*

$$\dot{\mathbf{y}} = \mathbf{g}(\mathbf{y}), \quad \text{where} \quad \mathbf{g} = (g_1, \ldots, g_m)^T$$

for every solution $\mathbf{x}$ *of* (1). *We call* m *the dimension of the lumping, and we will refer to* $\mathbf{y}$ *as* macro-variables.

In other words, a lumping of dimension m is a linear change of variables from n to m variables such that the new variables satisfy a self-contained ODE system. In the language of differential geometry, a lumping is a linear map φ from the state space such that there exists a vector field on the range of the map which is φ-related to $\mathbf{f}$ (see [33, Definition 1.54]).

Example 1. Consider the following differential system:

$$\begin{cases} \dot{x}_1 = \dfrac{x_2^2 + 4x_2 x_3 + 4x_3^2}{x_1^3 - x_2 - 2x_3}, \\[2mm] \dot{x}_2 = \dfrac{4x_3 - 2x_1}{x_1 + x_2 + 2x_3}, \\[2mm] \dot{x}_3 = \dfrac{x_1 + x_2}{x_1 + x_2 + 2x_3}, \end{cases}$$

Then the matrix

$$\begin{pmatrix} 1 & 0 & 0 \\ 0 & 1 & 2 \end{pmatrix},$$

is a lumping of dimension 2, since:

$$\begin{cases} \dot{y}_1 = \dot{x}_1 = \dfrac{(x_2 + 2x_3)^2}{x_1^3 - x_2 - 2x_3} = \dfrac{y_2^2}{y_1^3 - y_2} \\ \dot{y}_2 = \dot{x}_2 + 2\dot{x}_3 = \dfrac{2x_2 + 4x_3}{x_1 + x_2 + 2x_3} = \dfrac{2y_2}{y_1 + y_2}. \end{cases}$$

One way to force a lumping to keep the information of interest is to fix a set of observables to be preserved. This leads to the notion of *constrained lumping*.

Definition 2 (Constrained lumping). *Let* $\mathbf{x}_{obs}$ *be a vector of linear forms in* $\mathbf{x}$ *(i.e., there is a matrix* $M \in \mathbb{R}^{p \times n}$ *with* $\mathbf{x}_{obs} = M\mathbf{x}$*). We say that a lumping* L *of* $\dot{\mathbf{x}} = \mathbf{f}(\mathbf{x})$ *is a constrained lumping with observables* $\mathbf{x}_{obs}$ *if each entry of* $\mathbf{x}_{obs}$ *can be expressed as a linear combination of* $\mathbf{y} = L\mathbf{x}$*.*

Remark 1 (On the constrained and partition-based lumpings). Software ERODE [9] can efficiently produce the minimal (in the sense of the dimension) lumping, in which the macro-variables correspond to a partition of the state variables. This means that $y_1 = \sum_{i \in S_1} x_i, \ldots, y_m = \sum_{i \in S_m} x_i$ such that $\{1, \ldots, n\} = S_1 \bigsqcup S_2 \bigsqcup \ldots \bigsqcup S_m$.

In this case, we always have $y_1 + \ldots + y_m = x_1 + \ldots + x_n$, hence a lumping found by ERODE will be always a constrained lumping with the sum of the state variables being the only observable. Therefore, an algorithm (like the one proposed in this paper) for computing a constrained linear lumping of the smallest possible dimension will be always able to find either the lumping found by ERODE or even lump it further. To be fair, we would like to point out that ERODE is typically faster than our algorithm.

Remark 2 (Lumping via polynomialization). It is known that, by introducing new variables, a rational dynamical system can always be embedded into a polynomial one. Therefore, a natural approach to finding constrained linear lumpings for (1) would be to combine such an embedding with any algorithm applicable to polynomial dynamics (e.g., CLUE). However, it has been demonstrated in [11, p. 149] that such an embedding not only increases the dimension of the ambient space but also may miss some of the reductions of original model.

2.2 Overview of the CLUE Algorithm for Polynomial Dynamics [25]

The algorithm for computing a constrained lumping of the minimal dimension for a given system $\dot{\mathbf{x}} = \mathbf{f}(\mathbf{x})$ with polynomial right-hand side and observables $\mathbf{x}_{obs}$, designed and implemented in [25], can be summarized as follows:

Algorithm 1: Finding constrained linear lumping (polynomial case)

Input: a *polynomial* ODE system $\dot{\mathbf{x}} = \mathbf{f}(\mathbf{x})$ of dimension n; a list of observables $\mathbf{x}_{obs} = A\mathbf{x}$ for $A \in \mathbb{R}^{s \times n}$.

Output: minimal lumping L containing $\mathbf{x}_{obs}$.

(1) Compute the Jacobian matrix $J(\mathbf{x})$ of $\mathbf{f}$;

(2) Write $J(\mathbf{x})$ as $J_1 m_1 + \ldots + J_N m_N$, where $m_1, \ldots, m_N$ are distinct monomials in $\mathbf{x}$ and $J_1, \ldots, J_N$ are constant matrices;

(3) Compute the minimal subspace V of the space of linear forms in $\mathbf{x}$ containing $\mathbf{x}_{obs}$ and invariant under $J_1, \ldots, J_N$ (using [25, Alg. 3 or 4]);

(4) Return matrix L with rows being basis vectors of V.

This algorithm is based on the criterion from [20], which states that a matrix L is a lumping for the system $\dot{\mathbf{x}} = \mathbf{f}(\mathbf{x})$ if an only if the row space of L is invariant under $J(\mathbf{x})$ for every value of $\mathbf{x}$. In order to use this criterion, it was shown in [25, Supplementary Materials, Lemma I.1], that L is a lumping for $\dot{x} = \mathbf{f}(\mathbf{x})$ if and only if the row space of L is invariant under J_i for $1 \leqslant i \leqslant N$ (as defined in Step (2) of Algorithm 1). This reduces the problem of finding the lumping to the one solved in Step (3) of Algorithm 1.

3 Algorithm for Rational Dynamical Systems

In this section, we will use the following "finite" version of the Jacobian-based criterion from [20] which is proved in the Appendix.

Lemma 1. *Consider the rational dynamical system $\dot{\mathbf{x}} = \mathbf{f}(\mathbf{x})$ and $J(\mathbf{x})$ the Jacobian matrix of $\mathbf{f}(\mathbf{x})$. Let $\mathcal{B}$ be any set of matrices spanning the vector space $\langle J(\mathbf{x}) \mid \mathbf{x} \in \mathbb{R}^n$ and $J(\mathbf{x})$ is well-defined$\rangle$. Then $L \in \mathbb{R}^{r \times n}$ is a lumping if and only if the row space of L is invariant with respect to all $J \in \mathcal{B}$.*

Remark 3 (Lemma 1 and the polynomial case). In the context of Algorithm 1, each value of $J(\mathbf{x})$ is a linear combination of $J_1, \ldots, J_N$, so $\mathcal{B}$ can be taken to be $\{J_1, \ldots, J_N\}$. Thus Lemma 1 implies the correctness of Algorithm 1.

3.1 Straightforward Extension of Algorithm 1

In the case of rational dynamics, the Jacobian matrix $J(\mathbf{x})$ has only rational function entries, so we can compute the common denominator $q(\mathbf{x})$ such that $J(\mathbf{x}) = \frac{B(\mathbf{x})}{q(\mathbf{x})}$, where $B(\mathbf{x})$ is a matrix with polynomial entries. Let $B(\mathbf{x}) = B_1 m_1 + \ldots + B_N m_N$ be a decomposition where $B_1, \ldots, B_N$ are constant matrices and $m_1, \ldots, m_N$ are distinct monomials appearing in $B(\mathbf{x})$ (compare with Step (2) of Algorithm 1). Then, for each value of $\mathbf{x}$ not annihilating q, $J(\mathbf{x})$ is a linear combination of $B_1, \ldots, B_N$, so one can take $\mathcal{B} = \{B_1, \ldots, B_N\}$ in Lemma 1. This yields the following algorithm for rational case:

Algorithm 2: Finding constrained linear lumping (rational case)

Input: a *rational* ODE system $\dot{\mathbf{x}} = \mathbf{f}(\mathbf{x})$ of dimension n; a list of observables $\mathbf{x}_{obs} = A\mathbf{x}$ for $A \in \mathbb{R}^{s \times n}$.

Output: minimal lumping L containing $\mathbf{x}_{obs}$.

(1) Compute $J(\mathbf{x})$, the Jacobian of $f(\mathbf{x})$;

(2.1) Compute $p(\mathbf{x})$, the common denominator of the entries of $J(\mathbf{x})$

(2.2) Set $B(\mathbf{x}) := p(\mathbf{x}) \cdot J(\mathbf{x})$

(2.3) Write $B(\mathbf{x})$ as $B_1 m_1 + \ldots + B_N m_N$, where $m_1, \ldots, m_N$ are distinct monomials and $B_1, \ldots, B_N$ are constant matrices.

(3) Compute the minimal subspace V of the space of linear forms in $\mathbf{x}$ containing $\mathbf{x}_{obs}$ and invariant under $B_1, \ldots, B_N$ (using [25, Alg. 3 or 4]);

(4) Return matrix L with rows being basis vectors of V;

However, the size of the expressions involved after bringing everything to the common denominator and differentiation can be prohibitively large. The following toy example illustrates this phenomenon (for a comparison with the more refined Algorithm 3, see Sect. 4.2).

Example 2. Consider the following differential system:

$$\dot{x} = \frac{y - z}{x - y}, \qquad \dot{y} = \frac{x + z}{x + y}, \qquad \dot{z} = \frac{x + y + z}{z - x - y}.$$

Here is the Jacobian matrix:

$$J(x, y, z) = \begin{pmatrix} -\frac{y-z}{(x-y)^2} & \frac{y-z}{(x+y)^2} & \frac{2z}{(x+y-z)^2} \\ -\frac{x-z}{(x-y)^2} & -\frac{x+z}{(x+y)^2} & \frac{2z}{(x+y-z)^2} \\ -\frac{1}{x-y} & \frac{1}{x+y} & \frac{-2(x+y)}{(x+y-z)^2} \end{pmatrix}$$

The polynomial $p(\mathbf{x})$ from Algorithm 2 will be equal to $(x - y)^2 (x + y)^2 (x + y - z)^2$. Then the matrix $B(x, y, z)$ will satisfy

$$J(x, y, z) = \frac{1}{(x - y)^2 (x + y)^2 (x + y - z)^2} B(x, y, z).$$

In this example, the matrix $B(x, y, z)$ has all entries of degree 5 that we need to expand which is substantially more complicated than the original expressions.

3.2 The Main Algorithm Based on Evaluation-Interpolation

Our main algorithm is built upon the following observation: we can evaluate $J(\mathbf{x})$ efficiently at a given point without writing down the symbolic matrix explicitly (which would be quite large even in small examples such as Example 2), using automatic differentiation techniques. Then we can use sufficiently many such evaluations to span the space $\langle J(\mathbf{x}) \mid \mathbf{x} \in \mathbb{R}^n$ and $J(\mathbf{x})$ is well-defined$\rangle$ from

Lemma 1. The resulting Algorithm 3 is shown below. It relies on Algorithm 4 for generating "sufficiently many" evaluations which we present in Sect. 3.3.

Algorithm 3: Finding constrained linear lumping (probabilistic)

Input: a rational ODE system $\dot{\mathbf{x}} = \mathbf{f}(\mathbf{x})$ of dimension n; a list of observables
   $\mathbf{x}_{obs} = A\mathbf{x}$ for $A \in \mathbb{R}^{s \times n}$; and a real number $\varepsilon \in (0, 1)$.
Output: minimal lumping L containing $\mathbf{x}_{obs}$.
   The result is correct with probability at least $1 - \varepsilon$.

(1) Compute $J(\mathbf{x})$, the Jacobian of $f(\mathbf{x})$.
(2) Compute points $\mathbf{x}_1, \ldots, \mathbf{x}_M \in \mathbb{Q}^n$ such that $J(\mathbf{x}_1), \ldots, J(\mathbf{x}_M)$ span
   $\langle J(\mathbf{x}) \mid \mathbf{x} \in \mathbb{R}^n$ and $J(\mathbf{x})$ is well-defined$\rangle$ with probability at least $1 - \varepsilon$
   (using Algorithm 4).
(3) Compute the minimal subspace V of linear forms in $\mathbf{x}$ containing $\mathbf{x}_{obs}$ and
   invariant under $J(\mathbf{x}_1), \ldots, J(\mathbf{x}_M)$ (using [25, Alg. 3 or 4]);
(4) Return matrix L with rows being basis vectors of V;

Remark 4 (On probabilities in algorithms). The outputs of Algorithms 3 and 4 are guaranteed to be correct with a user-specified probability. This should be understood as follows. The algorithm makes some random choices, that is, draws a point from a probability space. The specification of the algorithm means that, for the fixed input, the probability of the output being correct (with respect to the probability space above) is at least $1-\varepsilon$.

   In the highly unlikely case (the used probability bounds are quite conservative) the computation at Step **(2)** of Algorithm 3 was incorrect, the computed space V will be a subspace of the "true" V, so the resulting matrix will not provide a lumping, and the reduced model returned by the software will be incorrect. The returned model can be checked using a direct substitution. This substitution is not performed in our implementation by default, but we offer a method to gain full confidence in the result.

Proposition 1. *Algorithm 3 is correct, that is the returned matrix L is a minimal lumping with probability at least $1 - \varepsilon$ (see Remark 4).*

Proof. Assume that we have found points $\mathbf{x}_1, \ldots, \mathbf{x}_M \in \mathbb{R}^n$ such that the evaluations $J(\mathbf{x}_1), \ldots, J(\mathbf{x}_M)$ span $\langle J(\mathbf{x}) \mid \mathbf{x} \in \mathbb{R}^n$ and $J(\mathbf{x})$ is well-defined$\rangle$, and consider L, a matrix with the rows being basis vectors of V, computed using [25, Algorithm 3 or Algorithm 4]. Then it will be a lumping by Lemma 1.

   If that is not the case, then it means that $J(\mathbf{x}_1), \ldots, J(\mathbf{x}_M)$ do not span $\langle J(\mathbf{x}) \mid \mathbf{x} \in \mathbb{R}^n$ and $J(\mathbf{x})$ is well-defined$\rangle$. This event only happens with probability at most ε. Hence the algorithm is correct with at least probability $1-\varepsilon$.

3.3   Generating "Sufficiently Many" Evaluations

In order to complete Algorithm 3, we will present in this section a procedure (Algorithm 4) for sampling values of $J(\mathbf{x})$ spanning the whole $\langle J(\mathbf{x}) \mid \mathbf{x} \in$

$\mathbb{R}^n$ and $J(\mathbf{x})$ is well-defined⟩ with high probability. The main theoretical tool to achieve the desired probability is the following proposition (proved in the appendix) based on the Schwartz-Zippel lemma [36, Proposition 98].

Proposition 2. *Let* $\mathbf{f} = (f_1, \ldots, f_n)^T$ *be a vector of elements of* $\mathbb{R}(\mathbf{x})$, *where* $\mathbf{x} = (x_1, \ldots, x_n)^T$, *and* $\varepsilon \in (0,1)$ *be a real number. Let* D_d *(resp.,* D_n*) be an integer such that the degree of the denominator (resp., numerator) of* f_i *does not exceed* D_d *(resp.,* D_n*) for every* $1 \leqslant i \leqslant n$.

Let $J(\mathbf{x})$ *be the Jacobian matrix of* $\mathbf{f}$ *and* $\mathbf{x}_1, \ldots, \mathbf{x}_m$ *be points such that* $J(\mathbf{x}_1), \ldots, J(\mathbf{x}_m)$ *do not span* ⟨$J(\mathbf{x}) \mid \mathbf{x} \in \mathbb{R}^n$ *and* $J(\mathbf{x})$ *is well-defined⟩. Consider a point* $\mathbf{x}_{m+1}$ *with each coordinate being an integer sampled uniformly at random from* $\{1, 2, \ldots, N\}$ *where*

$$N > \frac{D_n + (2m+1)D_d}{\varepsilon} + nD_d.$$

Then we have

$$\mathbb{P}[J(\mathbf{x}_{m+1}) \in ⟨J(\mathbf{x}_1), \ldots, J(\mathbf{x}_m)⟩ \mid J(\mathbf{x}_{m+1}) \text{ is well-defined}] < \varepsilon.$$

Proposition 2 tells us that, when the points are sampled from a large enough range, if a value of $J(\mathbf{x})$ belongs to the space spanned by the previous evaluations, then these evaluations span, with high probability, the whole space ⟨$J(\mathbf{x}) \mid \mathbf{x} \in \mathbb{R}^n$ and $J(\mathbf{x})$ is well-defined⟩. This yields the following algorithm.

Algorithm 4: Sampling the values of the Jacobian

Input:
- n-dimensional vector $\mathbf{f}$ of rational functions in $\mathbf{x} = (x_1, \ldots, x_n)$;
- real number $\varepsilon \in (0,1)$.

Output:] Points $\mathbf{x}_1, \ldots, \mathbf{x}_M \in \mathbb{Q}^n$ such that

$$⟨J(\mathbf{x}_i) \mid 1 \leqslant i \leqslant n⟩ = ⟨J(\mathbf{x}) \mid \mathbf{x} \in \mathbb{R}^n \text{ and } J(\mathbf{x}) \text{ is well-defined}⟩, \qquad (2)$$

with probability at least $1 - \varepsilon$, where $J(\mathbf{x})$ is the Jacobian matrix of $\mathbf{f}$.

(1) Compute D_d (resp., D_n) as the maximum of the degrees of denominators (resp., numerators) of entries of $\mathbf{f}$, respectively.
(2) $M \leftarrow 0$
(3) Repeat
    (a) Compute $\mathbf{x}_{M+1}$ by sampling each coordinate uniformly at random from $\{1, 2, \ldots, N\}$, where

$$N = \left[\frac{D_n + (2M+1)D_d}{\varepsilon} + nD_d \right] + 1.$$

    Repeat sampling until none of the denominators of $\mathbf{f}$ vanishes at $\mathbf{x}_{M+1}$.
    (b) Compute $J(\mathbf{x}_{M+1})$ using automatic differentiation (see Section 3.4).
    (c) If $J(\mathbf{x}_{M+1}) \in ⟨J(\mathbf{x}_i) \mid 1 \leqslant i \leqslant M⟩$, return $\mathbf{x}_1, \ldots, \mathbf{x}_M$.
    (d) $M \leftarrow M + 1$

Proposition 3. *Algorithm 4 is correct, that is, for the returned points* $\mathbf{x}_1, \ldots, \mathbf{x}_M$ *the relation (2) holds with probability at least* $1 - \varepsilon$ *(see Remark 4).*

Proof. Assume that the output condition (2) for Algorithm 4 does not hold. Since the algorithm has returned, the value $J(\mathbf{x}_{M+1})$ belonged to the span of $J(\mathbf{x}_1), \ldots, J(\mathbf{x}_M)$. Then Proposition 2 implies that the probability of this event is at most ε, so the output of the algorithm is correct with a probability of at least $1 - \varepsilon$.

3.4 Improving the Efficiency of Algorithm 3

Evaluating the Jacobian. An attractive feature of Algorithm 3 is that it does not require a symbolic expression for the Jacobian $J(\mathbf{x})$ but only a way to efficiently evaluate it at chosen points. These evaluations can be preformed efficiently using exact automatic differentiation [4,15,18,35].

More precisely, our implementation uses a forward algorithm for automatic differentiation by computing with the extended dual numbers. Extended dual numbers are tuples $(a_0, a_1, \ldots, a_n)$ of real numbers (where n is the dimension of the model) with the following arithmetic rules:

$$(a_0, a_1, \ldots, a_n) + (b_0, b_1, \ldots, b_n) = (a_0 + b_0, \ldots, a_n + b_n),$$
$$(a_0, a_1, \ldots, a_n)(b_0, b_1, \ldots, b_n) = (a_0 b_0, a_1 b_0 + a_0 b_1, \ldots, a_0 b_n + a_n b_0).$$

If one evaluates a rational function $f(\mathbf{x})$ at the point

$$((x_1, 1, 0, \ldots, 0), (x_2, 0, 1, \ldots, 0), \ldots, (x_n, 0, 0, \ldots, 1)), \qquad (3)$$

the output will be (see [18, Section 4]) precisely the vector

$$\left(f(\mathbf{x}), \frac{\partial f}{\partial x_1}(\mathbf{x}), \ldots, \frac{\partial f}{\partial x_n}(\mathbf{x}) \right).$$

Thus, we can evaluate $J(\mathbf{x})$ by evaluating $f_1, \ldots, f_n$ at (3). The complexity of such evaluation is directly related to the cost of evaluating $f_1, \ldots, f_n$. If evaluating each of $f_i(\mathbf{x})$ costs A operations, then evaluating the whole $J(\mathbf{x})$ has cost $\mathcal{O}(n^2 A)$. There are techniques (based on backward mode automatic differentiation) that could compute these evaluations within $\mathcal{O}(nA)$ operations [3]. For the moment, we decided to stick to the forward mode due to its simplicity and generalizability.

Selecting starting evaluation points. The sampling strategy for the evaluation points $\mathbf{x}_1, \ldots, \mathbf{x}_M$ at Step (3) of Algorithm 4 derived from Proposition 2 is used to ensure (with the prescribed probability) that we do not stop sampling too early. Therefore, before going into Step (3), we can start with choosing several evaluation points $\mathbf{x}_1, \ldots, \mathbf{x}_\ell$ in a different way than it is described in Step (3), and the probability of correctness will be preserved.

More precisely, before starting to sample evaluation points as in Step (3), our implementation of Algorithm 4 proceeds as follows:

1. *Sparse points.* For every $1 \leqslant i \leqslant n$, we do the following. Let $\mathbf{e}_1$ be the i-th standard basis vector in $\mathbb{R}^n$. We consider the point $\mathbf{e}_i$ and analyze the variables that need to be made different from zero so that the Jacobian $J(\mathbf{x})$ will be well-defined. We do this modifications in $\mathbf{e}_i$ and obtain (typically sparse) evaluation point.
2. *"Small Points".* Then we consider randomly sampled evaluation points but sample them from a small range until we detect a linear dependence between the evaluations of the Jacobian at already sampled points.

These evaluation points yield simpler evaluations of the Jacobian $J(\mathbf{x})$ which simplifies further computations. The optimization has led to significant speedup, see Table 1 (the models used in the table are described in Sect. 4.1).

Table 1. Speedup through a refined choice of the evaluation points

Model	Speedup	Model	Speedup
BIOMD0000000013	4.72	*Section* 5.1 $(n = 4)$	5.31
BIOMD0000000023	2.35	*Section* 5.1 $(n = 5)$	31.45
BIOMD0000000033	1.92	*Section* 5.1 $(n = 6)$	> 100
MODEL1502270000	8.98		

4 Implementation and Performance

We implement both our algorithms described in Sect. 3 in the software CLUE [25]. This software is written in Python and before it allowed to compute optimal constrained lumpings for any polynomial dynamical system. We have extended the functionality to the case of rational dynamics (starting with version 1.5). CLUE is available on the following GitHub repository:

  https://github.com/pogudingleb/CLUE

and can be installed using `pip` directly from GitHub using the command line

  `pip install git+https://github.com/pogudingleb/CLUE`

We refer to the README and tutorial in the repository for further details on how to use the software.

4.1 Performance of Algorithm 3

In this section, we report the runtimes for our implementation of our main algorithm, Algorithm 3. The timings reported below were measured on a laptop with Intel i7-9850H, 16 GB RAM, and Python 3.8.10. The runtimes are average through 5 independent executions on each model. We have measured runtimes for two sets of models

- Several models with rational dynamics from the BioModels database [22] in Table 2. For each model, the table contains the name in the database and a reference to the related paper. The observables in Table 2 were chosen as the states yielding nontrivial reductions, further analysis of these reductions is a question of future research.
- The models we use as examples in this paper (see Sect. 5) in Table 3.

In these examples, we have fixed the probability bound $\varepsilon = 0.01$. Changing this value to 0.001 may affect the timings measured, although its effect very moderate (less than 3% in all models). In both tables, we also include the observable used in the lumping and the change of the dimension (in the format $before \rightarrow after$).

Table 2. Execution time for Algorithm 3 for models from BioModels

Name	Reference	Obs.	Size	Time (min.)
BIOMD0000000013	[28]	x_CO2	28 → 25	2.19
BIOMD0000000023	[30]	Fru	13 › 11	0.04
BIOMD0000000113	[14]	Y	20 → 12	0.01
BIOMD0000000182	[23]	AC_cyto_mem	45 → 14	44.1
BIOMD0000000313	[29]	IL13_DecoyR	35 → 5	0.09
BIOMD0000000448	[7]	mTORC1a	67 → 48	1.37
BIOMD0000000526	[19]	DISC	32 → 19	0.1
MODEL1502270000	[34]	rmr	46 → 45	56.44

Table 3. Execution time for Algorithm 3 for the examples from Sect. 5

Example	Obs.	Size	Time (min.)
Section 5.2	freeEGFReceptor	80 → 5	0.25
Section 5.1, $n = 6$	x_1	6 → 2	0.02
Section 5.1, $n = 7$	x_1	7 → 2	0.04
Section 5.1, $n = 8$	x_1	8 → 2	0.17
Section 5.1, $n = 9$	x_1	9 → 2	0.33
Section 5.1, $n = 10$	x_1	10 → 2	3.74

For all the models in Table 2, the optimal reduction found by the algorithm is not of the type that could be found by ERODE, so the reduction by ERODE would be of higher dimension (but faster to compute). For example, the reduction found by ERODE for BIOMD0000000182 would be of dimension 33 while we reduce further to 14. We leave a more detailed comparison of CLUE and ERODE for future research.

From the timings above one can see that the complexity of the model for our algorithm is not solely determined by its order but rather by its structure: compare, for example, BIOMD0000000013 and BIOMD0000000448.

We have analyzed the breakdown of the total runtimes for these examples and observed that the most time-consuming part is Algorithm 4 while reading the models and computing the actual lumping typically take less than 1% of the full computation time. Since the sampled matrices depend only on the model, not on the observables, we enhanced CLUE with caching of the matrices so that one can reuse them if wants to check several different sets of observables for a single model. In this case, every subsequent computation will be much faster (~ 100 times) than reported in the table.

4.2   Comparing Algorithm 2 and Algorithm 3

In our implementation, both Algorithms 2 and 3 are available. For rational dynamical systems, the latter (probabilistic) is used by default. For the polynomial systems, the original algorithm from CLUE is used.

Table 4 contains the runtimes of Algorithms 2 and 3 for several biological models and, for each model, we count the number of distinct denominators appearing in the right-hand side of the ODE system. We separate the polynomial systems that have been taken from the paper [25] from the rational systems we studied in this paper. Note that, for the polynomial systems, Algorithm 3 is essentially the original algorithm from CLUE. One can observe that, if there is only a couple of denominators, then Algorithm 2 may be much faster but once the number of denominators grows, Algorithm 3 outperforms it substantially. In the future, it would be interesting to determine an algorithm to use on the fly.

Table 4. Execution times for Algorithms 2 and 4 (in minutes)

Polynomial models	Reference	# denoms	Algorithm 2	Algorithm 3
Barua	[2]	0	1.1913	> 30
OrderedPhosphorylation	[6]	0	0.0131	> 30
MODEL1001150000	[26]	0	0.0116	> 30
MODEL8262229752	[21]	0	0.0005	0.0262
fceri_fi	[16]	0	0.0494	> 30
ProteinPhosphorylation (7)	[32]	0	3.433	> 30
Rational models				
BIOMD0000000526	[19]	1	0.0222	0.0972
BIOMD0000000448	[7]	2	0.0387	1.3719
BIOMD0000000313	[29]	2	0.0111	0.0881
BIOMD0000000113	[14]	4	0.4914	0.0073
BIOMD0000000013	[28]	10	> 30	2.1906
BIOMD0000000023	[30]	11	> 30	0.0435
BIOMD0000000033	[8]	22	> 30	2.3012

5   Examples

5.1   Michaelis-Menten Kinetics with Competing Substrates

Consider an enzymatic reaction with a single enzyme E and multiple competing substrates $S_1, \ldots, S_n$ (and the corresponding products $P_1, \ldots, P_n$). The reaction can be described by the following chemical reaction network

$$S_i + E \underset{k_{i,-1}}{\overset{k_{i,1}}{\rightleftharpoons}} SE \overset{k_{i,2}}{\longrightarrow} P_i \quad \text{for } i = 1, \ldots, n.$$

Then quasi-steady-state approximation can be used to deduce a system of ODEs for the substrate concentrations only. For n competing substrates, using the general approach due to [12] (for competing substrates, see also [31, Section 3]), we obtain the following ODE system

$$\dot{x}_i = \frac{a_i x_i}{1 + \sum\limits_{j=1}^{n} \frac{x_j}{K_j}} \quad \text{for } i = 1, \ldots, n, \tag{4}$$

where x_i is the concentration of S_i, $K_i = \frac{k_{i,-1} + k_{i,2}}{k_{i,1}}$, $a_i = \frac{k_{i,2} E_0}{K_i}$, and E_0 is the total enzyme concentration. Assume that we are interested in the dynamics of a particular substrate, say S_1. We will now analyze possible constrained lumpings of the system (4) depending on the relations between the parameters (K_i's and a_i's). If a_i's are arbitrary distinct numbers or distinct symbolic parameters, our algorithm shows that there are no nontrivial reductions. However, if some of a_i's are equal, the situation becomes more interesting.

- *Simplest case: $a_2 = a_3 = \ldots = a_n$.* In this case, independently from n (checked for $n \leqslant 10$), the algorithm produces the system

$$\begin{cases} y_1' = \frac{a_1 y_1}{1 + \frac{y_1}{K_1} + y_2}, \\ y_2' = \frac{a_2 y_2}{1 + \frac{y_1}{K_1} + y_2}, \end{cases} \quad \text{where} \quad \begin{cases} y_1 = x_1, \\ y_2 = \sum\limits_{i=2}^{n} \frac{y_i}{K_i}. \end{cases}$$

- *More general case: some of a_i's are equal.* For example, if $n = 6$ and we have $a_2 = a_3$ and $a_4 = a_5 = a_6$, the optimal reduction produced by the algorithm will be

$$\begin{cases} y_1' = \frac{a_1 y_1}{1 + \frac{y_1}{K_1} + y_2 + y_3}, \\ y_2' = \frac{a_2 y_2}{1 + \frac{y_1}{K_1} + y_2 + y_3}, \\ y_3' = \frac{a_4 y_3}{1 + \frac{y_1}{K_1} + y_2 + y_3}, \end{cases} \quad \text{where} \quad \begin{cases} y_1 = x_1, \\ y_2 = \frac{x_2}{K_2} + \frac{x_3}{K_3}, = \frac{x_4}{K_4} + \frac{x_5}{K_5} + \frac{x_6}{K_6}. \end{cases}$$

More generally, if several of a_i's are equal, the corresponding x_i's are lumped together with the coefficients $\frac{1}{K_i}$, and the reduced model defines again the Michaelis-Menten kinetics for competing substrates. Interestingly, the substrates can be lumped together if the corresponding a_i's are equal but not all the $k_{i,1}, k_{i,-1}, k_{i,2}$ (cf. the scaling transformation in [31, p. 161]).

The reduction above cannot be found by ERODE [9] unless the $K_i = K_j$ whenever $a_i = a_j$ since the coefficients are not only ones and zeros. Note also that we treat all the parameters symbolically (instead simulating them as states with zero derivatives) so that they can appear in the coefficients of the lumping. To the best of our knowledge, CLUE is the only lumping software with this feature.

Table 3 in Sect. 4 reports the runtimes for different values of n.

5.2 Nerve Growth Factor Signaling

Motivated by the study of differentiation of neuronal cells, Brown et al. [8] considered a model describing the actions of nerve growth factor (NGF) and mitogenic epidermal growth factor (EGF) in rat pheochromocytoma (PC12) cells. In the model, these factors stimulated extracellular regulated kinase (Erk) phosphorylation with distinct dynamical profiles via a network of intermediate signaling proteins, the network is shown in Fig. 1. Each intermediate protein (and Erk) is modeled using two species: active and inactive states. The resulting model is described by a system of 32 differential equations with 48 parameters, the full system can be found in [8, Supplementary materials] or as BIOMD0000000033 in the BioModels database [22]. The exact reduction of this model has been earlier studied in [27, Sect. 5.4] using ERODE.

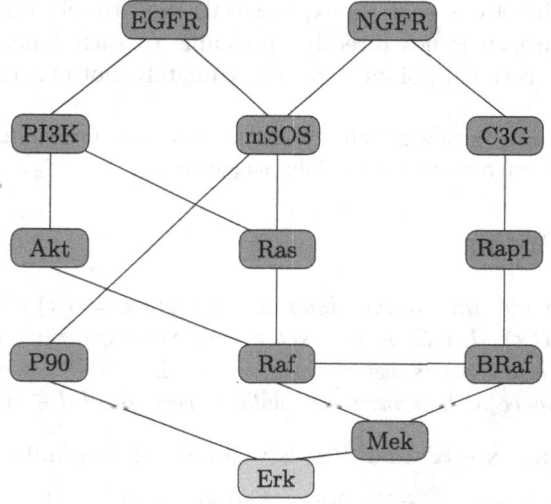

Fig. 1. A network diagram describing the action of NGF and EGR on Erk (a simplified version of [8, Fig. 1])

We have applied our algorithm to the model with the observable being the sum of all species (see Remark 1). The resulting reduction agrees with the one from [27, Sect. 5.4], that is, the macro-variables are the following

- concentrations of free and bound EGF and NGF and the corresponding receptors (EGFR and NGFR) remain separate variables;
- a single variable with constant dynamics equal to the sum of the concentrations of all other species (intermediate proteins and Erk) is introduced;
- only 4 out of 48 parameters remain in the reduced model.

Therefore, the reduced model is defined by 7 variables and 4 parameters and captures exactly the dynamics of EGF and NGF.

6   Conclusions

We have presented the first (to the best of our knowledge) algorithms for finding optimal constrained linear lumping of a rational dynamical system which non trivially extends the existing algorithm for the polynomial case.

While being based on the Jacobian-invariance criterion going back to [20] and used in the polynomial case [25], our main algorithm approaches the key step of [25], turning a nonconstant Jacobian matrix into a finite collection of constant matrices, from a different angle, via automatic differentiation and randomized evaluation. We implement our algorithms, report runtimes for them on a set of benchmarks, and demonstrate how they can be applied to models from the literature.

Directions for future research include extending the algorithm to models involving other functions such as exponential, logarithmic, and trigonometric. Our current approach is not directly applicable to such functions as one can evaluate them at rational points only approximately, not exactly.

Acknowledgement. We are grateful to Mirco Tribastone for helpful discussions. We are grateful to the referees for their helpful suggestions.

Appendix: Proofs

Lemma 1. *Consider the rational dynamical system* $\dot{\mathbf{x}} = \mathbf{f}(\mathbf{x})$ *and* $J(\mathbf{x})$ *the Jacobian matrix of* $\mathbf{f}(\mathbf{x})$. *Let* $\mathcal{B}$ *be any set of matrices spanning the vector space* $\langle J(\mathbf{x}) \mid \mathbf{x} \in \mathbb{R}^n$ *and* $J(\mathbf{x})$ *is well-defined*$\rangle$. *Then* $L \in \mathbb{R}^{r \times n}$ *is a lumping if and only if the row space of* L *is invariant with respect to all* $J \in \mathcal{B}$.

Proof. Since $\langle J(\mathbf{x}) \mid \mathbf{x} \in \mathbb{R}^n$ and $J(\mathbf{x})$ is well-defined$\rangle$ has finite dimension, then:

- There are points $\mathbf{x}_1, \ldots, \mathbf{x}_N \in \mathbb{R}^n$ such that $J(\mathbf{x}_1), \ldots, J(\mathbf{x}_N)$ is a basis.
- There are matrices $J_1, \ldots, J_N \in \mathcal{B}$ that form also a basis.

Hence, there is a invertible matrix $C \in \mathbb{R}^{N \times N}$ such that (the $n \times n$ matrices are considered as n^2-dimensional vectors)

$$\begin{pmatrix} J_1 \\ \vdots \\ J_N \end{pmatrix} = C \begin{pmatrix} J(\mathbf{x}_1) \\ \vdots \\ J(\mathbf{x}_N) \end{pmatrix}.$$

Assume that L is a lumping, then [25, Proposition II.1] implies that there is $A(\mathbf{x})$ such that $A(\mathbf{x})L = LJ(\mathbf{x})$. Hence, for the matrices defined by

$$\begin{pmatrix} A_1 \\ \vdots \\ A_N \end{pmatrix} = C \begin{pmatrix} A(\mathbf{x}_1) \\ \vdots \\ A(\mathbf{x}_N) \end{pmatrix},$$

we obtain $A_i L = LJ_i$. So for all $J \in \mathcal{B}$ there is A_J such that $A_J L = LJ$.

On the other hand, if for all $J \in \mathcal{B}$ there is A_J such that $A_J L = LJ$, then we can check that, if $J(\mathbf{x}) = \sum_{i=1}^{N} \alpha_i(\mathbf{x})J(\mathbf{x}_i)$, then

$$LJ(\mathbf{x}) = \left[(\alpha_1(\mathbf{x}), \ldots, \alpha_N(\mathbf{x}))C^{-1} \begin{pmatrix} A_{J_1} \\ \vdots \\ A_{J_N} \end{pmatrix} \right] L,$$

so the row space of L is invariant under $J(\mathbf{x})$ and L is a lumping.

Proposition 2. *Let* $\mathbf{f} = (f_1, \ldots, f_n)^T$ *be a vector of elements of* $\mathbb{R}(\mathbf{x})$, *where* $\mathbf{x} = (x_1, \ldots, x_n)^T$, *and* $\varepsilon \in (0, 1)$ *be a real number. Let* D_d *(resp.,* D_n*) be an integer such that the degree of the denominator (resp., numerator) of* f_i *does not exceed* D_d *(resp.,* D_n*) for every* $1 \leqslant i \leqslant n$.

Let $J(\mathbf{x})$ *be the Jacobian matrix of* $\mathbf{f}$ *and* $\mathbf{x}_1, \ldots, \mathbf{x}_m$ *be points such that* $J(\mathbf{x}_1), \ldots, J(\mathbf{x}_m)$ *do not span* $\langle J(\mathbf{x}) \mid \mathbf{x} \in \mathbb{R}^n$ *and* $J(\mathbf{x})$ *is well-defined*$\rangle$. *Consider a point* $\mathbf{x}_{m+1}$ *with each coordinate being an integer sampled uniformly at random from* $\{1, 2, \ldots, N\}$ *where*

$$N > \frac{D_n + (2m+1)D_d}{\varepsilon} + nD_d.$$

Then we have

$$\mathbb{P}[J(\mathbf{x}_{m+1}) \in \langle J(\mathbf{x}_1), \ldots, J(\mathbf{x}_m) \rangle \mid J(\mathbf{x}_{m+1}) \text{ is well-defined}] < \varepsilon.$$

Proof. Define $\mathbf{v}_1, \ldots \mathbf{v}_m \in \mathbb{R}^{n^2}$ as the vector representations of the matrices $J(\mathbf{x}_1), \ldots, J(\mathbf{x}_m)$ and $\mathbf{w}(x)$ be the corresponding vector representation for the matrix $J(\mathbf{x})$ with the entries being rational functions. By removing the corresponding evaluation points if necessary, we will further assume that $\mathbf{v}_1, \ldots, \mathbf{v}_m$ are linearly independent. The fact that $\langle J(\mathbf{x}_i) \mid 1 \leqslant i \leqslant m \rangle \neq \langle J(\mathbf{x}) \mid \mathbf{x} \in \mathbb{R}^n \rangle$ implies that

$$\operatorname{rank} E = m + 1, \quad \text{where} \quad E := (\mathbf{v}_1 \mid \mathbf{v}_2 \mid \ldots \mid \mathbf{v}_m \mid \mathbf{w}(\mathbf{x})).$$

Each maximal minor of E is a linear combination of the entries of $\mathbf{w}(\mathbf{x})$. Since $\operatorname{rank} E = m + 1$, at least one of these minors is a nonzero rational function, we write it as $\frac{A(\mathbf{x})}{B(\mathbf{x})}$. Note the degrees of the denominator and numerator of each entry of $\mathbf{w}(\mathbf{x})$ are bounded by $2D_d$ and $D_d + D_n$, respectively. Therefore, we have

$$\deg(A(\mathbf{x})) \leqslant D_n + (2m+1)D_d, \qquad \deg(B(\mathbf{x})) \leqslant 2(m+1)D_d.$$

Let $q(\mathbf{x})$ be the product of the denominators of $f_1, \ldots, f_n$. Then $\deg q(\mathbf{x}) \leqslant n D_d$.

Let $\mathbf{x}_{m+1} \in \{1, \ldots N\}^n$ as in the statement of the proposition. We would like to find an upper bound for

$$\mathbb{P}[J(\mathbf{x}_{m+1}) \in \langle J(\mathbf{x}_1), \ldots, J(\mathbf{x}_m) \rangle \mid J(\mathbf{x}_{m+1}) \text{ is well-defined}]. \qquad (5)$$

We write this as

$$\frac{\mathbb{P}[J(\mathbf{x}_{m+1}) \in \langle J(\mathbf{x}_1), \ldots, J(\mathbf{x}_m) \rangle \text{ and } J(\mathbf{x}_{m+1}) \text{ is well-defined}]}{\mathbb{P}[J(\mathbf{x}_{m+1}) \text{ is well-defined}]}.$$

For the numerator:

$$\mathbb{P}[J(\mathbf{x}_{m+1}) \in \langle J(\mathbf{x}_1), \ldots, J(\mathbf{x}_m) \rangle \text{ and } J(\mathbf{x}_{m+1}) \text{ is well-defined}] \leqslant$$

$$\leqslant \mathbb{P}[A(\mathbf{x}_{m+1}) = 0] \leqslant \frac{D_n + (2m+1)D_d}{N},$$

where the latter inequality follows from the Schwartz-Zippel lemma [36, Proposition 98]. Using this lemma again, we bound the probability of the denominator:

$$\mathbb{P}[J(\mathbf{x}_{m+1}) \text{ is well-defined}] = \mathbb{P}[q(\mathbf{x}) \neq 0] \geqslant 1 - \frac{n D_d}{N}.$$

Putting everything together, we obtain

$$\mathbb{P}[J(\mathbf{x}_{m+1}) \in \langle J(\mathbf{x}_1), \ldots, J(\mathbf{x}_m) \rangle \mid J(\mathbf{x}_{m+1}) \text{ is well-defined}] \leqslant \frac{D_n + (2m+1)D_d}{N - n D_d}.$$

Now a direct computation shows that

$$\frac{D_n + (2m+1)D_d}{N - n D_d} < \varepsilon \iff N > \frac{D_n + (2m+1)D_d}{\varepsilon} + n D_d.$$

References

1. Antoulas, A.: Approximation of large-scale dynamical systems. Advance in Design and Control, SIAM (2005)
2. Barua, D., Faeder, J.R., Haugh, J.M.: A bipolar clamp mechanism for activation of Jak-family protein tyrosine kinases. PLoS Comput. Biol. **5**(4), e1000364 (2009). https://doi.org/10.1371/journal.pcbi.1000364
3. Baur, W., Strassen, V.: The complexity of partial derivatives. Theoret. Comput. Sci. **22**(3), 317–330 (1983). https://doi.org/10.1016/0304-3975(83)90110-X
4. Baydin, A.G., Pearlmutter, B.A., Radul, A.A., Siskind, J.M.: Automatic differentiation in machine learning: a survey. J. Mach. Learn. Res. **18**(153), 1–43 (2018). https://jmlr.org/papers/v18/17-468.html
5. Borisov, N., Markevich, N., Hoek, J., Kholodenko, B.: Signaling through receptors and scaffolds: independent interactions reduce combinatorial complexity. Biophys. J. **89**(2), 951–966 (2005). https://doi.org/10.1529/biophysj.105.060533
6. Borisov, N., Kholodenko, B., Faeder, J., Chistopolsky, A.: Domain-oriented reduction of rule-based network models. IET Syst. Biol. **2**(5), 342–351 (2008). https://doi.org/10.1049/iet-syb:20070081

7. Brännmark, C., et al.: Insulin signaling in type 2 diabetes. J. Biol. Chem. **288**(14), 9867–9880 (2013). https://doi.org/10.1074/jbc.m112.432062
8. Brown, K.S., et al.: The statistical mechanics of complex signaling networks: nerve growth factor signaling. Phys. Biol. **1**(3), 184–195 (2004). https://doi.org/10.1088/1478-3967/1/3/006
9. Cardelli, L., Tribastone, M., Tschaikowski, M., Vandin, A.: ERODE: a tool for the evaluation and reduction of ordinary differential equations. In: Legay, A., Margaria, T. (eds.) TACAS 2017. LNCS, vol. 10206, pp. 310–328. Springer, Heidelberg (2017). https://doi.org/10.1007/978-3-662-54580-5_19
10. Cardelli, L., Tribastone, M., Tschaikowski, M., Vandin, A.: Maximal aggregation of polynomial dynamical systems. Proc. National Acad. Sci. **114**(38), 10029–10034 (2017). https://www.pnas.org/content/114/38/10029
11. Cardelli, L., Tribastone, M., Tschaikowski, M., Vandin, A.: Symbolic computation of differential equivalences. Theoret. Comput. Sci. **777**, 132–154 (2019). https://doi.org/10.1016/j.tcs.2019.03.018
12. Chou, T.C., Talalay, P.: A simple generalized equation for the analysis of multiple inhibitions of Michaelis-Menten kinetic systems. J. Biol. Chem. **252**, 6438–6442 (1977). https://www.jbc.org/article/S0021-9258(17)39978--7/pdf
13. Conzelmann, H., Fey, D., Gilles, E.: Exact model reduction of combinatorial reaction networks. BMC Syst. Biol. **2**(1), 78 (2008). https://doi.org/10.1186/1752-0509-2-78
14. Dupont, G., Goldbeter, A.: Protein phosphorylation driven by intracellular calcium oscillations: a kinetic analysis. Biophys. Chem. **42**(3), 257–270 (1992). https://doi.org/10.1016/0301-4622(92)80018-z
15. Elliott, C.: Beautiful differentiation. In: International Conference on Functional Programming (ICFP) (2009). http://conal.net/papers/beautiful-differentiation
16. Faeder, J.R., et al.: Investigation of early events in fcεRI-mediated signaling using a detailed mathematical model. J. Immunol. **170**(7), 3769–3781 (2003). https://doi.org/10.4049/jimmunol.170.7.3769
17. Feret, J., Danos, V., Krivine, J., Harmer, R., Fontana, W.: Internal coarse-graining of molecular systems. Proc. National Acad. Sci. **106**(16), 6453–6458 (2009). https://doi.org/10.1073/pnas.0809908106
18. Hoffmann, P.H.: A Hitchhiker's guide to automatic differentiation. Numer. Algorithms **72**, 775–811 (2016). https://doi.org/10.1007/s11075-015-0067-6
19. Kallenberger, S.M., et al.: Intra- and interdimeric caspase-8 self-cleavage controls strength and timing of CD95-induced apoptosis. Sci. Signal. **7**(316), ra23 (2014). https://doi.org/10.1126/scisignal.2004738
20. Li, G., Rabitz, H.: A general analysis of exact lumping in chemical kinetics. Chem. Eng. Sci. **44**(6), 1413–1430 (1989). https://doi.org/10.1016/0009-2509(89)85014-6
21. Li, J., et al.: A stochastic model ofEscherichia coliAI-2 quorum signal circuit reveals alternative synthesis pathways. Mol. Syst. Biol. **2**(1), 67 (2006). https://doi.org/10.1038/msb4100107
22. Malik-Sheriff, R.S., et al.: BioModels - 15 years of sharing computational models in life science. Nucleic Acids Res. **48**(D1), D407–D415 (2020). https://doi.org/10.1093/nar/gkz1055
23. Neves, S.R., et al.: Cell shape and negative links in regulatory motifs together control spatial information flow in signaling networks. Cell **133**(4), 666–680 (2008). https://doi.org/10.1016/j.cell.2008.04.025
24. Okino, M., Mavrovouniotis, M.: Simplification of mathematical models of chemical reaction systems. Chem. Rev. **2**(98), 391–408 (1998). https://doi.org/10.1021/cr9502231

25. Ovchinnikov, A., Pérez Verona, I., Pogudin, G., Tribastone, M.: CLUE: exact maximal reduction of kinetic models by constrained lumping of differential equations. Bioinformatics **37**(19), 3385–3385 (2021). https://doi.org/10.1093/bioinformatics/btab258

26. Pepke, S., Kinzer-Ursem, T., Mihalas, S., Kennedy, M.B.: A dynamic model of interactions of Ca^{2+}, calmodulin, and catalytic subunits of Ca^{2+}/calmodulin-dependent protein kinase II. PLoS Comput. Biol. **6**(2), e1000675 (2010). https://doi.org/10.1371/journal.pcbi.1000675

27. Perez-Verona, I.C., Tribastone, M., Vandin, A.: A large-scale assessment of exact lumping of quantitative models in the BioModels repository. Theoret. Comput. Sci. **893**, 41–59 (2021). https://doi.org/10.1016/j.tcs.2021.06.026

28. Poolman, M.G., Assmus, H.E., Fell, D.A.: Applications of metabolic modelling to plant metabolism. J. Exp. Bot. **55**(400), 1177–1186 (2004). https://doi.org/10.1093/jxb/erh090

29. Raia, V., et al.: Dynamic mathematical modeling of IL13-induced signaling in hodgkin and primary mediastinal b-cell lymphoma allows prediction of therapeutic targets. Can. Res. **71**(3), 693–704 (2010). https://doi.org/10.1158/0008-5472.can-10-2987

30. Rohwer, J.M., Botha, F.C.: Analysis of sucrose accumulation in the sugar cane culm on the basis of in vitro kinetic data. Biochem. J. **358**(2), 437–445 (2001). https://doi.org/10.1042/bj3580437

31. Schnell, S., Mendoza, C.: Enzyme kinetics of multiple alternative substrates. J. Math. Chem. **27**(1/2), 155–170 (2000). https://doi.org/10.1023/a:1019139423811

32. Sneddon, M.W., Faeder, J.R., Emonet, T.: Efficient modeling, simulation and coarse-graining of biological complexity with NFsim. Nat. Meth. **8**(2), 177–183 (2010). https://doi.org/10.1038/nmeth.1546

33. Warner, F.W.: Foundations of Differentiable Manifolds and Lie Groups. Springer, New York, NY (1983). https://doi.org/10.1007/978-1-4757-1799-0

34. Weiße, A.Y., Oyarzún, D.A., Danos, V., Swain, P.S.: Mechanistic links between cellular trade-offs, gene expression, and growth. Proc. Natl. Acad. Sci. **112**(9), E1038–E1047 (2015). https://doi.org/10.1073/pnas.1416533112

35. Wengert, R.E.: A simple automatic derivative evaluation program. Commun. ACM **7**(8), 463–464 (1964). https://doi.org/10.1145/355586.364791

36. Zippel, R.: Effective Polynomial Computation. Springer (1993). https://doi.org/10.1007/978-1-4615-3188-3

Limit Cycle Analysis of a Class of Hybrid Gene Regulatory Networks

Honglu Sun[1]([✉])[iD], Maxime Folschette[2][iD], and Morgan Magnin[1][iD]

[1] Nantes Université, École Centrale Nantes, CNRS, LS2N, UMR 6004,
44000 Nantes, France
`honglu.sun@ls2n.fr`
[2] Univ. Lille, CNRS, Centrale Lille, UMR 9189 CRIStAL, 59000 Lille, France

Abstract. Many gene regulatory networks have periodic behavior, for instance the cell cycle or the circadian clock. Therefore, the study of formal methods to analyze limit cycles in mathematical models of gene regulatory networks is of interest. In this work, we study a pre-existing hybrid modeling framework (HGRN) which extends René Thomas' widespread discrete modeling. We propose a new formal method to find all limit cycles that are simple and deterministic, and analyze their stability, that is, the ability of the model to converge back to the cycle after a small perturbation. Up to now, only limit cycles in two dimensions (with two genes) have been studied; our work fills this gap by proposing a generic approach applicable in higher dimensions. For this, the hybrid states are abstracted to consider only their borders, in order to enumerate all simple abstract cycles containing possible concrete trajectories. Then, a Poincaré map is used, based on the notion of transition matrix of the concrete continuous dynamics inside these abstract paths. We successfully applied this method on existing models: three HGRNs of negative feedback loops with 3 components, and a HGRN of the cell cycle with 5 components.

Keywords: Hybrid modeling · Celerity · Transition matrix · Limit cycle · Gene regulatory networks · Poincaré map

1 Introduction

Using mathematical models to study the dynamics of gene regulatory networks is fundamental because of the complex nature of biological systems. Two widely used formalisms are discrete models (like Boolean networks [29]) and continuous models (differential equations [5], stochastic models [28]). The dynamics of discrete models are easy to analyze but sometimes not precise enough (for example, it is hard to identify damped oscillation in discrete models). Continuous models are more precise but their dynamics are sometimes hard to analyze. To make a bridge between discrete and continuous models, hybrid models were proposed

Supported by China Scholarship Council.

[13,14,34]: they can be seen as a simplification of the continuous models or an extension of the discrete model. These hybrid models contain both continuous and discrete components.

In this work, we study a class of hybrid models: *hybrid gene regulatory networks* (HGRN) [6,15] which is an extension of Thomas' discrete modeling framework [35,36]. HGRNs have been used to model the circadian clock [15] and the cell cycle [6]. In HGRNs, the state space is separated into several discrete states, as for discrete models, and in each discrete state, the temporal derivative of the system is described by a constant vector making the system evolve continuously over time, as for differential equations. The most important property of HGRNs is that the sliding mode is allowed, which means that when a trajectory reaches a black wall (a boundary of the discrete state which cannot be crossed by trajectories) it is forced to move along the black wall.

Previous studies of HGRNs mostly focused on parameters identification [7]. Another important aspect is the dynamical analysis of HGRNs, such as the location of the attractors and their nature (fixed point, limit cycle, etc.). Dynamical properties can be used for model verification and for the discovery of new possible biological behaviors. For now, few results about analysis of HGRNs have been published. We can mainly cite [15] which discussed necessary and sufficient conditions of the existence of a limit cycle of a HGRN in two dimensions, that is, containing two genes. In higher dimensions, limit cycles of HGRNs are more complex. However, no result about the analysis of limit cycles of HGRNs in N dimensions has been published yet, although many genetic oscillators contain several genes. In this work, we seek to fill this gap by studying limit cycles of HGRNs in N dimensions. The main contribution of this work is a new formal method to find all simple limit cycles that do not visit several times the same discrete state in one loop, in a HGRN of N dimensions, and to analyze their stability. The limitations are: we do not consider trajectories that reach several borders simultaneously, which is a very particular case, and we do not consider trajectories containing states which can potentially reach several discrete states and would introduce non-determinism.

The main idea of this new method is based on the notion of *Poincaré map*. The Poincaré map was initially proposed to study periodic orbits of nonlinear dynamical system and has also been used later to study limit cycles of hybrid systems [8,20]. The Poincaré map describes the intersection of a periodic orbit of the system with a lower dimension subspace which is called the Poincaré section. In other words, the Poincaré map allows to witness the shift made by an oscillatory trajectory in a chosen hyperplane of the state space. Thus, by using a Poincaré map, the study of the limit cycle in the original system is transformed into the study of the related fixed point in another system in lower dimensions. One major problem of the application of the Poincaré map to study limit cycles of hybrid systems is that the computation of the Poincaré map can be difficult and the shape of a Poincaré map can be complex, making it hard to analyze. In HGRNs, the shape of the Poincaré map is a simple affine map, but its calculation is still complex because of the existence of sliding modes (two different trajectories that cross the same sequence of discrete states have different Poincaré maps if they have different sliding modes). It is the major difficulty of this method, and to deal with this problem, a new abstraction based on the new concept of *discrete domain* is proposed. Relying on such discrete domains, we also (re)define the notions of *discrete trajectory, transition matrix* and *compatible zone* to calculate the Poincaré map. After the Poincaré map is obtained, the fixed point of the Poincaré map is computed to find the limit cycle and an eigenanalysis is applied to analyze the stability of the limit cycle found.

Most of the works about dynamical analyses of hybrid systems focus on reachability analyses [16]. Among these analysis methods, our method is most similar to the discrete abstraction method [1,2] of which the main purpose is to obtain a finite state transition system from a hybrid automaton. The study of periodic orbits in hybrid system is also a lively field: we can cite, for example, works [11,12,33] based on the Poincaré-Bendixson theorem for systems in two dimensions, and works [21,23,26,27,37] based on the Poincaré map for hybrid systems in N dimensions.

Even though few works exist about limit cycles analysis in HGRNs, limit cycles were studied in other hybrid models of gene regulatory networks. Most of these works are also based on the Poincaré map. In [8,20], the Poincaré map is used to study the limit cycle of simple piecewise affine systems in two dimensions. In these works, since the system is planar, it is easy to compute and analyze the Poincaré map. In [17,18,30], methods are proposed to find and analyze limit cycles in higher dimensions of piecewise affine system with a uniform decay rate. The hypothesis of a uniform decay rate in these works makes it always possible to calculate a Poincaré map because they have a simple shape. However, for a general piecewise affine system, it is difficult to prove theoretically the existence of limit cycle except for some particular examples such as negative loops [10,19].

Compared to previous works about limit cycles in hybrid system, our work has two major novelties: (1) We consider limit cycles with sliding modes, and (2) We use an abstraction method in order to find cycles of discrete regions, which might contain limit cycles (in other words, cycles of discrete regions which contain at least one continuous trajectory).

It is worth mentioning that our work is similar to [17,18,30]. These works use similar methods based on the Poincaré map but apply it to a different kind of hybrid framework, in which temporal derivatives are affine functions in each discrete region, while in HGRNs they are constants. Following these works, [24] proposes a framework that is close to ours, in which temporal derivatives are constants in each discrete region, but is not focused on enumerating limit cycles. Our work also uses a similar approach than [7] to compute trajectories based on constraints, which constitutes one of the steps our method.

There are other works which are based on piecewise constant derivative systems, which are similar to HGRNs, but study different problems, for example, the decidability of the reachability problem [3,4], the stability of fixed point [32].

This paper is organized as follows. In Sect. 2, we define HGRNs. In Sect. 3, we use a simple example to describe our method to find and analyze limit cycles of

HGRNs. In Sect. 4, we apply the method on three HGRNs of negative feedback loops in 3 dimensions and a HGRN of the cell cycle in 5 dimensions. And finally in Sect. 5, we make a conclusion by discussing the merits and limit of this method and our future work.

2 Hybrid Gene Regulatory Networks

This section defines Hybrid Gene Regulatory Networks (HGRNs). Compared to the original paper regarding HGRNs [6], we introduce here some new notions about HGRNs, including (closed) trajectory and (input/output/attractive/neutral) boundary.

Consider a gene regulatory network with N genes, the ith gene has $n_i + 1$ discrete levels which are represented by integers: $\{0, 1, 2, ..., n_i\}$. A discrete state s is obtained by attributing a valuation for each gene among its discrete levels. We denote d_s the integer vector which describes the discrete levels of all genes in s in order; in the following, for simplicity, we also call d_s a discrete state. The set of all discrete states is $E_d = \{d_s \in \mathbb{N}^N \mid \forall i \in \{1, 2, ..., N\}, d_s^i \in \{0, 1, ..., n_i\}\}$, where d_s^i is the ith component of d_s.

Based on the notion of discrete state, HGRNs are defined as follows:

Definition 1 (Hybrid gene regulatory network (HGRN)). *A hybrid gene regulatory network (HGRN) is noted $\mathcal{H} = (E_d, c)$. E_d is the set of all discrete states. c is a function from E_d to $\mathbb{R}^N$. For each $d_s \in E_d$, $c(s)$, also noted c_s, is called the* celerity *of discrete state d_s and describes the temporal derivative of the system in d_s.*

In HGRNs, a hybrid state is used to fully describe the state of the system: it contains the discrete state in which the system currently is, and a fractional part that represents the (normalized) position of each variable inside this discrete state.

Definition 2 (Hybrid state of HGRN). *A* hybrid state *of a HGRN is a couple $h = (\pi, d_s)$ containing a fractional part π, which is a real vector $[0, 1]^N$, and a discrete state d_s in E_d. E_h is the set of all hybrid states.*

In the following, we will use simply *state* to denote a hybrid state. Based on this notion of state, a trajectory and a boundary are defined as follow.

Definition 3 (Trajectory). *A* trajectory τ *is a function from a time interval $[0, t_0]$ to $E_\tau = E_h \cup E_{sh}$, where $t_0 \in \mathbb{R}^+ \cup \{\infty\}$, E_h is the set of all states, and E_{sh} is the set of all finite sequences of states ($E_{sh} = \{(h_0, h_1, ..., h_m) \in (E_h)^{m+1} \mid m \in \mathbb{N}\}$).*

A trajectory represents a simulation of the system over time. Consider a trajectory τ on $[0, t_0]$. For any $t \in [0, t_0]$, if $\tau(t) \in E_{sh}$, this means that there is a sequence of instant transitions at t; otherwise, if $\tau(t) \in E_h$, then the trajectory in t is made of a regular point. A trajectory τ defined on $[0, \infty[$ is called a *closed trajectory* if $\exists T > 0, \forall t \in [0, \infty[, \tau(t) = \tau(t + T)$. The smallest T is the period of τ.

Definition 4 (Boundary). *A boundary in a discrete state d_s is a set of states defined by $e(i, \pi_0, d_s) = \{(\pi, d_s) \in E_h \mid \pi^i = \pi_0, \}$, where $i \in \{1, 2, ..., N\}$, $d_s \in E_d$ and $\pi_0 \in \{0, 1\}$. The boundary $e(i, \pi_0, d_s)$ is inside the discrete state d_s. In the rest of this paper, we simply use e to represent a boundary.*

A toy example of HGRN, not based on any real-world biological system, is shown in Fig. 1. This example is related to a negative feedback loop with two genes: A (first dimension) and B (second dimension), where A activates B and B inhibits A. Each gene has two discrete levels, so there are four discrete states in this system. In the right part of the figure representing the model's dynamics, black arrows represent the celerities of each discrete state and red arrows represent a possible trajectory of this system, which happens to be a closed trajectory.

The state $h_M = ((\pi_M^1, 1)^T, (1, 1)^T)$ of point M belongs to the upper boundary e_1 in the second dimension of the discrete state $(1, 1)^T$. Since there is no other discrete state on the other side of e_1, the trajectory from h_M cannot cross e_1 and has to slide along e_1. Boundaries like e_1, which can be reached by trajectories but cannot be crossed, are defined as *attractive boundaries*. If there was another discrete state on the other side of e_1, in which the celerity is negative in the second dimension (towards the boundary), then the trajectory from h_M could still not cross it, and in this case e_1 would also be an attractive boundary.

The state $h_P = ((\pi_P^1, 0)^T, (0, 1)^T)$ of point P belongs to the lower boundary e_2 in the second dimension of the discrete state $(0, 1)^T$. The trajectory from h_P will reach instantly $h_Q = ((\pi_Q^1, 1)^T, (0, 0)^T)$, which belongs to the upper boundary e_3 in the second dimension of $(0, 0)^T$, because the celerities on both sides allow this (instant) discrete transition. e_2 is defined as an *output boundary* of $(0, 1)^T$ and e_3 is defined as an *input boundary* of $(0, 0)^T$.

When a trajectory reaches several output boundaries at same time, a priori, it is non-deterministic because it can cross any of them. In this work, Constraint 1 is proposed to make HGRNs deterministic in any state. This is convenient for simulation purposes.

Formal details about the simulation of general HGRNs are presented as follows. Consider a state $h = (\pi, d_s)$ and a trajectory τ which reaches h at $t > 0$.

- If h does not belong to any boundary, then $\frac{d\tau(t)}{dt} = c_s$ (the temporal derivative of a hybrid state $h = (\pi, d_s)$ is defined as $\frac{dh}{dt} = \frac{d\pi}{dt}$).
- If h only belongs to one boundary e, let us consider that e is the upper boundary in i^{th} dimension (the result is easily adapted when e is the lower boundary). In case d_s^i is not the maximal discrete level of ith gene, the discrete state on the other side of e is noted as d_r, where $d_s^k = d_r^k$ for all $k \neq i$, and $d_s^i + 1 = d_r^i$. There are four possible cases:
  - If $c_s^i < 0$, then the trajectory from the current state will enter the interior of the current discrete state. e called an *input boundary* of d_s. $\frac{d\tau(t^+)}{dt} = c_s$, $\frac{d\tau(t^-)}{dt} = c_r$ and $\tau(t) = ((\pi', d_r), (\pi, d_s))$, where $\pi'^k = \pi^k$ for all $k \neq i$, and $\pi'^i = 0$, which means that there is an instant transition from (π', d_r) to (π, d_s) at t.

- If $c_s^i = 0$, then the trajectory from the current state will slide along the boundary e, which is then called a *neutral boundary* of d_s. $\frac{d\tau(t^+)}{dt} = \frac{d\tau(t^-)}{dt} = c_s$ and $\tau(t) = (\pi, d_s)$.
- If $c_s^i > 0$, and either d_s^i is the maximal discrete level of the ith gene, or d_s^i is not the maximal discrete level of the ith gene but the ith component of c_r is negative, then the trajectory from the current state will slide along the boundary e, which is called an *attractive boundary* of d_s. If τ reaches e at t, then: $\frac{d\tau(t^+)}{dt}^k = c_s^k$ for all $k \neq i$, $\frac{d\tau(t^+)}{dt}^i = 0$, $\frac{d\tau(t^-)}{dt} = c_s$ and $\tau(t) = (\pi, d_s)$. If τ reaches e at $t_0 < t$, then: $\frac{d\tau(t)}{dt}^k = c_s^k$ for all $k \neq i$, $\frac{d\tau(t)}{dt}^i = 0$, and $\tau(t) = (\pi, d_s)$.
- If $c_s^i > 0$, d_s^i is not the maximal discrete level of the ith gene, and the ith component of c_r is positive, then the trajectory from the current state will cross instantly the boundary e and enter the discrete state d_r. e is called an *output boundary* of d_s. $\frac{d\tau(t^+)}{dt} = c_r$, $\frac{d\tau(t^-)}{dt} = c_s$ and $\tau(t) = ((\pi, d_s), (\pi', d_r))$, where $\pi'^k = \pi^k$ for all if $k \neq i$, and $\pi'^i = 0$.
- If h belongs to several boundaries, then the previous cases can be mixed:
  - If in these boundaries there is no output boundary, then the trajectory from the current state will exit all input boundaries and slide along all attractive or neutral boundaries.
  - If in these boundaries there is only one output boundary, then the trajectory from the current state will cross this output boundary.
  - If in these boundaries there are several output boundaries, then the trajectory from the current state will cross one of them following Constraint 1.

Constraint 1. *If a state of an HGRN is on several output boundaries of dimensions $dim_1, dim_2, ..., dim_m$, where dim_i is the gene number, such that $dim_1 < dim_2 < ... < dim_m$, then from this state the trajectory will only cross the output boundary of dimension dim_1 (the dimension of lowest value).*

An attractive boundary can also be considered as a black wall which is a boundary that attracts neighbor trajectories and cannot be crossed. In general hybrid systems, the behavior on black wall is not easy to define because the derivatives might be different on the different sides of a black wall. In HGRNs, by using hybrid states, a black wall is separated into two boundaries, therefore the system can have different derivatives on the different sides of the wall. There exist other methods to define behaviors of the system on a black wall [25,31].

3   Limit Cycle Analysis

This section presents new methods to find closed trajectories (potential limit cycles) (Sect. 3.1) and to analyze their stability (Sect. 3.2).

In this paper, we make two assumptions about limit cycles in HGRNs.

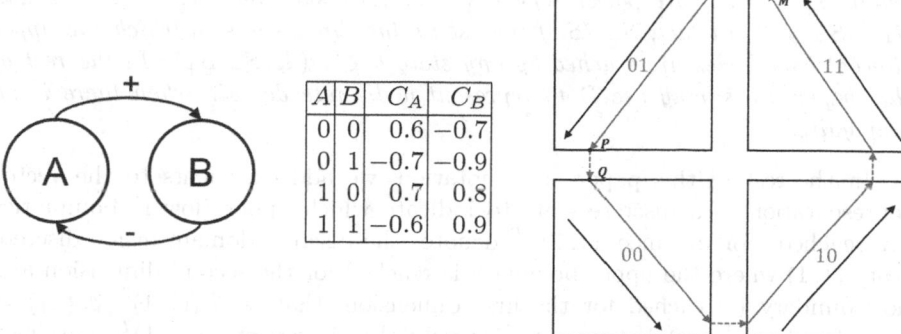

A	B	C_A	C_B
0	0	0.6	−0.7
0	1	−0.7	−0.9
1	0	0.7	0.8
1	1	−0.6	0.9

Fig. 1. Example of HGRN in 2 dimensions. Left: Influence graph (negative feedback loop with 2 genes). Middle: Example of corresponding parameters (celerities). Right: Corresponding example of dynamics; abscissa represents gene A and ordinate represents gene B.

Assumption 1. *Any non instant transition on closed trajectory does not reach more than one new boundary at the same time.*

Assumption 2. *For any instant transition on the closed trajectory (from state h_i to state h_j), there is at most one output boundary to which h_j belongs.*

For Assumption 1, in real-life systems, it is indeed very unlikely for parameters to be that constrained due to measurement noise. Assumption 2 can be satisfied if we assume that a threshold of one gene only influences at most one another gene. A counter example for Assumption 1 is: $((0.3, 0.7)^T, (a, b)^T) \rightarrow ((1, 1)^T, (a, b)^T)$, for any values of a and b, which is a non instant transition reaching two new boundaries at the same time. A counter example for Assumption 2 is: $((1, 1, 1)^T, (a, b, c)^T) \rightarrow ((0, 1, 1)^T, (a + 1, b, c)^T)$, for any values of a, b and c, where the upper boundaries in the second and third dimensions of $(a + 1, b, c)^T$ are output boundaries.

3.1 Identification of Closed Trajectories

In this section, we describe our method to find closed trajectory using the example in Fig. 1. This method has three steps which are described in order.

(1) Abstract the HGRN with discrete domains
First, the HGRN is transformed into a graph of discrete domains. A discrete domain is a new concept proposed in this work which is defined as follows.

Definition 5 (Discrete domain). *A discrete domain $\mathcal{D}(d_s, S_-, S_+)$ is a set of states inside one discrete state d_s, defined by:*

$$\mathcal{D}(d_s, S_-, S_+) = \{(\pi, d_s) \mid \forall i \in \{1, 2, ..., N\}, \pi^i \in \begin{cases} \{1\} & \text{if } i \in S_+ \\ \{0\} & \text{if } i \in S_- \\ ]0, 1[& \text{if } i \notin S_- \cup S_+ \end{cases} \}$$

where S_+ and S_- are power sets of $\{1, 2, ..., N\}$ such that $S_+ \cap S_- = \emptyset$ and $S_+ \cup S_- \neq \emptyset$. In fact, S_+ (S_-) represents the dimensions in which the upper (lower) boundaries are reached by any state $h \in \mathcal{D}(d_s, S_-, S_+)$. In the rest of this paper, we simply use $\mathcal{D}$ to represent a discrete domain when there is no ambiguity.

In the rest of this paper, as a notation, we add exponents to the vector representation of a discrete state to indicate which upper (lower) boundaries are reached. For instance, $(1, 1^+)^T$ denotes the discrete domain inside discrete state $(1, 1)$ where the upper boundary is reached for the second dimension and no boundary is reached for the first dimension, that is: $\mathcal{D}((1, 1)^T, \emptyset, \{2\}) = \{(\pi, (1, 1)^T) \mid \pi^1 \in \,]0, 1[\, \wedge \pi^2 = 1\}$ Actually, the discrete state $(1, 1)^T$ contains 8 discrete domains: $(1^-, 1^-)^T$, $(1, 1^-)^T$, $(1^+, 1^-)^T$, $(1^-, 1^+)^T$, $(1, 1^+)^T$, $(1^+, 1^+)^T$, $(1^-, 1)^T$ and $(1^+, 1)^T$.

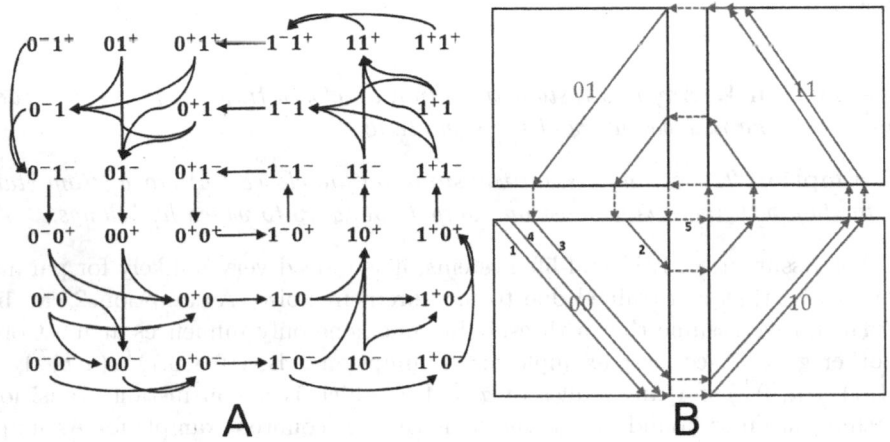

Fig. 2. A: Graph of discrete domains of the HGRN of Fig. 2. B: Examples of trajectories inside different discrete trajectories.

It is then possible to build the *graph of discrete domains* of a HGRN, such as in Fig. 2 A, where the nodes are the discrete domains and the edges are computed by considering only the signs of the celerities. In this graph, a discrete domain $\mathcal{D}_j$ is a successor of discrete domain $\mathcal{D}_i$ if:

- There are instant transitions from $\mathcal{D}_i$ to $\mathcal{D}_j$, which means that trajectories from $\mathcal{D}_i$ will cross a boundary and instantly reach $\mathcal{D}_j$; see for example $(0^+, 0)^T$ and $(1^-, 0)^T$ in Fig. 2 A.
- Only considering the sign of celerities, it is possible that there is a trajectory which begins from $\mathcal{D}_i$ and reaches $\mathcal{D}_j$ without going through another boundary; see for example $(0, 0^+)^T$ and $(0^+, 0)^T$ in Fig. 2 A: since the celerity of $(0, 0)^T$ is positive in the first dimension and negative in the second, it is possible that there is trajectory from $(0, 0^+)^T$ which reaches $(0^+, 0)^T$. We exclude

cases where two new boundaries are reached at the same time; for instance, there is no edge between $(0, 0^+)^T$ and $(0^+, 0^-)^T$.

(2) Find the closed discrete trajectories

Based on the graph of discrete domains, we consider a sequence of discrete domains $\mathcal{T} = (\mathcal{D}_0, \mathcal{D}_1, ... \mathcal{D}_p)$ which is a walk in the graph of discrete domains. A trajectory is said to be *inside* such a sequence of discrete domains if it begins from the first discrete domain and reaches by order all discrete domains in the sequence. Based on this, we define two new notions on such a sequence: the transition matrix, which allows to compute the final state of a trajectory inside a given sequence of discrete domains, when it exists, and the compatible zone, which is the set of initial states so that such a trajectory exists.

Definition 6 (Transition matrix). *Consider two different discrete domains $\mathcal{D}_i$ and $\mathcal{D}_j$ such that there exists a sequence of discrete domains $\mathcal{T}$ from $\mathcal{D}_i$ to $\mathcal{D}_j$. If there exists a state $h_i = (\pi_i, d_{s_i})$ in $\mathcal{D}_i$ so that from h_i there is a trajectory τ (defined on $[0, t_0]$) which is inside $\mathcal{T}$ and reaches $\mathcal{D}_j$ on $h_j = (\pi_j, d_{s_j})$ at t_0, then there exists a transition matrix M which describes the relation between π_i and π_j, that is: $\pi_j = s^{-1}(Ms(\pi_i))$, where s is a function that adds an extra dimension and the value in the extra dimension is always 1: $s((a_1, a_2, ..., a_N)^T) = (a_1, a_2, ..., a_N, 1)^T$. The transition matrix M only depends on $\mathcal{T}$.*

Considering the HGRN in Fig. 1, the transition matrix of $((0, 0^+)^T, (0^+, 0)^T)$ is $\begin{bmatrix} 0 & 0 & 1 \\ -\frac{c_{00}^2}{c_{00}^1} & 1 & \frac{c_{00}^2}{c_{00}^1} \\ 0 & 0 & 1 \end{bmatrix}$ and the transition matrix of $((0^+, 0)^T, (1^-, 0)^T)$ is $\begin{bmatrix} 1 & 0 & -1 \\ 0 & 1 & 0 \\ 0 & 0 & 1 \end{bmatrix}$.

Definition 7 (Compatible zone). *Consider a sequence of discrete domains $\mathcal{T} = (\mathcal{D}_0, \mathcal{D}_1, \mathcal{D}_2, ..., \mathcal{D}_m)$. The compatible zone $\mathcal{S}$ is the maximal subset of $\mathcal{D}_0$ such that any trajectory starting from $\mathcal{S}$ contains a sub-trajectory that is inside $\mathcal{T}$. More formally, for any state $h \in \mathcal{S}$, if τ is the trajectory defined on $[0, \infty]$ and beginning from h, then there exists t_0 such that the restriction of τ on $[0, t_0]$ is a trajectory inside $\mathcal{T}$.*

The compatible zone $\mathcal{S}$ of a sequence of discrete domains $\mathcal{T} = (\mathcal{D}_0, \mathcal{D}_1, \mathcal{D}_2, ..., \mathcal{D}_m)$ can be expressed with linear inequalities: $\mathcal{S} = \{(\pi, d_{s_0})^T \mid (\pi, d_{s_0}) \in \mathcal{D}_0 \land A\pi < b\}$ where A is a square matrix and b a vector. The idea to calculate compatible zone is based on Theorem 1.

Theorem 1. *A state $h = (\pi, d_{s_0})$ belongs to the compatible zone $\mathcal{S}$ of $\mathcal{T} = (\mathcal{D}_0, \mathcal{D}_1, \mathcal{D}_2, ..., \mathcal{D}_m)$ if and only if $(\pi, d_{s_0}) \in \mathcal{D}_0$, $(s^{-1}(M_{(\mathcal{D}_0, \mathcal{D}_1)}s(\pi)), d_{s_1}) \in \mathcal{D}_1$, $(s^{-1}(M_{(\mathcal{D}_0, \mathcal{D}_1, \mathcal{D}_2)}s(\pi)), d_{s_2}) \in \mathcal{D}_2$, ..., $(s^{-1}(M_{(\mathcal{D}_0, \mathcal{D}_1, ..., \mathcal{D}_{m-1})}s(\pi)), d_{s_{m-1}}) \in \mathcal{D}_{m-1}$ and $(s^{-1}(M_{(\mathcal{D}_0, \mathcal{D}_1, ..., \mathcal{D}_m)}s(\pi)), d_{s_m}) \in \mathcal{D}_m$, where $M_{(\mathcal{D}_0, \mathcal{D}_1, ..., \mathcal{D}_i)}$ is the transition matrix of $(\mathcal{D}_0, \mathcal{D}_1, ..., \mathcal{D}_i)$ and $\mathcal{D}_i$ is inside discrete state d_{s_i} ($i \in \{0, 1, ..., m\}$).*

Proof. Proof of sufficient condition: We can easily see that if $h = (\pi, d_{s_0})$ belongs to the compatible zone $\mathcal{S}$ of $\mathcal{T} = (\mathcal{D}_0, \mathcal{D}_1, \mathcal{D}_2, ..., \mathcal{D}_m)$, then $\forall i \in \{1, 2, ..., m\}$, h also belongs to the compatible zone of $(\mathcal{D}_0, \mathcal{D}_1, \mathcal{D}_2, ..., \mathcal{D}_i)$, so $(s^{-1}(M_{(\mathcal{D}_0, \mathcal{D}_1, ..., \mathcal{D}_i)} s(\pi)), d_{s_i}) \in \mathcal{D}_i$.

Proof of necessary condition: By induction. Consider a sequence of discrete domains of length 2: $(\mathcal{D}_0, \mathcal{D}_1)$, $(s^{-1}(M_{(\mathcal{D}_0, \mathcal{D}_1)} s(\pi)), d_{s_1}) \in \mathcal{D}_1$ means that, when the trajectory from h reaches the new boundary $\mathcal{D}_1$ in all dimensions in which an attractive boundary is not reached and which are not part of the boundaries related to $\mathcal{D}_1$, the fractional parts are all strictly between 0 and 1, so h belongs to the compatible zone of $(\mathcal{D}_0, \mathcal{D}_1)$. Now suppose that it is true for any sequence of discrete domains of length $k + 1$, and consider a sequence of discrete domains of length $k + 2$: $(\mathcal{D}_0, \mathcal{D}_1, \mathcal{D}_2, ..., \mathcal{D}_{k+1})$. Since it is true for a sequence of discrete domains of length $k + 1$, h belongs to the compatible zone of $(\mathcal{D}_0, \mathcal{D}_1, \mathcal{D}_2, ..., \mathcal{D}_k)$, so the trajectory from h will stay inside $(\mathcal{D}_0, \mathcal{D}_1, \mathcal{D}_2, ..., \mathcal{D}_k)$ and will reach $\mathcal{D}_k$ at $h_k = (s^{-1}(M_{(\mathcal{D}_0, \mathcal{D}_1, ..., \mathcal{D}_k)} s(\pi)), d_{s_k})$. Let $h_k = (\pi_k, d_{s_k})$. We can easily see that $s^{-1}(M_{(\mathcal{D}_0, \mathcal{D}_1, ..., \mathcal{D}_{k+1})} s(\pi)) = s^{-1}(M_{(\mathcal{D}_k, \mathcal{D}_{k+1})} s(\pi_k))$. So we have $(s^{-1}(M_{(\mathcal{D}_k, \mathcal{D}_{k+1})} s(\pi_k)), d_{s_{k+1}}) \subset \mathcal{D}_{k+1}$. Similarly to the case of length 2, h_k belongs to the sable zone of $(\mathcal{D}_k, \mathcal{D}_{k+1})$. Therefore, h belongs to the compatible zone of $(\mathcal{D}_0, \mathcal{D}_1, \mathcal{D}_2, ..., \mathcal{D}_{k+1})$. □

A sequence of discrete domains $\mathcal{T}$ is called a *discrete trajectory* if the compatible zone of $\mathcal{T}$ is not empty. A discrete trajectory $\mathcal{T} = (\mathcal{D}_1, \mathcal{D}_2, ... \mathcal{D}_m)$ is said *closed* if $\mathcal{D}_1 = \mathcal{D}_m$.

In order to find closed discrete trajectories, we use a depth first algorithm. For this, we rely on the notion of *Poincaré section* which, in our case, is a boundary of dimension $N - 1$ that a given closed trajectory always crosses. We first choose one or several input boundaries of discrete states as Poincaré sections by studying the cycles in the transition graph of discrete states, and then on each discrete domain on the Poincaré section, we apply this depth first algorithm. In each step of this depth first algorithm the compatible zone is calculated and the search will continue if the compatible zone is not empty. This algorithm finds all discrete trajectories which begin from a discrete domain and return to the initial discrete state without crossing the same discrete state more than once. An execution of this algorithm on discrete domain $(0, 0^+)^T$ is illustrated in Fig. 3. Among these discrete trajectories, we can easily find the closed ones.

Consider the HGRN in Fig. 1, we can easily see that there is only one cycle of discrete states in this system, which is:

$$(0, 0)^T \rightarrow (1, 0)^T \rightarrow (1, 1)^T \rightarrow (0, 1)^T \rightarrow (0, 0)^T$$

Therefore, for this system, we only need one Poincaré section and any boundary in this cycle can take this role. Let us choose for instance the input boundary of discrete state $(0, 0)^T$ from $(0, 1)^T$ as Poincaré section, that is, the union of the three discrete domains $(0^-, 0^+)^T$, $(0, 0^+)^T$ and $(0^+, 0^+)^T$. We thus apply the depth first algorithm on each of these three discrete domains. As a result, we can find 5 discrete trajectories which begin from the Poincaré section and returns to the initial discrete state:

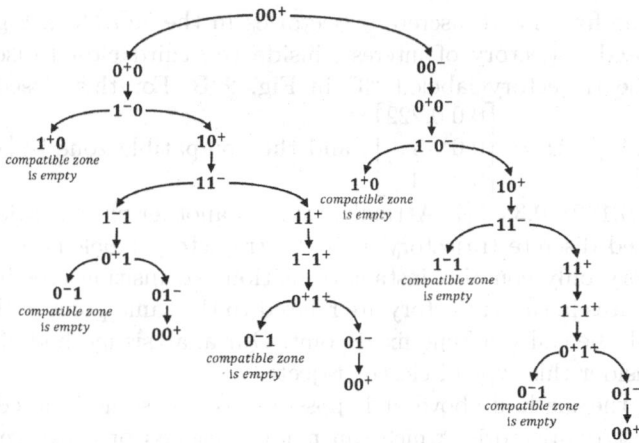

Fig. 3. Illustration of the depth first algorithm on discrete domain $(0, 0^+)^T$

$1 : (0^-, 0^+)^T \rightarrow (0, 0^-)^T \rightarrow (0^+, 0^-)^T \rightarrow (1^-, 0^-)^T \rightarrow (1, 0^+)^T \rightarrow (1, 1^-)^T \rightarrow (1, 1^+)^T \rightarrow (1^-, 1^+)^T \rightarrow (0^+, 1^+)^T \rightarrow (0, 1^-)^T \rightarrow (0, 0^+)^T$

$2 : (0, 0^+)^T \rightarrow (0^+, 0)^T \rightarrow (1^-, 0)^T \rightarrow (1, 0^+)^T \rightarrow (1, 1^-)^T \rightarrow (1^-, 1)^T \rightarrow (0^+, 1)^T \rightarrow (0, 1^-)^T \rightarrow (0, 0^+)^T$

$3 : (0, 0^+)^T \rightarrow (0^+, 0)^T \rightarrow (1^-, 0)^T \rightarrow (1, 0^+)^T \rightarrow (1, 1^-)^T \rightarrow (1, 1^+)^T \rightarrow (1^-, 1^+)^T \rightarrow (0^+, 1^+)^T \rightarrow (0, 1^-)^T \rightarrow (0, 0^+)^T$

$4 : (0, 0^+)^T \rightarrow (0, 0^-)^T \rightarrow (0^+, 0^-)^T \rightarrow (1^-, 0^-)^T \rightarrow (1, 0^+)^T \rightarrow (1, 1^-)^T \rightarrow (1, 1^+)^T \rightarrow (1^-, 1^+)^T \rightarrow (0^+, 1^+)^T \rightarrow (0, 1^-)^T \rightarrow (0, 0^+)^T$

$5 : (0^+, 0^+)^T \rightarrow (1^-, 0^+)^T \rightarrow (1^-, 1^-)^T \rightarrow (0^+, 1^-)^T \rightarrow (0^+, 0^+)^T$

Examples of trajectories inside each of these 5 discrete trajectories are shown in Fig. 2 B. We note that there always exists at least one trajectory inside a discrete trajectory since, by definition, its compatible zone is not empty. Among the 5 discrete trajectories above, only the first one is not closed.

(3) Find a closed trajectory inside each closed discrete trajectory
Consider a closed discrete trajectory $\mathcal{T} = (\mathcal{D}_0, \mathcal{D}_1, ...\mathcal{D}_m, \mathcal{D}_0)$. A *closed trajectory inside* $\mathcal{T}$ is a looping trajectory inside $\mathcal{T}$, that is, a trajectory which begins from a state $h \in \mathcal{D}_0$, reaches by order all discrete domains of $\mathcal{T}$ and finally reaches back state h. To check if there is a closed trajectory inside $\mathcal{T} = (\mathcal{D}_0, \mathcal{D}_1, ...\mathcal{D}_m, \mathcal{D}_0)$, we only need to verify the two following properties:

- $\exists (\pi_0, d_{s_0}) \in \mathcal{D}_0$ such that $s^{-1}(M_\mathcal{T} s(\pi_0)) = \pi_0$, and
- (π_0, d_{s_0}) belongs to the compatible zone of $\mathcal{T}$.

Then (π_0, d_{s_0}) is called a fixed point of $\mathcal{T}$.

Under Assumption 1, any closed trajectory which crosses the Poincaré section must be inside one of the closed discrete trajectories found by the depth first algorithm. Meanwhile, if a closed trajectory reaches more than one new boundary at the same time, then it is not inside any closed discrete trajectories found by the algorithm.

Among the five closed discrete trajectories in the HGRN in Fig. 1, we find only one closed trajectory of interest, inside the third closed discrete trajectory; it is the trajectory labeled "3" in Fig. 2 B. For this closed trajectory,

$$\pi_0 = (0.222, 1)^T, \; M = \begin{bmatrix} 0 & 0 & 0.222 \\ 0 & 0 & 1 \\ 0 & 0 & 1 \end{bmatrix}, \text{ and the compatible zone is } \{(\pi, (0,0)^T) \mid$$

$\pi^2 = 1, \pi^1 \in]0.1428, 0.3469[\}$. Actually, there is another closed trajectory inside the fifth closed discrete trajectory; it is the trajectory labeled "5" in Fig. 2 B. This trajectory only contains instant transitions (transition crossing a boundary), so all states in this trajectory are related to the same point in the euclidean space. It could be called a Zeno fixed point. Our analysis method of limit cycles does not consider this type of closed trajectory.

By using the method above, it is possible to find some isolated closed trajectories (closed trajectories which can not be reached or converged to by any trajectory) which are not limit cycles. They can be identified by the analysis method proposed in the next section.

3.2   Stability Analysis

Before introducing our stability analysis method of limit cycles in HGRNs, firstly we define the stability of limit cycles in HGRNs.

Definition 8 (Neighborhood in the same discrete state). *The neighborhood in the same discrete state of a state* $h = (\pi_0, d_s)^T$ *is a set of states defined as:* $N_d(h, r) = \{(\pi, d_s) \mid d(\pi, \pi_0) < r, \pi \in [0,1]^N\}$, *with* $r > 0$ *the radius of this neighborhood, and* d *the maximum norm between vectors:* $d(\pi, \pi_0) = \max_{i \in \{1,2,...,N\}} |\pi^i - \pi_0^i|$.

Definition 9 (Stability of limit cycles in HGRNs). *A limit cycle* C_τ *is stable if, for any state* h *on* C_τ, *there exists a neighborhood in the same discrete state of radius* r *such that any trajectory* τ_0 *that begins from this neighborhood* $N_d(h, r)$ *satisfies:* $\lim_{t \to \infty} (Dis_{min}(\tau_0(t), C_\tau)) = 0$ *where* $Dis_{min}(h', C_\tau)$ *is defined as* $Dis_{min}(h', C_\tau) = \min_{h_0 \in C_\tau} d(x(h'), x(h_0))$, *with* $h' \in E_h$, $x(h')$ *the sum (dimension by dimension) of the fractional part and the discrete state of state* h', *and* d *the maximum norm.*

It is noteworthy that in most cases, a value of t high enough is sufficient to obtain $Dis_{min}(\tau_0(t), C_\tau) = 0$, without needing a limit computation.

In the following, we call *neighborhood of a trajectory* a union of neighborhood in the same discrete state of all the states in this trajectory. A limit cycle is said to respect the *continuity of neighborhood* if there exists a neighborhood of this cycle that is small enough so that all trajectories starting from this neighborhood remain in this neighborhood. When a limit cycle does not have continuity of neighborhood, some trajectories in the neighborhood may undergo a "disruption" by touching another boundary and thus follow another sequence of discrete states. Without Assumption 1 and Assumption 2, some neighborhoods of a limit cycle might not respect this continuity, no matter how small they are. For example, consider a limit cycle that contains a state $((1,1)^T, (a,b)^T)$, for given values of a and b, where the upper boundaries in the first and second dimensions are both output boundaries. According to Constraint 1, the trajectory from this state crosses the boundary in the first dimension at first. However, in the neighborhood of $((1,1)^T, (a,b)^T)$, no matter how small it is, we can always find a state which reaches the boundary in the second dimension at first, and as it will reach a different discrete state, it might never return to the neighborhood of the limit cycle. We claim that Assumption 1 and Assumption 2 together are sufficient conditions for the continuity of neighborhood of any limit cycle in HGRNs, although we do not show a proof of this. In the following, the continuity of neighborhood is thus assumed for any limit cycle.

Now we present the method to analyze the stability of limit cycle. Consider a closed trajectory τ inside the closed discrete trajectory $\mathcal{T} = (\mathcal{D}_1, \mathcal{D}_2, ...\mathcal{D}_m, \mathcal{D}_1)$. τ begins from $h = (\pi, d_{s_1}) \in \mathcal{D}_1$. By definition of a closed trajectory, we have:

$$\pi = s^{-1}(M_{\mathcal{T}} s(\pi)) \tag{1}$$

For π, there might be some dimensions in which the values are 0 or 1 because in these dimensions the upper or lower boundaries are reached. If we only consider the dimensions in which the boundaries are not reached, Eq. 1 becomes:

$$x = Ax + b \tag{2}$$

where x is a reduction of π which only contains the dimensions in which the boundaries are not reached. The matrix A is called the reduction matrix of $\mathcal{T}$ and vector b is called the constant vector of $\mathcal{T}$.

The stability analysis method of the limit cycle is based Theorem 2.

Theorem 2. *Consider a limit cycle τ inside the closed discrete trajectory $\mathcal{T} = (\mathcal{D}_1, \mathcal{D}_2, ...\mathcal{D}_m, \mathcal{D}_1)$, and $\lambda_1, \lambda_2, ..., \lambda_p$ the eigenvalues of the reduction matrix A of $\mathcal{T}$. If $\max_{i \in \{1,2,...,p\}} |\lambda_i| < 1$ then τ is stable, otherwise τ is not stable.*

Proof. For this proof, we define the *neighborhood in the same discrete domain* of a state $h = (\pi_0, d_{s_0})^T$ as the set of states: $N_{\mathcal{D}}(h, r) = \{(\pi, d_{s_0}) \mid d(\pi, \pi_0) < r \land (\pi, d_{s_0}) \in \mathcal{D}_0\}$, where $\mathcal{D}_0$ is the discrete domain which includes h.

Consider a closed trajectory $\mathcal{C}_\tau$ that exists inside a closed discrete trajectory $\mathcal{T}$. The intersection of $\mathcal{C}_\tau$ with the Poincaré section e is $h_0 = (\pi_0, d_{s_0})$. The Poincaré map in the compatible zone of $\mathcal{T}$ is noted as $x^{k+1} = Ax^k + b$, where x is the reduction of the fractional part considering only the dimensions in which the boundaries are not reached (the reduction of π_0 is x_0). The stability of the fixed point(s) of the system $x^{k+1} = Ax^k + b$ depends on the eigenvalues of A.

If the absolute values of all eigenvalues of A are less than 1, then x_0 is asymptotically stable for the system $x^{k+1} = Ax^k + b$. And since the neighborhood of $\mathcal{C}_\tau$ is continuous, we can find a neighborhood in the same discrete domain of h_0: $N_{\mathcal{D}}(h_0, r_0)$, such that any trajectory τ from $N_{\mathcal{D}}(h_0, r_0)$ stays inside the neighborhood of C_τ and converges asymptotically to or reaches C_τ. Also, based on the fact that the neighborhood of C_τ is continuous, for any state h' on C_τ, we can find a neighborhood in the same discrete state of h': $N_d(h', r)$, such that any trajectory from $N_d(h', r)$ reaches $N_{\mathcal{D}}(h_0, r_0)$. Thus, for any trajectory τ from $N_d(h', r)$, we have: $\lim_{t \to \infty} Dis_{min}(\tau(t), C_\tau) = 0$, which proves that C_τ is a stable limit cycle.

If the maximum absolute value of all eigenvalues of A equals to or is greater than 1, then x_0 is marginally stable or unstable for system $x^{k+1} = Ax^k + b$; in both cases we cannot guarantee that any trajectory from a small neighborhood in the same discrete domain of h_0 converges to or reaches C_τ. Therefore, C_τ is not stable. $\qquad \square$

For the HGRN in Fig. 1, the reduction matrix of the third closed discrete trajectory is $[0]$, so the closed trajectory inside this closed discrete trajectory is a stable limit cycle. Consider for example the fourth trajectory in Fig. 2 B which is a trajectory from the neighborhood of the limit cycle: we can see it finally reaches the limit cycle. In fact, in this HGRN the basin of attraction of this limit cycle is the set of all states of the system.

In fact, if all eigenvalues of A are equal to 1, then the relevant closed trajectory is an isolated closed trajectory, that is, a closed trajectory that can not be reached or converged to by any trajectory, which is not a limit cycle.

4   Application

In this section, we apply our proposed limit cycle analysis method on three HGRNs of negative feedback loop in 3 dimensions and one HGRN of cell cycle in 5 dimensions. The negative feedback loop in 3 dimensions can be used to describe real biological oscillators, for example the p53 system [22]. The signs of the celerities in these three HGRNs are determined by the influence graph (positive for an activation and negative for an inhibition) and their absolute values of celerities are randomly selected. The parameters of the HGRN in 5 dimensions are generated randomly respecting the constraints in Table 3 of [6]. The influence graphs of both systems can be found in Fig. 4. Details about implementation can be found at https://doi.org/10.5281/zenodo.6524936.

4.1   HGRNs of Negative Feedback Loop in 3 Dimensions

The parameters of these three HGRNs of negative feedback loop in 3 dimensions are shown in Table 1. The signs of celerities in these three models are the same so they have the same graph of discrete states. There is only one cycle of discrete states in each of these systems, which is:
$(1,1,1)^T \to (0,1,1)^T \to (0,1,0)^T \to (0,0,0)^T \to (1,0,0)^T \to (1,0,1)^T \to (1,1,1)^T$
Therefore, for these three models, we choose the input boundary e of $(0,0,0)^T$ in the cycle as the Poincaré section. Simulations depicting the convergence to the

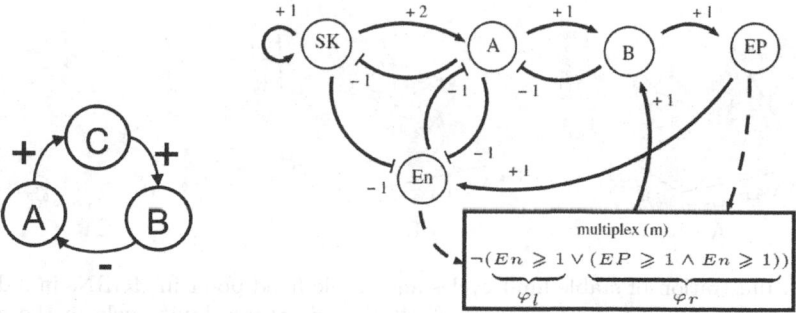

Fig. 4. Left: Influence graph of a negative feedback loop with 3 genes, used to build three models given in Table 1. Right: Influence graph of a cell cycle model with 5 genes from [6]; the multiplex (m) expresses constrains on the joint activation of En and Ep on B.

Table 1. Parameters of the three HGRNs of negative feedback loop in 3 dimensions. Left: First model. Middle: Second model. Right: third model.

A	B	C	C_A	C_B	C_C
0	0	0	1	-0.6	-0.7
0	0	1	1	0.7	-0.9
0	1	0	-0.8	-0.8	-0.7
0	1	1	-0.8	0.6	-0.9
1	0	0	0.7	-0.6	0.6
1	0	1	0.7	0.7	0.5
1	1	0	-0.9	-0.8	0.6
1	1	1	-0.9	0.6	0.5

A	B	C	C_A	C_B	C_C
0	0	0	3	-0.6	-0.7
0	0	1	3	0.7	-2.9
0	1	0	-2.8	-0.8	-0.7
0	1	1	-2.8	0.6	-2.9
1	0	0	2.7	-0.6	2.6
1	0	1	2.7	0.7	0.5
1	1	0	-2.9	-0.8	2.6
1	1	1	-2.9	0.6	0.5

A	B	C	C_A	C_B	C_C
0	0	0	3	-0.6	-0.7
0	0	1	3	0.7	-2.9
0	1	0	-0.8	-0.8	-0.7
0	1	1	-0.8	0.6	-2.9
1	0	0	0.7	-0.6	2.6
1	0	1	0.7	0.7	0.5
1	1	0	-2.9	-0.8	2.6
1	1	1	-2.9	0.6	0.5

stable cycle or to the fixed point (see below) in these three HGRNs are shown in Fig. 5.

In the first HGRN, by using our limit cycle analysis method, we find one stable limit cycle and one closed trajectory which only contains instant transitions (that we call a fixed point). Regarding the stable limit cycle, the fixed point of this limit cycle in discrete domain $(0^-, 0^+, 0)^T$ is $((0, 1, 0.125)^T$,

$(0, 0, 0)^T)$, the transition matrix is $\begin{bmatrix} 0 & 0 & 0 & 0 \\ 0 & 0 & 0 & 1 \\ 0 & 0 & 0 & 0.125 \\ 0 & 0 & 0 & 1 \end{bmatrix}$, the compatible zone is

$\{(\pi, (0, 0, 0)^T) \mid \pi^1 = 0, \pi^2 = 1, \pi^3 \in]0, 0.7[\}$ and the reduction matrix is $[0]$, therefore trajectories from the neighborhood of this limit cycle will reach this limit cycle very quickly (less than one turn if the neighborhood is small enough).

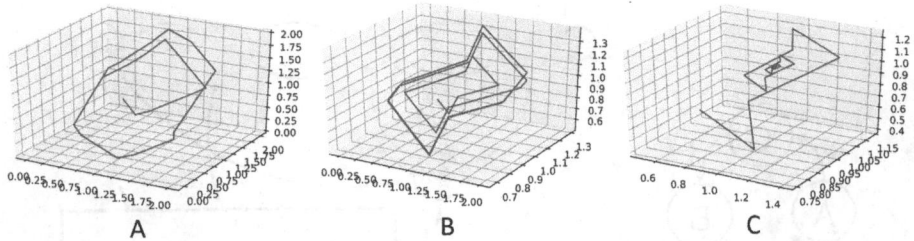

Fig. 5. Illustration of stable limit cycles and stable fixed point in HGRNs in 3 dimensions. A: Stable limit cycle in the first HGRN. B: Stable limit cycle in the second HGRN. C: Stable fixed point in the third HGRN.

In this HGRN we can also prove that all trajectories will reach this limit cycle except trajectories which can reach the fixed point of the system. All discrete trajectories which begin from the Poincaré section and return to the Poincaré section in this HGRN are shown in Fig. 6 A. Since in this HGRN there is only one cycle of discrete states which is also a global attractor, any trajectory from the Poincaré section must return to the Poincaré section and it must begin from the compatible zone or the boundary of the compatible zone of one of the discrete trajectories in Fig. 6 A. We see that all discrete trajectories which are not closed will finally reach closed discrete trajectories (31, 32, 33, 9, 10). Discrete trajectories 31, 32 and 33 have the same transition matrix and their reduction matrix is $[0]$ so any trajectory from $(0^-, 0^+, 0)^T$ will reach the limit cycle. For the discrete trajectory 10, the two eigenvalues of the reduction matrix are 7.0306 and 0.0368, so trajectories inside discrete trajectory 10 will finally leave the compatible zone and reach $(0^-, 0^+, 0)^T$. From here, we can see that any trajectories from the Poincaré section will reach the limit cycle except the trajectories inside discrete trajectory 9 which are related to a fixed point. As any trajectory in this system will finally reach this Poincaré section, all trajectories will reach this limit cycle except trajectories which can reach the fixed point.

For the second HGRN, by using our method, we can also find one stable limit cycle and one fixed point. Unlike the first HGRN, trajectories from the neighborhood of the limit cycle converge asymptotically to the limit cycle: The limit cycle is inside the discrete trajectory which begins from $(0^-, 0^+, 0)^T$ and the reduction matrix of the limit cycle is $[0.0298]$. We can also prove that all trajectories will converge to this limit cycle except trajectories which can reach the fixed point by using the same method as for the first HGRN.

Contrary to the first and the second HGRN, we cannot find a limit cycle in the third HGRN but only a fixed point which is related to the discrete trajectory 2 in Fig. 6 C. By analyzing the eigenvalues and the fixed points of discrete trajectories 3 and 4, we can prove that all trajectories in this system will converge to the fixed point.

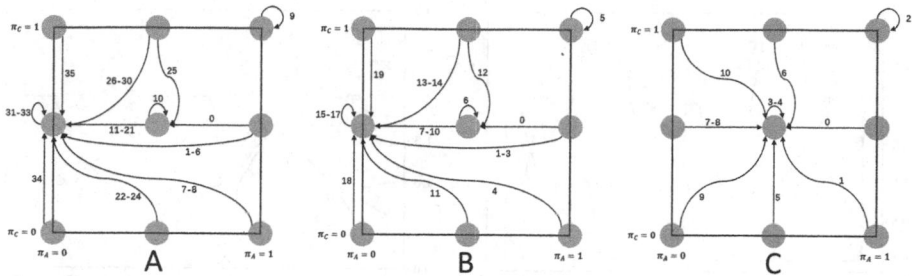

Fig. 6. Abstracted representations of the chosen Poincaré sections in the HGRNs in 3 dimensions, illustrating all possible discrete trajectories which start from and return to this Poincaré section. The blue dots represent the discrete domains and each arrow depicts one or several different discrete trajectories (each following a unique sequence of discrete domains). A: First HGRN. B: Second HGRN. C: Third HGRN. (Color figure online)

4.2 HGRN of Cell Cycle in 5 Dimensions

For the HGRN in 5 dimensions, the transition graph of discrete states is more complex. By using a depth first algorithm, we find that there are 1104 cycles of discrete states in which 930 cycles contain the discrete transition $(0, 1, 0, 1, 0)^T \rightarrow (0, 1, 0, 1, 1)^T$, 94 cycles contain $(0, 0, 0, 1, 1)^T \rightarrow (0, 1, 0, 1, 1)^T$ and all the rest contain $(0, 0, 1, 1, 0)^T \rightarrow (0, 0, 0, 1, 0)^T$. Therefore, for this model, we use the three input boundaries crossed by these transition as Poincaré sections, and perform as many analyses. Our method exhibits one stable limit cycle and one unstable limit cycle. The stable one is the same one studied in [6] to calculate the constraints of parameters. The simulations of both cycles are shown in Fig. 7 A and B. We need to mention that for now we have not identified any biological behavior related to this unstable limit cycle yet.

For the stable limit cycle of cell cycle model, the fixed point of this limit cycle in the discrete domain $(0, 1^+, 0, 1^+, 1^-)^T$ is $((0.3714, 1, 0.8581, 1, 0)^T,$ $(0, 1, 0, 1, 1)^T)$, and the reduction matrix is $\begin{bmatrix} 0 & 0 \\ 0 & 0 \end{bmatrix}$.

For the unstable limit cycle of cell cycle model, the fixed point of this limit cycle in the discrete domain $(0, 0, 0^|, 1, 0^-)^T$ is $((0.6375,$ $0.2552, 1, 0.3472, 0)^T, (0, 0, 0, 1, 0)^T)$, and the reduction matrix A is $\begin{bmatrix} 4.95359512 \cdot 10^3 & 0 & 1.37489884 \cdot 10^{-13} \\ -5.25996267 \cdot 10^2 & 0 & -1.45993292 \cdot 10^{-14} \\ -7.15619779 \cdot 10^2 & 0 & -1.98624389 \cdot 10^{-14} \end{bmatrix}$. The eigenvalues of A are 0, $4.95359512 \cdot 10^3$ and $3.15544362 \cdot 10^{-30}$, making it unstable.

This current naive implementation in Python reaches its limits w.r.t. execution time when the size of the system increases: finding the limit cycles above takes less than one minute for the HGRNs in 3 dimensions, and 8 h for the HGRN

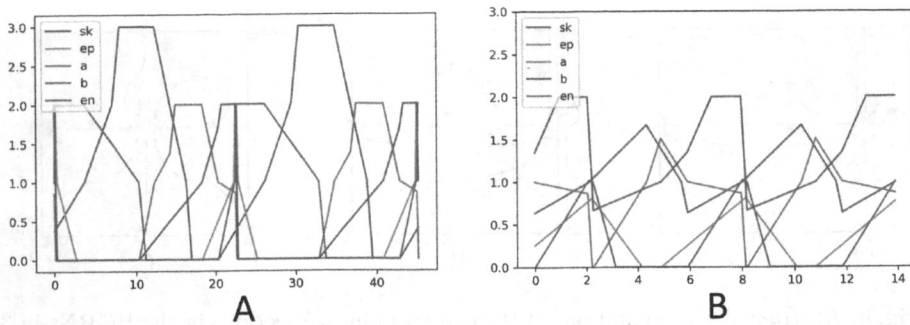

Fig. 7. Simulation of the two limit cycles found in the HGRN of 5 dimensions. A: Stable limit cycle. B: Unstable limit cycle.

in 5 dimensions[1]. In future works, we plan to make adjustments to improve the implementation performance.

5   Conclusion

In this work, we proposed a formal method to find all limit cycles of HGRNs with some minor restrictions, mainly to remove non-deterministic behaviors and complex loops, and to analyze their stability. To our knowledge, this method is the first one to find and analyze limit cycles of HGRNs in N dimensions. We showed the merits of this method on random generated HGRNs of a negative feedback loop with 3 components and a HGRN of the cell cycle with 5 components taken from the literature.

As stated above, a first limitation of this method is that we do not handle non-determinism, and we might thus miss some complex closed trajectories, consisting of a composition of several loops using states with a non-deterministic future. Considering closed trajectories inside more complex attractors by assessing non-determinism is thus an interesting continuation. Another limitation in the application of this method is that we first need to construct a HGRN of a specific gene regulatory network; however, the observation of real biological systems is limited and it is not always possible to determine all parameters. In some cases, some parameters can only be described by constraints, or remain unknown. Thus, considering extensions of this method that are parameterized or that take into account a set of constraints on parameters is also of interest.

Finally, in future works, we will also focus on the application of this method on the problem of the control of gene regulatory networks. Similar works have been done with other classes of hybrid models, for instance [8] for the control of oscillations and [9] for the control of bistable switches.

---

[1] Computations were performed on a standard laptop computer, with an Intel Core I7-8550U 1.80 GHz processor and 16.0 GB RAM.

Acknowledgements. We would like to thank Gilles Bernot and Jean-Paul Comet for their fruitful discussions.

References

1. Alur, R., Dang, T., Ivančić, F.: Counter-example guided predicate abstraction of hybrid systems. In: Caravel, H., Hatcliff, J. (eds.) TACAS 2003. LNCS, vol. 2619, pp. 208–223. Springer, Heidelberg (2003). https://doi.org/10.1007/3-540-36577-X_15
2. Alur, R., Henzinger, T.A., Lafferriere, G., Pappas, G.J.: Discrete abstractions of hybrid systems. Proc. IEEE **88**(7), 971–984 (2000)
3. Asarin, E., Maler, O., Pnueli, A.: Reachability analysis of dynamical systems having piecewise-constant derivatives. Theor. Comput. Sci. **138**(1), 35–65 (1995)
4. Asarin, E., Mysore, V.P., Pnueli, A., Schneider, G.: Low dimensional hybrid systems-decidable, undecidable, don't know. Inf. Comput. **211**, 138–159 (2012)
5. Barik, D., Baumann, W.T., Paul, M.R., Novak, B., Tyson, J.J.: A model of yeast cell-cycle regulation based on multisite phosphorylation. Mol. Syst. Biol. **6**(1), 405 (2010)
6. Behaegel, J., Comet, J.P., Bernot, G., Cornillon, E., Delaunay, F.: A hybrid model of cell cycle in mammals. J. Bioinform. Comput. Biol. **14**(01), 1640001 (2016)
7. Behaegel, J., Comet, J.P., Folschette, M.: Constraint identification using modified Hoare logic on hybrid models of gene networks. In: 24th International Symposium on Temporal Representation and Reasoning (TIME 2017). Schloss Dagstuhl-Leibniz-Zentrum fuer Informatik (2017)
8. Belgacem, I., Gouzé, J.L., Edwards, R.: Control of negative feedback loops in genetic networks. In: 2020 59th IEEE Conference on Decision and Control (CDC), pp. 5098–5105. IEEE (2020)
9. Chaves, M., Gouzé, J.L.: Exact control of genetic networks in a qualitative framework: the bistable switch example. Automatica **47**(6), 1105–1112 (2011)
10. Chaves, M., Preto, M.: Hierarchy of models: from qualitative to quantitative analysis of circadian rhythms in cyanobacteria. Chaos: Interdisc. J. Nonlinear Sci. **23**(2), 025113 (2013)
11. Clark, W., Bloch, A.: A Poincaré-Bendixson theorem for hybrid dynamical systems on directed graphs. Math. Control Signals Syst. **32**(1), 1–18 (2020)
12. Clark, W., Bloch, A., Colombo, L.: A Poincaré-Bendixson theorem for hybrid systems. Math. Control Relat. Fields **10**(1), 27 (2020)
13. Comet, J.P., Bernot, G., Das, A., Diener, F., Massot, C., Cessieux, A.: Simplified models for the mammalian circadian clock. Procedia Comput. Sci. **11**, 127–138 (2012)
14. Comet, J.-P., Fromentin, J., Bernot, G., Roux, O.: A formal model for gene regulatory networks with time delays. In: Chan, J.H., Ong, Y.-S., Cho, S.-B. (eds.) CSBio 2010. CCIS, vol. 115, pp. 1–13. Springer, Heidelberg (2010). https://doi.org/10.1007/978-3-642-16750-8_1
15. Cornillon, E., Comet, J.P., Bernot, G., Enée, G.: Hybrid gene networks: a new framework and a software environment. In: Advances in Systems and Synthetic Biology (2016)
16. Doyen, L., Frehse, G., Pappas, G.J., Platzer, A.: Verification of hybrid systems. In: Clarke, E., Henzinger, T., Veith, H., Bloem, R. (eds.) Handbook of Model Checking, pp. 1047–1110. Springer, Cham (2018). https://doi.org/10.1007/978-3-319-10575-8_30

17. Edwards, R.: Analysis of continuous-time switching networks. Physica D **146**(1–4), 165–199 (2000)
18. Edwards, R., Glass, L.: A calculus for relating the dynamics and structure of complex biological networks. Adventures Chem. Phys.: A Special Volume of Advances in Chemical Physics **132**, 151–178 (2005)
19. Farcot, E., Gouzé, J.L.: Periodic solutions of piecewise affine gene network models with non uniform decay rates: the case of a negative feedback loop. Acta. Biotheor. **57**(4), 429–455 (2009)
20. Firippi, E., Chaves, M.: Topology-induced dynamics in a network of synthetic oscillators with piecewise affine approximation. Chaos: Interdisc. J. Nonlinear Sci. **30**(11), 113128 (2020)
21. Flieller, D., Riedinger, P., Louis, J.P.: Computation and stability of limit cycles in hybrid systems. Nonlinear Anal. Theory Methods Appl. **64**(2), 352–367 (2006)
22. Geva-Zatorsky, N., et al.: Oscillations and variability in the p53 system. Mol. Syst. Biol. **2**(1), 2006–0033 (2006)
23. Girard, A.: Computation and stability analysis of limit cycles in piecewise linear hybrid systems. IFAC Proc. Vol. **36**(6), 181–186 (2003)
24. Glass, L., Edwards, R.: Hybrid models of genetic networks: mathematical challenges and biological relevance. J. Theor. Biol. **458**, 111–118 (2018)
25. Gouzé, J.L., Sari, T.: A class of piecewise linear differential equations arising in biological models. Dyn. Syst. **17**(4), 299–316 (2002)
26. Hiskens, I.A.: Stability of hybrid system limit cycles: application to the compass gait biped robot. In: Proceedings of the 40th IEEE Conference on Decision and Control (Cat. No. 01CH37228), vol. 1, pp. 774–779. IEEE (2001)
27. Hiskens, I.A.: Stability of limit cycles in hybrid systems. In: Proceedings of the 34th Annual Hawaii International Conference on System Sciences, pp. 6-pp. IEEE (2001)
28. Karlebach, G., Shamir, R.: Modelling and analysis of gene regulatory networks. Nat. Rev. Mol. Cell Biol. **9**(10), 770–780 (2008)
29. Kauffman, S.A.: Metabolic stability and epigenesis in randomly constructed genetic nets. J. Theor. Biol. **22**(3), 437–467 (1969)
30. Mestl, T., Lemay, C., Glass, L.: Chaos in high-dimensional neural and gene networks. Physica D **98**(1), 33–52 (1996)
31. Plahte, E., Kjøglum, S.: Analysis and generic properties of gene regulatory networks with graded response functions. Physica D **201**(1–2), 150–176 (2005)
32. Prabhakar, P., Garcia Soto, M.: Abstraction based model-checking of stability of hybrid systems. In: Sharygina, N., Veith, H. (eds.) CAV 2013. LNCS, vol. 8044, pp. 280–295. Springer, Heidelberg (2013). https://doi.org/10.1007/978-3-642-39799-8_20
33. Simic, S.N., Sastry, S., Johansson, K.H., Lygeros, J.: Hybrid limit cycles and hybrid Poincaré-Bendixson. IFAC Proc. Vol. **35**(1), 197–202 (2002)
34. Sriram, K., Bernot, G., Képès, F.: Discrete delay model for the mammalian circadian clock. Complexus **3**(4), 185–199 (2006)
35. Thomas, R.: Boolean formalization of genetic control circuits. J. Theor. Biol. **42**(3), 563–585 (1973)
36. Thomas, R.: Regulatory networks seen as asynchronous automata: a logical description. J. Theor. Biol. **153**(1), 1–23 (1991)
37. Znegui, W., Gritli, H., Belghith, S., et al.: Design of an explicit expression of the Poincaré map for the passive dynamic walking of the compass-gait biped model. Chaos Solitons Fractals **130**(C) (2020)

Machine Learning

Bayesian Learning of Effective Chemical Master Equations in Crowded Intracellular Conditions

Svitlana Braichenko[1,2](✉) (ID), Ramon Grima[2] (ID), and Guido Sanguinetti[1,3] (ID)

[1] School of Informatics, University of Edinburgh, Edinburgh, UK
braichenko.svitlana@gmail.com
[2] School of Biological Sciences, University of Edinburgh, Edinburgh, UK
Ramon.Grima@ed.ac.uk
[3] SISSA, Trieste, Italy
gsanguin@sissa.it

Abstract. Biochemical reactions inside living cells often occur in the presence of crowders - molecules that do not participate in the reactions but influence the reaction rates through excluded volume effects. However the standard approach to modelling stochastic intracellular reaction kinetics is based on the chemical master equation (CME) whose propensities are derived assuming no crowding effects. Here, we propose a machine learning strategy based on Bayesian Optimisation utilising synthetic data obtained from spatial cellular automata (CA) simulations (that explicitly model volume-exclusion effects) to learn effective propensity functions for CMEs. The predictions from a small CA training data set can then be extended to the whole range of parameter space describing physiologically relevant levels of crowding by means of Gaussian Process regression. We demonstrate the method on an enzyme-catalyzed reaction and a genetic feedback loop, showing good agreement between the time-dependent distributions of molecule numbers predicted by the effective CME and CA simulations.

Keywords: Inference · Stochastic reactions · Crowding

1 Introduction

The empirical demonstration of stochasticity in gene expression [1] has had a profound impact both on experimental and computational biology. Experimentally, the last two decades have witnessed a flourishing of advanced technologies to measure stochastic effects in biology at unprecedented throughput and spatial/temporal resolution [2–4]. Computationally, considerable effort has gone towards developing algorithmic solutions to facilitate the *in silico* simulation of stochastic biological systems, and their calibration to observational data [5–10].

Supported by Leverhulme Trust.

The vast majority of stochastic modelling work operates within the framework of the classical Chemical Master Equation (CME), which describes the time evolution of the (single-time marginal) state probability distribution of a discrete state, continuous-time Markovian system [11]. In this framework, each reaction has associated with it a propensity function which is derived assuming molecules are point particles diffusing fast enough such that well-mixed conditions ensue [11–13]. The CME formulation provides a number of advantages, including a transparent and elegant mathematical formalism, and efficient simulation and approximation algorithms [8]. In particular, the existence of an exact Stochastic Simulation Algorithm (SSA) [5] has led to the wide use of this formulation. However, the assumption that reactants freely diffuse inside cells is clearly at odds with biological reality: the cellular environment is spatially highly structured, and, for every given reaction, it contains large numbers of particles that do not partake in the reaction, creating a crowding effect which can significantly affect the dynamics of biochemical processes inside the cell [14–17]. While the modelling community is keenly aware of this mismatch in assumptions, there have been only a few attempts at modifying the propensities of the SSA/CME to take into account crowding [18–20]. These pioneering studies have focused on simple biochemical systems but they do not provide a general recipe applicable to all intracellular reaction systems of interest. In contrast, particle-based algorithms (such as Brownian dynamics and cellular automata [18,21–27]) have been extensively used to study the effect of crowding; these approaches while naturally suited to study crowding, are computationally demanding since they model the movement of each particle (crowder or reactant) in the system.

In this paper, we devise a computational approach based on machine learning to adapt the efficiency of the CME approach to the reality of crowding. In a nutshell, the idea is to learn CME propensity functions that lead to particle number distributions which optimally match the ones resulting from particle-based algorithms. This leads to a general computational recipe for constructing effective CMEs that capture the stochastic dynamics of any biochemical system in crowded conditions.

2   Connecting Different Mathematical Descriptions of Stochastic Kinetics Using Bayesian Optimization

In this paper, we shall be concerned with two different descriptions of stochastic chemical kinetics: (i) Cellular automata (CA); (ii) Monte Carlo simulations using the SSA. We next describe each in detail and then show how a Bayesian optimization based procedure can be used to connect these different stochastic descriptions.

2.1   Cellular Automata

Cellular Automata (CA) can be characterized as a lattice of sites each holding a finite number of discrete states plus some rules which describe the evolution of the state of each site. Typically these update rules are a function of the states of the sites within a local neighbourhood. For a general introduction to CA in the context of biological and chemical modelling see [28, 29]. Many CA models that study the influence of crowding in biochemical reactions [21–23] have the following properties in common: (i) each lattice site is either occupied by a molecule or empty; (ii) at each time step, a molecule is selected at random and one of its neighbouring sites is also chosen at random; (iii) if the chosen neighbouring site is empty then the molecule moves to it otherwise a reaction is attempted (only if the site is occupied by a reactant). In Fig. 1(a) we illustrate a CA modelling a simple enzymatic reaction.

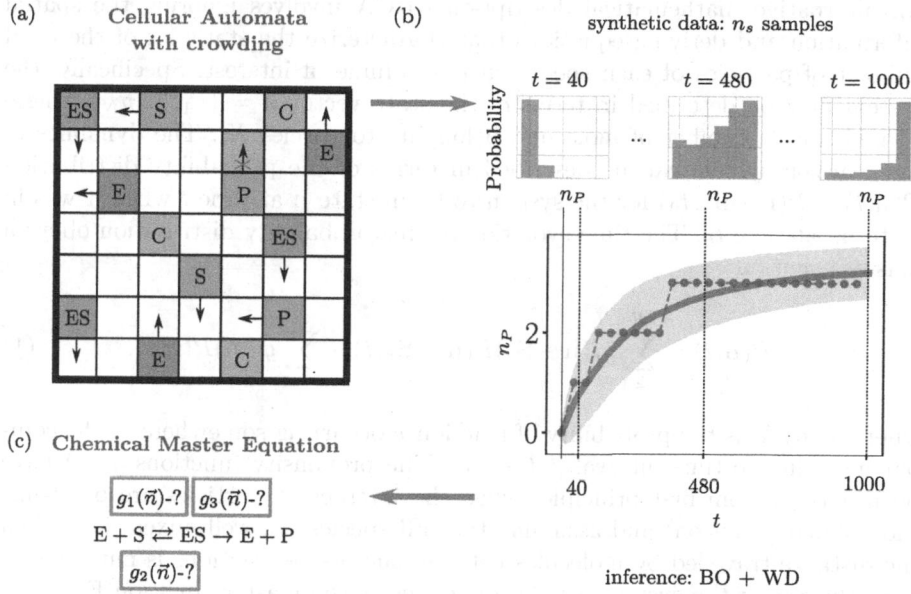

Fig. 1. Cartoon illustrating the learning procedure, exemplified by means of a simple enzyme reaction. (a) First, we generate synthetic data using cellular automata simulations of the enzyme reaction. We show reactants (substrate S in orange and enzyme E in teal), enzyme-substrate complexes (ES in purple), products (P in green), inert crowders (C in grey) and empty spaces (white). The particle movements are updated according to a set of rules (see Appendix B). In general, particles may move in any direction, but the arrows show a randomly chosen one; note that a particle can only move if an empty space is available. (b) We collect n_s CA sample trajectories of the numbers of molecules of E, S, ES and P in time (a typical trajectory for the number of P molecules is shown as red connected dots while the mean and standard deviation over the trajectories are shown by a solid blue line and a shaded blue area, respectively). The marginal distributions of the number of each species molecules sampled at a number of discrete time points constitute our synthetic data (we show only those for P). (c) Finally we use Bayesian optimization (BO) to minimize the Wasserstein distance (WD) between the time-dependent particle number distributions generated by the CA and the SSA; this leads to the propensities $g_i(\vec{n})$ of the effective CME. (Color figure online)

The main advantage of CAs is their simplicity, in particular it is easy to devise rules that mimic molecular movement and interaction in complex geometries. Their disadvantages are (i) each molecule, independent of type, occupies the same amount of space (equal to one lattice site); (ii) the regularity of the lattice can influence the simulated dynamics. These main disadvantages can be overcome by using Brownian dynamics, a lattice-free approach [18,24,25,30], however these simulations are much more computationally expensive and we do not consider them further here.

2.2   The Chemical Master Equation and the SSA

An alternative mathematical description to CA involves ignoring the spatial information and deriving equations that characterize the statistics of the total number of particles of each species in the volume of interest. Specifically, the system may be described in terms of the state vector $\mathbf{n} = (n_1, ..., n_N)$, where n_i indicates a number of molecules belonging to species X_i. The dynamics of the reaction system can be described in terms of the probability distribution $P(\mathbf{n}, t) = P(\mathbf{n}, t|\mathbf{n_0}, t_0)$ for the system to be in state $\mathbf{n}$ at time t when it was in state $\mathbf{n_0}$ at time t_0. The time evolution of this probability distribution obeys a master equation

$$\partial_t P(\mathbf{n}, t) = \sum_{r=1}^{R} g_r(\mathbf{n} - \mathbf{S_r})P(\mathbf{n} - \mathbf{S_r}, t) - \sum_{r=1}^{R} g_r(\mathbf{n})P(\mathbf{n}, t), \tag{1}$$

where $g_r(\mathbf{n})\Delta t$ is the probability of reaction r occurring somewhere in the compartment in the time interval $[t, t + \Delta t)$. The propensity functions $g_r(\mathbf{n})$ have been derived from first principles when the interacting particles are point-like (no volume exclusion) and assuming that all species are well-mixed, i.e., when the distance travelled by molecules between successive reactions is much larger than the size of the system [12,13]. In this case, the master equation Eq. (1) is known as the Chemical Master Equation. In the well-mixed limit, this is formally equivalent to the reaction-diffusion master equation [31] which has been shown to provide an accurate approximation of microscopic simulations that track point particle positions [32]. We note that there is no general closed-form solution to the CME and hence in practice, one uses the SSA [33] to estimate $P(\mathbf{n}, t)$ in a Monte Carlo setup. We refer the reader to [8,31] for comprehensive background on the CME and the various methods to approximate its solutions.

2.3   Bayesian Optimisation

The CME presents considerable computational advantages over CA simulations, however simply ignoring spatial effects will generally result in an inaccurate prediction of the stochastic dynamics of the total number of particles for each species in a compartment. Nevertheless, it is plausible that there exists a different CME parametrisation which leads to a time-dependent probability distribution of molecule numbers that well approximates the same distributions calculated from CA. The main purpose of this paper is to illustrate a machine-learning strategy that automates the task of finding appropriate propensity functions.

The task is akin to the inference of CME parameters from observations [8,34,35], however in this case the CME parameters are not determined in order to maximise a data likelihood, but rather to give rise to (transient and steady state) particle number distributions that match the distributions generated from CA. To do so, we use a machine learning technique called Bayesian optimisation (BO) [36,37]. BO is an efficient sequential algorithm to optimise objective functions which are very expensive to evaluate. In our case, the task is

$$\mathbf{x}^* = \arg\min_{\mathbf{x}} f(\mathbf{x}), \tag{2}$$

where $\mathbf{x} \in \mathbb{R}^d$ is a vector of CME parameters and the objective function $f(\mathbf{x})$ is a (rescaled) one-dimensional Wasserstein distance (WD) [38] between the empirical marginal distribution of CA and CME trajectories (1); see Appendix A for more details. BO works by using discrete evaluations of the expensive objective function $f(\mathbf{x})$ to construct a statistical surrogate (the *acquisition function*) which is easy to optimise and can be used to identify the next query point. Two subsequent queries are shown in the top and bottom of Fig. 2, respectively. Generally, Gaussian process (GP) regression [39] is used to construct the surrogate function. GP regression is a methodology to perform Bayesian inference over functional spaces; given some training points (in our case estimates of the objective function at some parameter values), one can obtain a posterior predictive distribution over function values everywhere in parameter space, in terms of a mean posterior function $\mu(\mathbf{x})$ and posterior variance function $\sigma(\mathbf{x})$, which provide a point-wise posterior distribution over the values of the expensive function $f(\mathbf{x})$. The acquisition function is then obtained as some analytical function of the posterior mean and variance. In our case, we use the expected improvement acquisition function

$$\alpha_{EI}(\mathbf{x}) = (\mu_n(\mathbf{x}) - \tau_n)\Phi\left(\frac{\mu_n(\mathbf{x}) - \tau_n}{\sigma_n(\mathbf{x})}\right) + \sigma_n(\mathbf{x})\phi\left(\frac{\mu_n(\mathbf{x}) - \tau_n}{\sigma_n(\mathbf{x})}\right), \tag{3}$$

where τ_n is the optimal value (minimum in the minimisation setup) of the estimated unknown function found after n steps (τ_4 and τ_5 on the left-hand side of Fig. 2), $\mu_n(\mathbf{x})$ and $\sigma_n(\mathbf{x})$ are the posterior mean and variance from GP regression after n function evaluations, Φ is the standard normal cumulative distribution function, and ϕ is the standard normal probability density function. The acquisition function is analytically computable and can be optimised using standard methods to provide an optimal query point for the next evaluation of the expensive function $f(\mathbf{x})$. We note that our choice of acquisition function is not unique; other possible choices are upper confidence bound, probability of improvement and knowledge gradient [36]. The expected improvement acquisition function that we use is guaranteed to find the optimum of the target function under mild assumptions [40].

A cartoon summarizing the overall learning procedure is shown in Fig. 1 and an illustration focusing on how BO works to minimize an objective function is shown in Fig. 2.

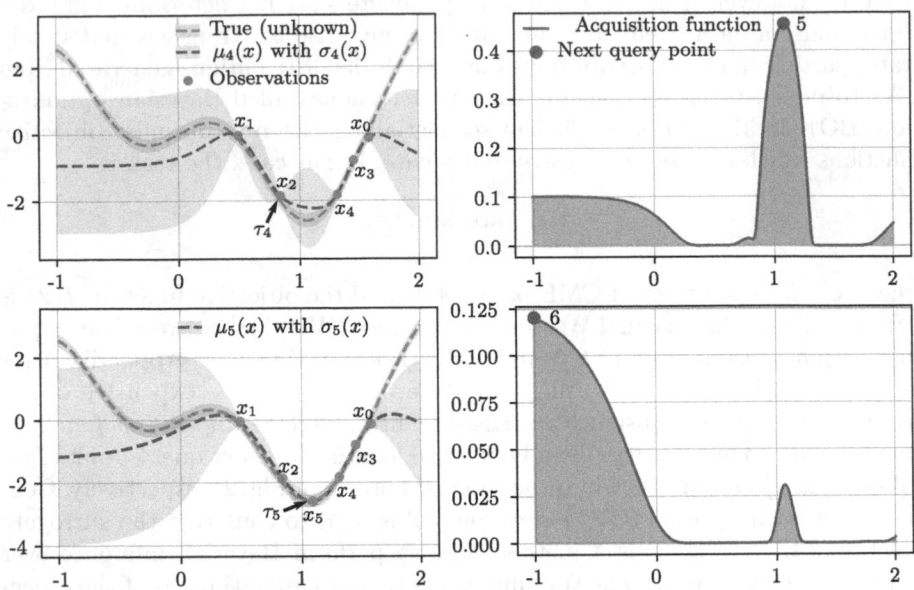

Fig. 2. Illustration of how BO works to find the minimum of an arbitrary function. The GP prediction is characterized by its mean (dashed green line) and variance (shaded green area). The true function of interest is shown using a red dashed line with shading denoting an uncertainty range. After observing the function for 5 times (red dots $x_0, \ldots, x_4$) – see top left figure – we select the next observation point 6 by calculating and maximizing the acquisition function estimated using τ_4, μ_4 and σ_4 in Eq. (3) and shown by the teal line – see top right figure. The bottom row of plots shows how the prediction approaches the true function with more observations. In this paper, the BO procedure is used to find the reaction rate values in the propensity functions of the CME (the parameter vector $\mathbf{x}$) which minimize the mean WD (the function) between the time-dependent distributions computed by CA and the CME. (Color figure online)

The computational recipe is as follows:

1. Generate a number n_s of CA trajectories for some chosen set of parameter values, and collect marginal probabilities of particle numbers by sampling them from the simulated molecule numbers at a set of time points (Fig. 1(b)).
2. Generate initial estimates of $f(\mathbf{x})$ by evaluating Wasserstein distances between CA distributions and CME distributions on a grid of CME parameters, and obtain the optimal initial value τ_0 as the minimum distance obtained.
3. Perform a sequential search which, at iteration n, selects a location $\mathbf{x}_n$ at which to query $f(\mathbf{x}_n)$ by maximising the expected improvement acquisition

function. Set $\tau_n = \min\{\tau_{n-1}, f(\mathbf{x}_n)\}$ and recompute the acquisition function by including $f(\mathbf{x}_n)$ among the training points.

4. Once the improvement in objective function becomes smaller than a pre-defined threshold, the algorithm makes a final recommendation $\mathbf{x}_p$, which represents the best estimate for the minimum of $f(\mathbf{x})$.

We use the python BO implementation in scikit-optimize and in parallel we confirm the results using the python package optuna. The BO procedure allows us to associate with one set of CA parameters a parametrisation of an effective CME which optimally matches (in a Wasserstein distance sense) the marginal probabilities of the CA. To extend this to *any* parametrisation of the CA, we once again appeal to smoothness and use CA/ CME pairs on a grid of CA parameters (each obtained via a separate BO procedure) to train a GP regression map to predict effective CME parametrisations also for unseen CA systems. This allows us to avoid expensive CA simulations and to provide effective CMEs for any setting of biochemical and crowding parameters in the CA system.

3   Applications

In this section, we apply our BO-based method to an enzyme system and a gene regulatory network. The code to recreate the results as well as the CA data are available at https://github.com/sb2g14/wasserstein_time_inference.

3.1   Michaelis-Menten Reaction in Crowded Conditions

Here we study the stochastic kinetics of Michaelis-Menten enzyme reaction system

$$E + S \underset{k_{-1}}{\overset{k_1}{\rightleftharpoons}} ES \overset{k_2}{\longrightarrow} E + P, \tag{4}$$

where E, S, ES and P are enzyme, substrate, enzyme-substrate complex and protein, respectively; k_1, k_{-1} and k_2 are bimolecular, reverse and catalytic rates, respectively. Specifically, we consider this reaction in crowded conditions, where the crowders are assumed to be immobile because normally these molecules are large and inert [21,22]. The detailed set of rules for the CA simulations of this system is described in Appendix B.

Similar to [22], the reaction rates may be estimated from the CA simulations directly using the formulae

$$k_1 = \frac{d\gamma/dt}{[E][S]}, \tag{5}$$

$$k_{-1} = \frac{d\gamma/dt + d[S]/dt}{[ES]}, \tag{6}$$

$$k_2 = \frac{d[P]/dt}{[ES]}, \tag{7}$$

where $\gamma(t)$ is the average number of $E + S \rightarrow ES$ reactions that have occurred in the time interval $[0, t]$ divided by the number of grid points and $[X]$ is the concentration of species X, i.e. the average number of particles of species X divided by the number of grid points. Note that averages are here understood to be computed over an ensemble of independent CA simulations. The reactions occur on a 2D square lattice of size 100×100. The initial values of the molecule numbers of each species are distributed uniformly and the boundary conditions are periodic. The latter conditions are used since they typically give rise to small finite-size errors in Monte Carlo simulations [41].

In Fig. 3, we show the calculation of the effective bimolecular rate k_1 using Eq. (5) as a function of time and of the concentration of crowders ϕ in the system (dark grey points). Note that the effective rates of the other two reactions k_{-1} and k_2 (estimated using Eq. (6) and Eq. (7)) do not show any appreciable variation with crowding levels and hence we do not discuss them any further (the values of the estimated rates are in agreement with the probabilities of the associated reactions in the CA). Clearly, crowding induces a bimolecular rate that is monotonically decreasing with time – this is due to the increasing amounts of product (and the decreasing amounts of the substrate) which reduces the rate of encounter of substrate and enzyme molecules. We fit the time-dependent bimolecular rates using the Zipf-Mandelbrot law $k_1 = \frac{k_0}{(t+\tau)^h}$ with parameters k_0, τ and h obtained from the least-squares fit of the data estimated from Eqs. (5) – these are shown are orange lines in Fig. 3. Note that ϕ was limited to the range $0 - 0.4$ since this is the physiological range [22].

We next aim to learn the parameters k_0, τ and h that characterise the effective bimolecular reaction rate using BO. Specifically, we use BO to fit the time-dependent distributions of all species calculated from the CA with those obtained from SSA simulations where the propensity functions in the CME description (Eq. (1)) are:

$$g_1(\vec{n}) = \frac{k_0}{(t+\tau)^h} n_S n_E, \tag{8}$$

$$g_2(\vec{n}) = k_{-1} n_{ES}, \tag{9}$$

$$g_3(\vec{n}) = k_2 n_{ES}, \tag{10}$$

where n_X is the number of molecules of species X. Since some of the propensities have a time-dependent rate coefficient, the SSA simulations cannot be performed using the standard Gillespie algorithm; rather we use the exact Extrande algorithm [6] that takes into account time-dependent reaction rates. The objective function minimized by BO (see Eq. (2)) is given by

$$f = \sum_{i=1}^{N_t} f_i, \quad f_i = \frac{|\gamma_i^{ref} - \gamma_i^{SSA}|}{\gamma_i^{SSA}} + \sum_{j=1}^{N} \frac{\text{WD}(P_i^j, Q_i^j)}{\langle Q_i^j \rangle}, \tag{11}$$

where γ_i^{ref} and γ_i^{SSA} correspond to the sample averaged counter of bimolecular reactions in the system at time interval $[0, t_i]$ in the CA and SSA simulations,

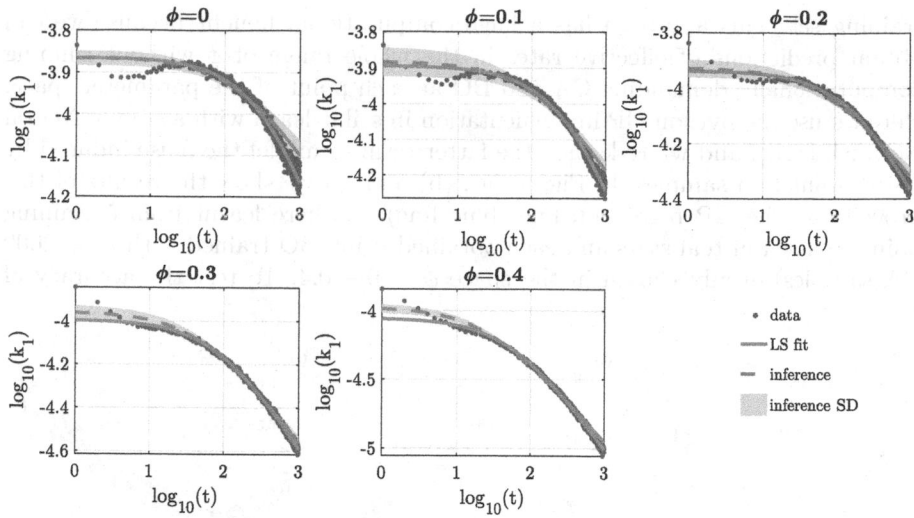

Fig. 3. Variation of the bimolecular reaction rate with time and the concentration of crowders (ϕ). The dark grey data points show the direct calculation of the rates using Eq. (5) where the concentrations and γ are calculated from an ensemble of 2000 trajectories generated using CA simulations. The time-dependent rates are fit to the functions $k_1 = \frac{k_0}{(t+\tau)^h}$ (following a Zipf-Mandelbrot law) using the method of least squares (red lines). They are also estimated using 10 BO runs (green dashed lines shows the average while the grey shaded area shows the standard deviation). The initial values of the molecule numbers are $N_S = 2000$, $N_E = 100$, $N_C = 0$ and $N_P = 0$. A detailed description of the CA simulations can be found in Appendix B. Note that time is in arbitrary units. (Color figure online)

respectively; t_i are discrete-time points that split simulation time into equally sized intervals; N_t is the number of time intervals; P_i^j and Q_i^j are the CA and SSA marginal distributions at time t_i for the number of molecules of species j, respectively; $\langle Q_i^j \rangle$ is the mean of the SSA number distribution Q_i^j; N is the total number of species. The first term in Eq. (11) helps to avoid parameters indistinguishably in the unimolecular reaction rates k_{-1} and k_2.

We repeat the BO-based estimation multiple times leading to a set of Zipf-Mandelbrot law curves – in Fig. 3, dashed green lines show the mean of these functions while the shaded areas show their standard deviation. These are in good agreement with the least square estimates (orange lines) calculated previously. In Table 1 (Appendix C) we show the inferred parameters and the objective function f for two different initial conditions.

To learn the functional dependence of the parameters of the effective bimolecular propensity $g_1(\vec{n})$ defined in Eq. (8) with the crowding level ϕ, we obtain effective parameters (k_0, τ, h) by minimisation of the objective function (11) for a set of training values of ϕ and then we use Gaussian Process (GP) regression to extend our predictions to a whole range of values of ϕ not covered by the training set. This approach has a huge computational benefit because we can obtain predictions of effective rates in the whole range of ϕ without running computationally demanding CA and BO at each point of the parameter space. Here we use the python GP implementation in scikit-learn with a sum of Neural Network kernel and white kernel, the latter term to model the noise induced by finite simulation samples. In Figs. 4 (a), (b) and (c) we show the results of this procedure. The GP regression line (blue line) was here learnt from 5 training points (shown in teal stars and each obtained using BO trained with $n_s = 2000$ CA samples) evenly chosen in the space $\phi = 0 - 0.4$. To test the accuracy of

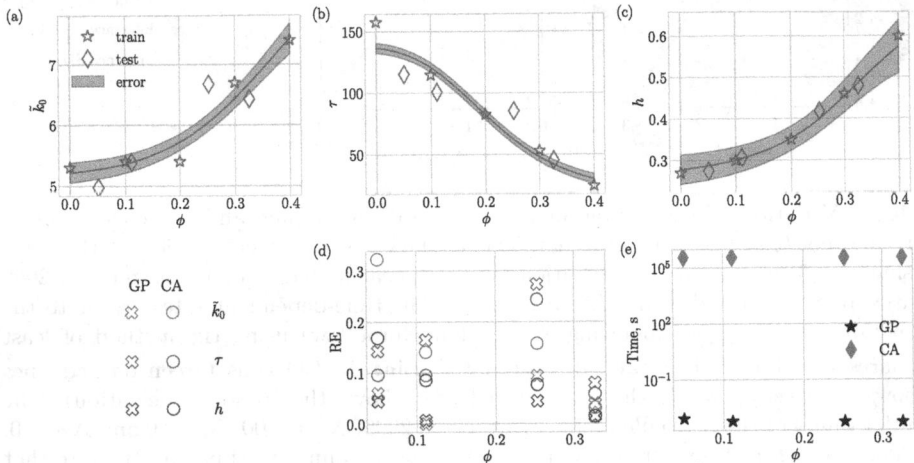

Fig. 4. Learning the effective CME parameters for an enzyme system (4) in crowded conditions. (a), (b) and (c) show the GP regression (blue line) of parameters of the Zipf-Mandelbrot law $\tilde{k}_0 = k_0 \Omega$ (here we rescale k_0 to avoid computational errors), τ and h, respectively; the prediction was built from 5 training points (teal stars) each obtained from BO trained with $n_s = 2000$ CA simulation samples shown in Fig. 3 and Table 1 (Appendix C). The shaded blue area shows the error in GP prediction. A number of test points (teal diamonds) each obtained from BO trained with $n_s = 6000$ CA simulation samples fall on or close to the GP regression line showing its accuracy. (d) Relative testing errors in the effective rate as estimated indirectly by GP regression or directly by BO for 4 different testing points. Error is computed relative to the ground truth rates evaluated by BO from 6000 CA samples. (e) CPU runtimes of 2000 CA samples (teal diamonds) in comparison to the CPU time of drawing a new sample from the function learnt with GP regression (black stars). (Color figure online)

GP regression, we calculated k_0, τ and h for another set of values of ϕ; these testing points are shown as teal diamonds and are close to the GP regression curve calculated from training data.

To further test the accuracy of the GP regression line, we calculated the relative errors between the GP's estimates of $\tilde{k}_0 = k_0 \Omega$, τ and h for 4 test points and a direct prediction from BO using 6000 CA samples at the same points (the ground truth). The results are shown by the open crosses in Fig. 4 (d) – the relative errors are relatively small showing the accuracy of the GP regression. We also compute the relative errors between the ground truth and the $\tilde{k}_0$, τ and h directly predicted from BO using the same number of CA samples as used to train the GP. We see that this relative error (shown by the open circles) is of comparable magnitude to the one obtained earlier for the GP prediction; in other words, GP predictions appear to be of a similar quality to ab initio re-learning of the effective rates from a new batch of CA simulations.

Finally, in Fig. 4 (e) we show the time of calculating a value of parameter k_1 for selected ϕ from GP regression predictions (black stars) in comparison to running 2000 CA samples (red diamonds) calculated on the single core. Note that while we previously found that the errors of GP prediction are comparable to the CA errors, the training time for a new GP sample is more than 6 orders of magnitude smaller. The BO training time for this system is $\sim 10^3$ s, thus, giving a significant computational advantage to our approach compared to evaluating CA samples for a range of ϕ. All the experiments were performed using a single core of Intel® Xeon® 3.5 GHz and 16 GB RAM.

In Fig. 5 we compare distributions drawn from CA simulations of the enzyme reaction system (4) (our ground truth; teal histograms) and the distributions generated with SSA using effective rates calculated using BO (orange outline histograms). We compare the marginal distributions of products P obtained in the enzyme reaction. The left column in Fig. 5 compares distributions sampled at the beginning of the experiment ($t = 40$) while the middle and the right columns compare the distributions in the middle ($t = 480$) and the end ($t = 1000$) of the experiment, respectively. The lower row of figures is for $\phi = 0$, the middle row is for $\phi = 0.2$ (middle row) and the top row is for $\phi = 0.4$. In the corner of each subplot, we show the WD between the marginal distributions (obtained from the CA and SSA+BO) and the value of f_i (as defined by Eq. (11)) for the respective time interval. Curiously, the distribution at $\phi = 0$ and $t = 480$, and the one at $\phi = 0.2$ and $t = 480$ look similar in the plot, but the difference in the WDs is significant at 2.3 vs 10.9. This is due to a systematic discrepancy between the CA and the SSA marginal distributions in the upper subplot that is not easily visible by eye but can be picked by zooming into the subplots. Also, we can see that the quality of the approximation degrades with an increase in ϕ (larger values of WD). This might happen because with an increase in crowding, the dynamics of the system become non-Markovian (time between successive reaction events is not any more exponential).

3.2   Gene Network with Negative Feedback

Next, we study the stochastic kinetics of a gene network with negative feedback in the presence of crowding (a well-mixed, non-crowded version of this system was studied in [42]).

The CA simulations for this system proceed via a set of rules and boundary conditions, akin to those used previously for enzyme kinetics. We consider a fictitious 2D cell defined by a square lattice of points (100×50) with periodic

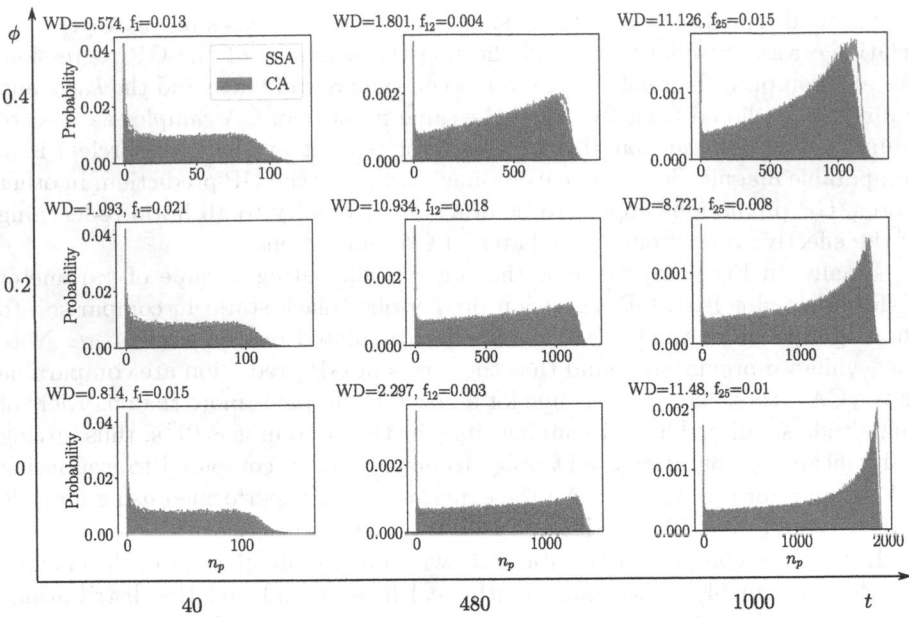

Fig. 5. Comparing the marginal distributions of products of the enzyme reaction generated with the SSA using rates learned with BO (orange lines obtained with $n_s = 500$) and data generated by CA (teal bars obtained with $n_s = 2000$) at $\phi = 0$, $\phi = 0.2$ and $\phi = 0.4$ for three different times $t = 40, 480, 1000$. The WDs between both distributions as well as the value of f_i (as given by Eq. (11)) are shown in the corner of each subplot. Note that f_i varies from f_1 to f_{25} because we divide the period $[0, 1000]$ into $N_t = 25$ subintervals. (Color figure online)

boundary conditions (to reduce finite-size effects). One (immobile) lattice point is the gene which can have one of three states: G (unbound to a protein), GP (bound to a protein) and GP_2 (bound to two proteins). The rest of the lattice points are either empty or occupied by an mRNA (M), a protein (P), a crowder (C), a free degrading enzyme (E) or a protein-enzyme complex (EP). All of these molecules are mobile, i.e. can jump to a neighbouring empty lattice point, except the crowders which are immobile at all times similarly to Sect. 3.1. M is produced when the gene is in state G or GP (transcription); subsequently, M can produce P (translation) or else it is removed from the system (mRNA degradation). P can bind to the enzyme E to form the complex EP which can then decay to E (protein degradation). P can also bind to the gene G to form GP and this can bind to another P to form GP_2. G and GP are assumed to produce M at the same rate but GP_2 is transcriptionally silent – hence this is a negative feedback loop since the gene-product (the protein) represses its own production. In all cases, the initial number of molecules of each species are $n_G = 1$, $n_{GP} = 0$, $n_{GP2} = 0$, $n_M = 0$, $n_P = 0$, $n_E = 100$, $n_{EP} = 0$. In Fig. 6 we show a cartoon of this system.

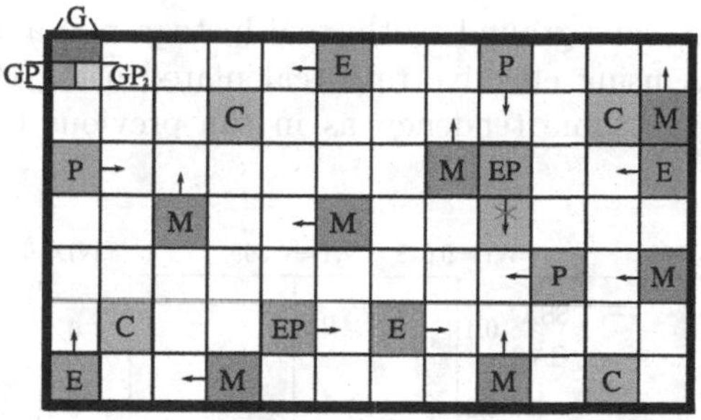

Fig. 6. Cartoon illustrating the setup for the CA simulations of a genetic negative feedback loop inside a cell. The colours represent the following: G (dark orange), GP (orange) and GP_2 (light orange) are different states of the gene with different numbers of protein-bound 0, 1, 2, respectively; protein P (green), mRNA M (teal), enzyme E (purple), complex EP (blue) and crowders C (grey). White represents empty space. All molecules can move except crowders and the gene which are immobile. The arrows show possible directions of movement of the molecules – movement is only possible if a neighbouring space is empty thus enforcing volume exclusion. For the possible reactions between molecules see the main text. (Color figure online)

The procedure leading to an effective CME, approximating the spatial CA dynamics, is the same as before. We use BO to fit CA generated time-dependent marginal distributions of all species to those generated by an SSA. In this case the reactions modelled by the SSA are given by

$$G \xrightarrow{k_0} G + M, \quad M \xrightarrow{k_s} P + M, \quad P + E \underset{k_{-3}}{\overset{k_3}{\rightleftharpoons}} EP \xrightarrow{k_4} E, \quad M \xrightarrow{k_{dM}} \varnothing,$$

$$P + G \underset{k_{-1}}{\overset{k_1}{\rightleftharpoons}} GP, \quad P + GP \underset{k_{-2}}{\overset{k_2}{\rightleftharpoons}} GP_2, \quad GP \xrightarrow{k_0} GP + M. \tag{12}$$

Note that the objective function minimized by BO is same as Eq. (11) but without the (first) γ dependent term; to lighten the computational burden, we choose to infer only the three bimolecular rates (k_1, k_2, k_3) and assume that the unimolecular reaction rates are fixed to the ones set in the CA simulations. This is a reasonable assumption since crowding tends to primarily affect bimolecular rates. Note that while the inferred bimolecular rates were time-dependent in the previous enzyme example, for the feedback loop they are found to quickly converge to a time-independent non-zero value and hence we do not need to assume a Zipf-Mandelbrot law for the rates – this is because for the feedback loop, in steady-state all species numbers fluctuate around a non-zero value.

The results of the parameter inference averaged over 5 BO runs are shown in Table 2 in Appendix C (here we also show the probabilities of the individual reactions in the CA). The standard deviation in the inferred parameters in most of the cases is smaller than 20% of the mean, therefore, we can conclude that the parameters are inferred fairly well. In Fig. 7 we compare distributions drawn from CA simulations (our ground truth; teal histograms) and the distributions generated with SSA using effective rates calculated using BO (orange outline histograms). Note the same tendency as in the previous example, where the

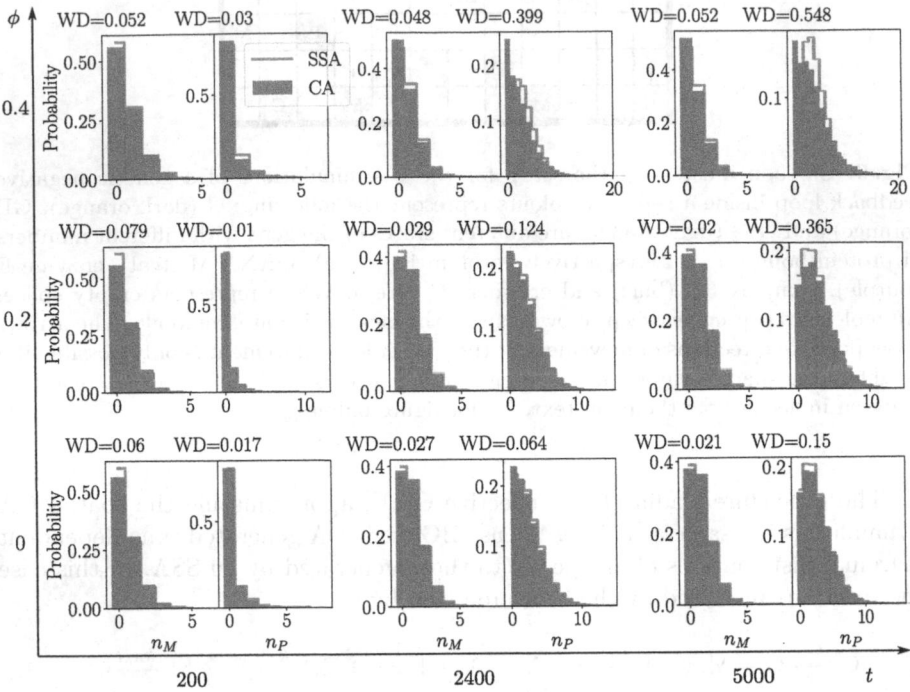

Fig. 7. Comparing the marginal distributions of mRNA and protein in a genetic feedback loop generated with the SSA using rates learned with BO (orange lines obtained with $n_s = 500$) and data generated by CA (teal bars obtained with $n_s = 2000$) at $\phi = 0$, $\phi = 0.2$ and $\phi = 0.4$ for three different times $t = 200, 2400, 5000$. The bimolecular reaction rates are inferred while the unimolecular rates are fixed at the same values as the CA (see Table 2 in Appendix C). The WDs between the distributions are shown on the top of each subplot. (Color figure online)

WDs increase with the level of crowding probably because of the breakdown of the Markovian assumption behind the CME. Interestingly, as shown in Fig. 8, the CME starts to fail as a good approximation of the CA for increased mRNA production rate k_0 even in the case where there is no crowding. Presumably, this happens because of increased mRNA production close to the gene which causes a large degree of volume exclusion due to self-crowding of mRNA molecules (the CA sample average of the fraction of occupied volume in steady-state is quite low at 0.023).

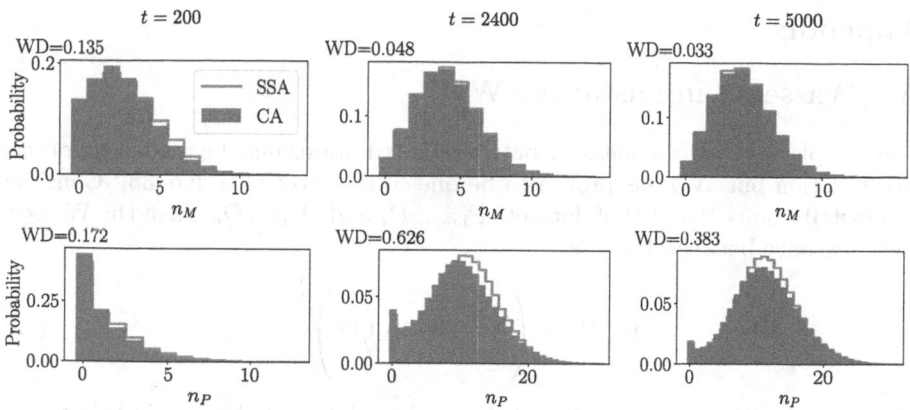

Fig. 8. Comparing the marginal distributions of mRNA and protein in a genetic feedback loop generated with the SSA using rates learned with BO (orange lines obtained with $n_s = 500$) and data generated by CA (teal bars obtained with $n_s = 2000$) at $\phi = 0$ for three different times $t = 200, 2400, 5000$. The inferred bimolecular reaction rates are $k_1 = 0.0018$, $k_2 = 0.0025$, $k_3 = 0.0002$, while the unimolecular rates are fixed; the latter have the same values as in Table 2) (Appendix C) except that $k_0 = 0.05$. (Color figure online)

4   Conclusions

As advances in measurement technology probe deeper into the spatial stochasticity of biochemical reactions, novel computational tools are needed to formulate quantitative theories of cellular function. Existing frameworks for modelling spatial stochastic effects such as Brownian Dynamics simulations and Cellular Automata (CA) inescapably suffer from a high computational load. This is further aggravated by the frequent need to explore a range of parametrisations for the models as biochemical parameters are seldom accessible, creating the need for even larger-scale simulation studies. In this paper, we propose an automatic approach to generate simpler effective CME models which can recapitulate the statistical behaviour of spatially crowded stochastic systems. Given a (limited) number of expensive spatial simulation runs, our approach can provide a fast CME-based simulator for *any* parametrisation of the spatial system which optimally matches its statistical properties. Our approach focussed on CA spatial systems, but in principle, the same procedure can be deployed for any spatial simulator.

As well as providing an efficient simulation tool, our approach opens potential new directions. As a first application, its computational efficiency would easily allow the analysis of 3D systems, as the scaling of the CME is clearly independent of the dimension of the space in which the reactions happen. Secondly, the availability of efficient simulation tools opens the way to the use of simulator-based inference tools to estimate the parameters of spatial crowded systems from data [43], therefore enabling a formal statistical link between computational methodology and experimental technology.

Appendix

A   Wasserstein Distance (WD)

In principle any distance measure between distributions may be used as an objective function but WD has proven to be one of the most effective [35]. Consider two distributions P and Q of datasets $P_1, ..., P_n$ and $Q_1, ..., Q_n$, then the Wasserstein distance between them is

$$\mathrm{WD}^{(p)} = \left(\sum_{i=1}^{n} ||P_i - Q_i||^p \right)^{\frac{1}{p}},\tag{13}$$

where $p \geq 1$ is the dimensionality of the original data distribution. In this paper, we always use $p = 1$.

B   CA Rules Modelling Enzyme Kinetics in Crowded Conditions

At the beginning of each simulation, the counter γ is reset to zero, and E, S and crowder molecules are randomly placed on the square lattice. At each time step, a "subject" molecule is randomly chosen and it is moved or participates in a reaction according to the following rules:

1. Choose randomly one of 4 nearest neighbouring "destination" sites.
2. If the destination site is empty and the "subject" molecule is E, S or P then move the molecule (simulates diffusion).
3. Otherwise:
   (a) If the "subject" molecule is E or S and the molecule occupying the "destination" site ("target" molecule) is, respectively, S or E then generate a uniform random number between 0 and 1. If this is lower than the reaction probability $P_1 = 1$, replace the "target" molecule with ES, remove "subject" molecule and increase the counter γ by one. This step models the reaction $E + S \rightarrow ES$.
   (b) If the "subject" molecule is ES, check if there are any molecules placed on the neighbouring sites. If at least one nearest neighbour site is empty, randomly choose a vacant "destination" site and generate a uniform random number between 0 and 1.
      i If the generated number is less than $P_{-1} = 0.02$, place E on "subject" site and S on "destination" site $(ES \rightarrow E + S)$.
      ii If the number is greater than P_{-1} but lower than $P_{-1} + P_2$, where $P_2 = 0.04$, then place E on the "subject" site and P on the "destination" site $(ES \rightarrow E + P)$.
      iii If the number is greater than $P_{-1} + P_2$ move ES to the "destination" site (only diffusion occurs).
4. Otherwise, if the "destination" site is occupied, reject the move (simulates volume exclusion effects).

C Supplementary Tables

Table 1. Parameters k_0, τ and h are estimated from the mean prediction of k_1 (as a function of time) over 6 or 10 BO runs for different ϕ in the CA simulations; k_{-1}, k_2 and f are mean predictions. The estimates are for two different initial conditions; the top table is for the data shown in Fig. 3.

Parameter	$\phi = 0$	$\phi = 0.1$	$\phi = 0.2$	$\phi = 0.3$	$\phi = 0.4$
	BO $t_{max} = 1000$, $N_S = 2000$, $N_E = 100$, $n_s = 2000$, estimated from the mean over 10 runs				
k_0	$5.3 \cdot 10^{-4}$	$5.4 \cdot 10^{-4}$	$5.4 \cdot 10^{-4}$	$6.7 \cdot 10^{-4}$	$7.4 \cdot 10^{-4}$
τ	157.6	114.5	82.5	53.0	24.7
h	0.27	0.3	0.35	0.46	0.6
k_{-1}	$1.9 \cdot 10^{-2}$	$2.0 \cdot 10^{-2}$	$1.9 \cdot 10^{-2}$	$1.9 \cdot 10^{-2}$	$1.9 \cdot 10^{-2}$
k_2	$4.0 \cdot 10^{-2}$	$4.0 \cdot 10^{-2}$	$3.9 \cdot 10^{-2}$	$3.9 \cdot 10^{-2}$	$3.7 \cdot 10^{-2}$
f	1.6	1.5	1.4	1.4	1.6
	BO $t_{max} = 4000$, $N_S = 3000$, $N_E = 20$, $n_s = 4000$, estimated from the mean over 10 runs				
k_0	$8.2 \cdot 10^{-4}$	$8.6 \cdot 10^{-4}$	$5.5 \cdot 10^{-4}$	$1.0 \cdot 10^{-3}$	$1.0 \cdot 10^{-3}$
τ	585.9	551.2	236.4	248.7	62.0
h	0.27	0.3	0.29	0.48	0.67
k_{-1}	$2.0 \cdot 10^{-2}$	$2.0 \cdot 10^{-2}$	$1.9 \cdot 10^{-2}$	$1.8 \cdot 10^{-2}$	$1.7 \cdot 10^{-2}$
k_2	$4.0 \cdot 10^{-2}$	$3.9 \cdot 10^{-2}$	$3.8 \cdot 10^{-2}$	$3.7 \cdot 10^{-2}$	$3.4 \cdot 10^{-2}$
f	1.4	1.4	1.8	1.7	1.6

Table 2. Parameters k_1, k_2 and k_3 are estimated from the mean predictions and standard deviation over 5 BO runs for different ϕ in the CA simulations, unimolecular rate are fixed at $k_0 = 0.01$, $k_s = 0.01$, $k_{-3} = 0.05$, $k_4 = 0.01$, $k_{dM} = 0.01$, $k_{-1} = 0.02$, $k_{-2} = 0.05$. Note that the probability of each first-order reaction in the CA is the same as the corresponding rate constant above; the probability of each bimolecular reaction is set to 1 for simplicity.

Parameter	$\phi = 0$	$\phi = 0.1$	$\phi = 0.2$	$\phi = 0.3$	$\phi = 0.4$
	BO $t_{max} = 5000$, $n_s = 2000$, estimated mean and standard deviation over 5 runs				
$k_1 \cdot 10^{-4}$	17.7 ± 3.0	23.3 ± 3.4	31.8 ± 3.1	44.6 ± 5.1	66.8 ± 3.8
$k_2 \cdot 10^{-4}$	15.9 ± 1.5	22.6 ± 3.8	32.3 ± 3.2	50.4 ± 5.6	97.8 ± 3.8
$k_3 \cdot 10^{-5}$	21.3 ± 2.4	21.5 ± 1.6	20.6 ± 0.7	16.4 ± 0.6	11.5 ± 0.4

References

1. Elowitz, M.B., Levine, A.J., Siggia, E.D., Swain, P.S.: Stochastic gene expression in a single cell. Science **297**(5584), 1183–1186 (2002)
2. Darzacq, X., et al.: Imaging transcription in living cells. Annu. Rev. Biophys. **38**, 173–196 (2009)
3. Shah, S., et al.: Dynamics and spatial genomics of the nascent transcriptome by intron seqFISH. Cell **174**(2), 363–376 (2018)
4. Larsson, A.J., et al.: Genomic encoding of transcriptional burst kinetics. Nature **565**(7738), 251–254 (2019)
5. Gillespie, D.T.: Exact stochastic simulation of coupled chemical reactions. J. Phys. Chem. **81**(25), 2340–2361 (1977)
6. Voliotis, M., Thomas, P., Grima, R., Bowsher, C.G.: Stochastic simulation of biomolecular networks in dynamic environments. PLoS Comput. Biol. **12**(6), e1004923 (2016)
7. Gillespie, D.T.: The chemical Langevin equation. J. Chem. Phys. **113**(1), 297–306 (2000)
8. Schnoerr, D., Sanguinetti, G., Grima, R.: Approximation and inference methods for stochastic biochemical kinetics—a tutorial review. J. Phys. A: Math. Theor. **50**(9), 093001 (2017)
9. Suter, D.M., Molina, N., Gatfield, D., Schneider, K., Schibler, U., Naef, F.: Mammalian genes are transcribed with widely different bursting kinetics. Science **332**(6028), 472–474 (2011)
10. Skinner, S.O., Xu, H., Nagarkar-Jaiswal, S., Freire, P.R., Zwaka, T.P., Golding, I.: Single-cell analysis of transcription kinetics across the cell cycle. Elife **5**, e12175 (2016)
11. Van Kampen, N.: Stochastic Processes in Physics and Chemistry, 3rd edn. North-Holland Personal Library. Elsevier, Amsterdam (2007)
12. Gillespie, D.T.: A rigorous derivation of the chemical master equation. Physica A **188**(1–3), 404–425 (1992)
13. Gillespie, D.T.: A diffusional bimolecular propensity function. J. Chem. Phys. **131**(16), 164109 (2009)
14. Van den Berg, B., Wain, R., Dobson, C.M., Ellis, R.J.: Macromolecular crowding perturbs protein refolding kinetics: implications for folding inside the cell. EMBO J. **19**(15), 3870–3875 (2000)
15. Zhou, H.X., Rivas, G., Minton, A.P.: Macromolecular crowding and confinement: biochemical, biophysical, and potential physiological consequences. Annu. Rev. Biophys. **37**, 375–397 (2008)
16. Tan, C., Saurabh, S., Bruchez, M.P., Schwartz, R., LeDuc, P.: Molecular crowding shapes gene expression in synthetic cellular nanosystems. Nat. Nanotechnol. **8**(8), 602–608 (2013)
17. Mourão, M.A., Hakim, J.B., Schnell, S.: Connecting the dots: the effects of macromolecular crowding on cell physiology. Biophys. J . **107**(12), 2761–2766 (2014)
18. Grima, R.: Intrinsic biochemical noise in crowded intracellular conditions. J. Chem. Phys. **132**(18), 05B604 (2010)
19. Cianci, C., Smith, S., Grima, R.: Molecular finite-size effects in stochastic models of equilibrium chemical systems. J. Chem. Phys. **144**(8), 084101 (2016)
20. Gillespie, D.T., Lampoudi, S., Petzold, L.R.: Effect of reactant size on discrete stochastic chemical kinetics. J. Chem. Phys. **126**(3), 034302 (2007)

21. Berry, H.: Monte Carlo simulations of enzyme reactions in two dimensions: fractal kinetics and spatial segregation. Biophys. J . **83**(4), 1891–1901 (2002)
22. Schnell, S., Turner, T.E.: Reaction kinetics in intracellular environments with macromolecular crowding: simulations and rate laws. Prog. Biophys. Mol. Biol. **85**(2), 235–260 (2004)
23. Grima, R., Schnell, S.: A systematic investigation of the rate laws valid in intracellular environments. Biophys. Chem. **124**(1), 1–10 (2006)
24. Smith, S., Grima, R.: Fast simulation of Brownian dynamics in a crowded environment. J. Chem. Phys. **146**(2), 024105 (2017)
25. Kim, J.S., Yethiraj, A.: Crowding effects on association reactions at membranes. Biophys. J . **98**(6), 951–958 (2010)
26. Chew, W.X., Kaizu, K., Watabe, M., Muniandy, S.V., Takahashi, K., Arjunan, S.N.: Reaction-diffusion kinetics on lattice at the microscopic scale. Phys. Rev. E **98**(3), 032418 (2018)
27. Andrews, S.S.: Smoldyn: particle-based simulation with rule-based modeling, improved molecular interaction and a library interface. Bioinformatics **33**(5), 710–717 (2017)
28. Deutsch, A., Dormann, S.: Mathematical Modeling of Biological Pattern Formation. Springer, Heidelberg (2005)
29. Wolf-Gladrow, D.A.: Lattice-Gas Cellular Automata and Lattice Boltzmann Models: An Introduction. Springer, Heidelberg (2004)
30. Wieczorek, G., Zielenkiewicz, P.: Influence of macromolecular crowding on protein-protein association rates—a Brownian dynamics study. Biophys. J . **95**(11), 5030–5036 (2008)
31. Gardiner, C.: Stochastic Methods: A Handbook for the Natural and Social Sciences. Springer Series in Synergetics, 4th edn. Springer, Heidelberg (2009)
32. Baras, F., Mansour, M.M.: Reaction-diffusion master equation: a comparison with microscopic simulations. Phys. Rev. E **54**(6), 6139 (1996)
33. Gillespie, D.T.: Stochastic simulation of chemical kinetics. Annu. Rev. Phys. Chem. **58**, 35–55 (2007)
34. Loskot, P., Atitey, K., Mihaylova, L.: Comprehensive review of models and methods for inferences in bio-chemical reaction networks. Front. Genet. **10**, 549 (2019)
35. Öcal, K., Grima, R., Sanguinetti, G.: Parameter estimation for biochemical reaction networks using Wasserstein distances. J. Phys. A: Math. Theor. **53**(3), 034002 (2019)
36. Shahriari, B., Swersky, K., Wang, Z., Adams, R.P., de Freitas, N.: Taking the human out of the loop: a review of Bayesian optimization. Proc. IEEE **104**(1), 148–175 (2016)
37. Brochu, E., Cora, V.M., de Freitas, N.: A tutorial on Bayesian optimization of expensive cost functions, with application to active user modeling and hierarchical reinforcement learning. arXiv:1012.2599 [cs] (2010)
38. Villani, C.: Optimal Transport: Old and New. Grundlehren Der Mathematischen Wissenschaften. Springer, Heidelberg (2009)
39. Rasmussen, C.E., Williams, C.K.I.: Gaussian Processes for Machine Learning. Adaptive Computation and Machine Learning. MIT Press, Cambridge (2006)
40. Vazquez, E., Bect, J.: Convergence properties of the expected improvement algorithm with fixed mean and covariance functions. J. Stat. Plann. Inference **140**(11), 3088–3095 (2010)
41. Allen, M.P., Tildesley, D.J.: Computer Simulation of Liquids. Oxford University Press (2017)

42. Thomas, P., Straube, A.V., Grima, R.: The slow-scale linear noise approxima-
    tion: an accurate, reduced stochastic description of biochemical networks under
    timescale separation conditions. BMC Syst. Biol. **6**(1), 39 (2012)
43. Cranmer, K., Brehmer, J., Louppe, G.: The frontier of simulation-based inference.
    Proc. Natl. Acad. Sci. **117**(48), 30055–30062 (2020)

Probabilistic Multivariate Early Warning Signals

Ville Laitinen$^{(\boxtimes)}$ (ID) and Leo Lahti (ID)

Department of Computing, University of Turku, Turku, Finland
{ville.laitinen,leo.lahti}@utu.fi

Abstract. A broad range of natural and social systems from human microbiome to financial markets can go through critical transitions, where the system suddenly collapses to another stable configuration. Anticipating such transition early and accurately can facilitate controlled system manipulation and mitigation of undesired outcomes. Generic data-driven indicators, such as autocorrelation and variance, have been shown to increase in the vicinity of an approaching tipping point, and statistical early warning signals have been reported across a range of systems. In practice, obtaining reliable predictions has proven to challenging, as the available methods deal with simplified one-dimensional representations of complex systems, and rely on the availability of large amounts of data. Here, we demonstrate that a probabilistic data aggregation strategy can provide new ways to improve early warning detection by more efficiently utilizing the available information from multivariate time series. In particular, we consider a probabilistic variant of a vector autoregression model as a novel early warning indicator and argue that it has certain advantages in model regularization, treatment of uncertainties, and parameter interpretation. We evaluate the performance against alternatives in a simulation benchmark and show improved sensitivity in warning signal detection in a common ecological model encompassing multiple interacting species.

Keywords: Early warning signals · Probabilistic programming · Complex systems

1 Introduction

The ability to anticipate and manage change plays a critical role in diverse domains from biomedicine to ecology, economics, or climate change [32]. Natural and social systems are complex arrangements of units and their interactions at various scales. Despite their complexity and size, such systems often exhibit remarkable stability where perturbations have only minor, and often temporary and reversible effects on the system. However, when the conditions are stretched far enough, a system may pass a critical threshold, a tipping point, leading to a rapid and potentially irreversible reorganization. These *critical transitions* can be observed across many different scenarios [23], including ecosystems [21,34], epidemics [27], and climate

© The Author(s), under exclusive license to Springer Nature Switzerland AG 2022
I. Petre and A. Păun (Eds.): CMSB 2022, LNBI 13447, pp. 259–274, 2022.
https://doi.org/10.1007/978-3-031-15034-0_13

[24]. As large transitions may have far-reaching consequences, the ability to anticipate them can provide valuable tools to manage change.

Generic early warning signals (EWS) have been introduced to detect signs that could alarm us about approaching tipping points, bifurcation points in the parameter space of a complex system after which transition to an alternative regime become inevitable [9,33]. A major challenge is that the sequence of events leading to critical transitions can be subtle and gradual, with little or no apparent changes in the observable system state [16]. However, the underlying system dynamics may change in ways that can be observed and quantified. For instance, measurable aspects of system resilience tend to decrease as a tipping point is approaching [33]. This is associated with *critical slowing down*, which can be indicated by certain statistical properties such as autocorrelation and variance. Increasing lag-1 autocorrelation and variance are some of the most robust and widely utilized EWS [8,12,13]. Autocorrelation tends to increase with slowing down, because the systemic rate of change decreases, and as a result the states at over short time intervals tend to become more and more alike. Similarly, changes in variance are arising as the lowering mean-reversion rate allows the state variable to fluctuate more freely. An important property of statistical early warning indicators is their generality. They provide data-driven quantification that can be informative even when our understanding of the data generating processes is limited. This facilitates the detection of early warnings even when accurate mechanistic modeling is infeasible due to the complexity of the phenomena and limitations in data collection, and makes the generic EWS indicators applicable across a broad range of different systems [23,32].

Despite the recent advances, the generic EWS often rely on the availability of relatively long and dense time series and manual parameter tuning. The ability to utilize the available data more efficiently and detect EWS from more limited time series would be important in many application fields such as ecology and human biomedical studies where the sample sizes can be remarkably low due to ethical, financial, or other constraints. Quantification of uncertainty is another key aspect in EWS analysis. Data is always limited, and may come with uneven observation times, measurement noise, or possible biases. The ability to quantify and control uncertainty is particularly relevant with limited sample sizes. The probabilistic framework provides tools to incorporate uncertainty and prior information into the models [14], and could lead to a more sensitive EWS detection. We recently demonstrated this by introducing a univariate probabilistic method for EWS detection [22], showing improvements in automated model selection and increased sensitivity in EWS detection performance. Here, we extend this earlier work into a multivariate context.

The univariate representation provides convenient, intuitively appealing, and robust ways to summarize changes in complex systems. Yet, the reliance on one-dimensional summaries may neglect important information that can be used to enhance the EWS detection. An enhanced use of multivariate observations in EWS design can provide improved sensitivity especially in shorter multivariate time series, where the information content of the data is more limited.

The advantages of probabilistic methods in treating uncertainties, and the potential for improving EWS detection with data aggregation strategies motivated us to investigate the possibility of extending our earlier work on probabilistic EWS into the multivariate domain.

Hence, we are in this work designing and investigating a novel probabilistic EWS indicator for multivariate systems. More specifically, we formulated and implemented a probabilistic variant of the time-varying vector autoregressive-1 model. A non-probabilistic version of a similar model was recently studied in [18]. The probabilistic version allows alternative ways to pool information across the multivariate time series and deal with uncertainties in the modeling. Moreover, the proposed method supports automated parameter inference, circumventing the need to manually select model parameters, such as sliding window size, which have posed problems in many EWS methods [12]. Besides these theoretically appealing properties, simulations based on an ecological model that has been commonly used in the EWS context demonstrate possible benefits against the currently available alternatives.

The work is structured as follows. In *Methods* we describe the novel approach along with a set of previously studied EWS indicators. We then compare these indicators in a simulation study in *Results* and, finally, conclude in *Discussion* with suggestions for further extension.

2 Methods

In this section, we provide a short overview of the currently available, related methods based on a recent review [40]. We formulate the probabilistic variant, tvPVAR(1), and describe the simulation model that is used to generate data for the experiments.

2.1 Autocorrelation Based EWS

Let us start by summarizing relevant methodology based a recent comparison between currently available indicators for detecting EWS in multivariate data [40]. In the present work we focus on autocorrelation-based indicators since these have shown robustness compared to the alternatives [13, 40], and can be naturally extended into the probabilistic framework that we explore in this study. More specifically, the methods detecting changes in lag-1 autocorrelation are based on the standard autoregressive-1 process, AR(1), defined by the recursion.

$$X_{t+1} = \phi X_t + \sigma \epsilon_t \tag{1}$$

where X_t is the centered (zero-mean) state variable at time t, ϕ the autoregressive parameter, ϵ_t a zero-mean, unit variance Gaussian random variable scaled with σ. The main interest here lies in the autoregressive parameter ϕ, which directly measures the lag-1 autocorrelation of the system.

Many variants and extensions of the AR(1) process have been studied in univariate context [12], and applications in multivariate data are also possible.

For instance, *maximum autocorrelation* (ac/max) [10] is based on fitting the AR(1) model separately to each node, or feature, of the system and then selecting the one with the highest autocorrelation as a proxy for the entire system. Other options include *average autocorrelation* (ac/mean) across the features, or *degenerate fingerprinting* which measures autocorrelation along the first principal component of the multivariate data [17]. Min/Max autocorrelation factors analysis (MAF) [39] is another method based on dimension reduction that aims to identify the subspace with the highest autocorrelation in a multidimensional system. This algorithm generates a set of vectors (MAFs) to project the multidimensional data onto a subspace where autocorrelation is maximized. Eigenvalues of the MAF subspace quantify autocorrelation in the respective directions, with lower values indicating higher autocorrelation. In addition to *MAF eigenvalues* (eigen/MAF), we used autocorrelation (ac/MAF) and variance (var/MAF) projected onto the 1st MAF as indicators. For a more detailed description of the these methods, see [39]. We did omit some of the methods considered in [40], such as information dissipation length [29] and time [28], since they require larger amounts of data, and our interest lies mainly in practically motivated situations where the sample sizes are modest.

The EWS detection based on these previously suggested indicators was carried out following standard procedures [12]. We estimated the early warning indicators in sliding windows along the time series, resulting in a trajectory of the indicator, which is the autoregressive parameter ϕ in our case. Except where otherwise noted, we set the sliding window to 50% of length of the time series, which is a common default choice in the EWS literature. In order to remove the effect of non-stationary trends in the data that could lead to spurious conclusions [12], we used Gaussian detrending (R function *stats::smooth*) as a preprocessing step before quantifying the indicator. We used a bandwidth of 10% of the total time series length, except where otherwise noted, which we chose based on visual assessment; the bandwidth length was chosen so that it removes long-term mean level variations unrelated to the short term correlation structure while aiming to avoid overfitting to short-term variations.

We then measured the strength of each estimated EWS by computing Kendall's rank correlation τ between the estimated autocorrelation trajectory and time. The rank correlation receives values in $[-1, 1]$, with $\tau = 1$ indicating a monotonously increasing trajectory, and strong a EWS, while $\tau \approx 0$ implies a negative finding. The rank correlation is defined as $\tau = (N_{\text{concordant pairs}} - N_{\text{disconcordant pairs}})/N_{\text{all pairs}}$, where N refers to the number of elements in the subscript set, and a pair $(t_i, \phi_i), (t_j, \phi_j)$ is said to be concordant if $t_i \leq t_j$ implies $\phi_i \leq \phi_j$ and disconcordant otherwise.

For hypothesis testing on these EWS indicators, we utilized the so-called surrogate data analysis methods [12]. This technique generates an approximate sampling distribution for Kendall's τ, which is then compared to the actual point estimate. The sampling distribution represents results that would be recovered under the null hypothesis that the indicator trend has arisen simply by change. We generated a collection of time series from the simulation model presented below in the Subsect. 2.3. We used constant parameters that produce data where the conditions remain constant and any estimated parameter trajectory is expected to have no correlation with time ($\tau = 0$). We generated 500 replicates of surrogate data for each experimental condition, and then estimated the EWS indicators and Kendall's rank correlations for these surrogate data sets, yielding approximate sampling distributions under the null hypothesis. P-values for a positive trend value were then computed as the proportion of the sampling distribution that exceeded (or were identical to) the actual point estimate.

2.2 The Probabilistic Time-Varying Vector Autoregressive-1 Process

We recently studied a probabilistic time-varying AR(1) process for detecting autocorrelation changes in univariate systems [22]. Here we investigate a multivariate extension of this model, the time-varying probabilistic vector AR(1) model, tvPVAR(1); for simplicity, we refer to this method as *ac/pooled* in the later comparisons. A non-probabilistic state space variant of this model was previously studied in [18]. Compared to the standard AR(1) process in Eq. 1, the time-varying model allows time-dependent variation in the model parameters.

The tvPVAR(1) model is defined as the recursion

$$X_{t+1} = \Phi_t \cdot X_t + \epsilon_t \tag{2}$$

where X_t is the D-dimensional centered state vector at time t, Φ_t the autoregressive matrix, ϵ_t the multivariate Gaussian random variable with covariance matrix Σ. The degrees of freedom grow rapidly as a function of dimensionality, which makes parameter estimation challenging especially when the sample size is low compared to the dimensionality of the data. We are hence making certain simplifying assumptions. First, we assume that Σ is diagonal and constant over time. Second, we assume that $\Phi_t = \phi_t I$, where ϕ_t is a real number for all t and I is the identity matrix. The latter assumption amounts to parameter pooling (whence the name ac/pooled), which means that a single parameter (ϕ_t) represents several units. Intuitively, this provides a measure for the average systemic autocorrelation.

The probabilistic formulation requires us to define the likelihood of the data, and the priors for the model parameters. Likelihood for the data X_t, $t = 1, \ldots, T$ is given by

$$\mathcal{L}(\Phi_t, \Sigma | X_t) = \prod_{t=1}^{T-1} \mathrm{MVN}(X_{t+1} | \phi_t I \cdot X_t, \Sigma), \tag{3}$$

where MVN refers to the multivariate normal distribution.

Regarding the prior distribution, we use a Gaussian process (GP) prior for ϕ_t. A Gaussian process $\mathcal{GP}(M, K)$ is formally defined as a collection of random variables where each finite set of these variables is multivariate normally distributed with mean M and covariance K [30]. GPs are a standard choice for Bayesian nonparametric regression, and incorporating them in the model forms a key difference compared to the autocorrelation metrics presented in Subsect. 2.1.

More specifically, GPs, as we utilize them, provide a means to regularize the posterior of ϕ_t towards areas of the parameter space that correspond to behaviour likely to be encountered in real systems, such as differentiability.

We utilize the Matèrn-3/2 covariance function that models the covariance between two random variables X_i and X_j as $k_{3/2}(\rho, \alpha) = \alpha^2 \left(1 + \frac{\sqrt{3}r}{\rho}\right) \exp(-\frac{\sqrt{3}r}{\rho})$, where α^2 is the process variance, ρ the length scale and $r = |X_i - X_j|$ [30,37]. In general, the Matèrn class is larger set of covariance functions characterized by a parameter ν which determines the level of differentiability of the output functions. We set $\nu = 3/2$, which restricts the posterior of ϕ_t to continuous and differentiable functions. This is a reasonable condition that allows flexibility in the model while avoiding overfitting to occassional large deviations. The length scale parameter ρ controls the dependence over time, while the process variance α^2 controls the average distance from the mean M. We set $M = 0$ and $\alpha = 1$, which restrains a majority of the prior values between -1 and 1. This is a justified choice as autoregressive-1 models are stationary if and only if the autoregressive parameter is within this interval. Length scale ρ was set, unless otherwise noted, to the length of the time series. We used the Cholesky factored parameterization of GPs for posterior sampling [20]. This models the process as a latent vector η which is mapped to the output space as $\phi_t = L\eta$ where L is the lower triangular matrix with positive diagonal from the Cholesky decomposition $k_{3/2} = LL^T$.

The fitting procedure is illustrated in Fig. 1. Hypothesis testing was carried out by first computing Kendall's τ for each posterior sample for ϕ_t. This provides a posterior distribution for τ, and the mass of this distribution on the positive side of the real line reflects the posterior probability of an increasing autocorrelation in the time series. The "Bayesian P-value" [14] can then be computed as the proportion on the negative side, facilitating comparison with non-probabilistic methods that generate frequentist P-values. While the Bayesian posterior proportion and frequentist P-value follow non-analogous logics and interpretations, they are comparable as metrics for statistical evidence, and below we use them interchangeably for a lack of choice. We set the EWS detection level at $P = 0.1$ and compare the methods in terms of the standard true positive rate (TPR) and true negative rate (TNR).

We implemented the tvPVAR(1) model in the probabilistic programming language Stan [36], utilizing the R interface RStan, and used its No-U-Turn variant of the Hamiltonian Monte Carlo algorithm with 2 chains, both with 2000 iterations for a given fit to sample the posterior. Sampling convergence was assessed with the $\hat{R}$ statistic which remained below the recommended limit 1.1 [15]. In addition, we encountered no divergent transitions indicating that the algorithm had converged and produced reliable estimates.

2.3   Simulation Model

We evaluate the performance of the EWS detection methods based on simulations from a well-studied ecological model [25]. The model characterizes systems with competition and mutualism, such as plant-pollinator interactions.

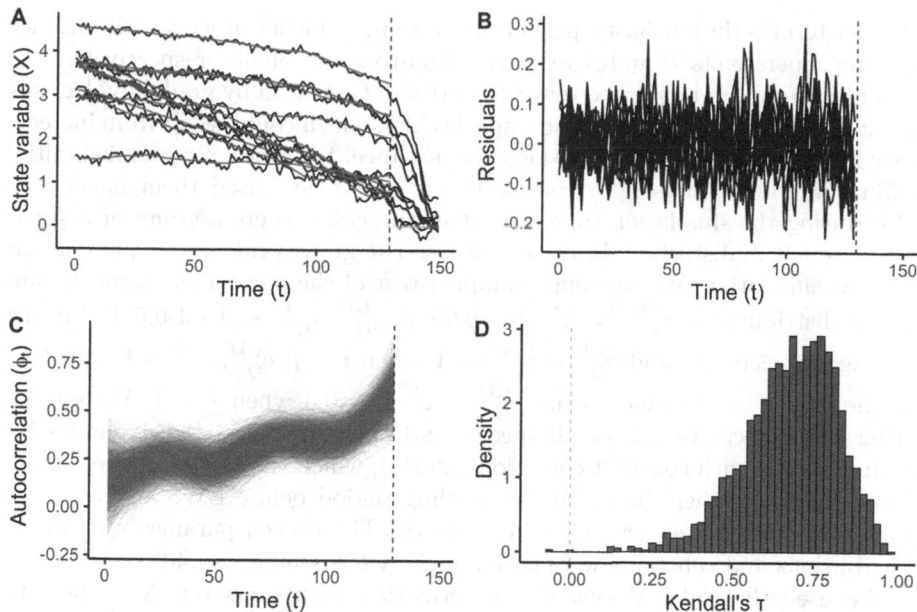

Fig. 1. Early warning signal detection with the proposed ac/pooled indicator (tvP-VAR(1) model) in simulated data. **A** The observed system state $X_i, i = 1, \ldots, N$ as a function of time (black) with estimated time series trends (red). The detrending is based on Gaussian kernel smoother applied separately on the individual features and removes mean-level variations unrelated autocorrelation. The system gradually approaches a tipping point before a system-wide collapse occurs starting at the vertical dashed line ($T = 134$). **B** Residuals from the detrending process are used to look for the EWS. **C** An increasing trend can be observed in the posterior samples of the autocorrelation parameter ϕ_t from tvPVAR(1). **D** Posterior of the Kendall's rank correlation for ϕ_t. More than 99.9% of the posterior mass is above 0 indicating strong evidence for an increasing average autocorrelation and an EWS before the observed state transition.

The deterministic part of the model consists of logistic growth which is stimulated by intergroup mutualism and limited by competition within the same group. The model is defined by the stochastic differential equation

$$dX_i^{(P)} = X_i^{(P)} \left(r_i^{(P)} + \frac{\Sigma_{k=1}^{S_P} \gamma_{ik}^{(P)} X_i^{(A)}}{1 + h^{(P)} \Sigma_{k=1}^{S_P} \gamma_{ik}^{(P)} X_i^{(A)}} - \Sigma_{l=1}^{S_A} c_{il}^{(P)} X_l^{(P)} \right) dt + \sigma_j^{(P)} dW$$

$$dX_j^{(A)} = X_j^{(A)} \left(r_j^{(A)} + \frac{\Sigma_{k=1}^{S_A} \gamma_{jk}^{(A)} X_j^{(P)}}{1 + h^{(A)} \Sigma_{k=1}^{S_A} \gamma_{jk}^{(A)} X_j^{(P)}} - \Sigma_{l=1}^{S_P} c_{jl}^{(A)} X_l^{(A)} \right) dt + \sigma_j^{(A)} dW$$

where X_i represents the abundance of species i, r the growth rate and h the half saturation constant, which was set to 0.5 for all species. The matrices γ_{ik} and c_{ij} represent the intergroup mutualism and interspecies competition, respectively.

The last term is the stochastic part of the system, a Wiener process with variance σ_i^2. The superscripts P and A refer to pollinators and plants, respectively.

The system can be pushed towards a critical transition by gradually decreasing the growth rate of the pollinator species [25], which could result from increasingly harsh environmental conditions, for instance. We randomly sampled initial pollinator growth rates $r^{(P)}$ from $\mathcal{N}(0, 0.1^2)$, and decreased them linearly to -1.5 during the simulation time, except in the cases where a group of pollinators were left undisturbed. In the latter case the growth rate was kept constant over the simulation. We randomly sampled rest of the parameters from the following distributions: $r_i^{(A)} \sim \mathcal{N}(-0.1, 0.05^2)$; $\gamma_{ij}^{(A)}, \gamma_{ij}^{(B)} \sim \text{Unif}(0.6, 1)$ for the off-diagonal elements and $\gamma_{ij}^{(A)} = \gamma_{ij}^{(B)} = 1$ when $i = j$; $c_{ij}^{(A)}, c_{ij}^{(B)} \sim \text{Unif}(0, 0.1)$ for the off-diagonal elements and $c_{ij}^{(A)} = c_{ij}^{(B)} = 0.3$ when $i = j$. We set the initial abundances to 2.5 for all species and then simulated the dynamics for 20 time points with constant conditions, during which the system settled into a stable state, and then discarded this settling period before EWS analysis. The stochastic noise σ was set to 0.1 in all cases. The chosen parameter sampling distributions and constants were based on previous studies [25,40].

We used the Euler-Maruyama discretization with time-step $\Delta t = 0.01$ to simulate the model and discarded all but every 100th observation, giving integer valued time points. For each replicate we required all species to be present after the settling period and defined a species to be present if its abundance is larger than 0.05. If this condition was not met, we repeated the parameter sampling process until a viable community emerged. We defined extinction to occur when any of the community species fell below 0.05. We only used the part preceding extinction in the EWS detection.

To assess the performance and robustness of the various EWS indicators in different settings, we generated data sets with varying data characteristics. For the first part of the experiments we simulated a community with $D = 10$ species in total, varied the number of perturbed pollinators from 1 to 5 and simulated $T = 150$ time points per replicate with no observation error. In the second part we simulated data with $D = 10$, $T = 150$ and included random Gaussian observation error, with standard deviations 0, 0.05, 0.1 and 0.2. In the third part we used time series lengths of $T = 50, 150, 250$, with $D = 10$ and no observation error. In the final part we varied the total number of features $D = 4, 10, 20$, with $T = 150$ and no observation error. In order to assess the indicators' specificity, we also generated data with corresponding data characteristics but where the conditions were kept constant, and no extinction took place. In each distinct set of data characteristics 50 replicates were generated, and in every simulation half of the community species were plants and half were pollinators.

3   Results

3.1   Simulation Benchmark

In this section we compare the performance of the alternative EWS indicators presented in *Methods* in ecological simulations. We limit our presentation here to the five top-performing autocorrelation-based methods, based on average TPR over all of our experiments. This filtering process excluded the degenerate finger-printing and MAF eigenvalue indicators, and retained the ac/pooled, ac/mean, ac/max, ac/MAF and var/MAF.

We studied the effect of four different data characteristics on the detection performance: the number of perturbed features when the full dimensionality is kept constant (at 10), Gaussian additive observations error with four levels of standard deviations, time series length, and the total dimensionality of the system. Figure 2 provides a graphical presentation of the results.

The true positive rate (TPR) in EWS detection increased with the number of affected features (the number of different pollinator species). No clear differences between the alternative indicators were observed when only 1–2 features were affected. However, when a larger fraction of the system, or a higher number of features was affected, a clear distinction between the methods emerges and ac/pooled achieves superior classification accuracy.

Regarding observation error, the performance for all indicators decreased at the error level 0.1 or higher, compared to the noise-free case. At the lower error levels we observed mixed results, with increasing accuracy in some cases. Random variation may explain these differences (ANOVA $F = 0.33$, $P = 0.86$ between error levels 0 and 0.05).

Increasing time series length led to a better accuracy. At $T = 50$ ac/pooled achieved a TPR of approximately 0.5, whereas the other models performed only slightly above the theoretical level for random guess, 0.1. We observed a large further improvement in EWS detection with the longer $T = 150$ set. Difference between 150 and 250 time points did not amount to a large improvement in TPR. This would either imply that the accuracy began to saturate, or that substantially larger amounts of samples are needed for further improvements in the TPR. We have omitted the analysis of longer time series in the present work because the longer time series are increasingly slow to model, and because our analysis is primary motivated by the practically important set of biomedical and ecological applications where the availability of longitudinal observations is limited to a few dozen time points.

By varying the total number of features in the data we observed, perhaps surprisingly, that EWS detection accuracy was best at the lowest-dimensional system $D = 4$. The dimensionalities of $D = 10$ and $D = 20$ had reduced, and approximately similar performance. The better performance at the lowest dimension level might be explained by a lower level of mutualistic links, which in turn causes transients to be short lived. This could be detected visually from the time series, as the lower dimension cases experienced a more sudden col-lapse compared to the higher dimensional cases that collapsed in a more gradual fashion.

In summary, our proposed indicator ac/pooled achieved the best performance (average TPR over all experiments 0.71), compared to ac/mean (0.51), var/MAF (0.4), ac/MAF (0.34), ac/max (0.31). One-way analysis of variance (ANOVA) indicated statistically significant differences between these methods ($F = 10.2$,

$P = 1.4 \cdot 10^{-6}$). We also looked at all of the aforementioned aspects in data with constant conditions and no EWS signal. Here, we found no meaningful differences between any of the methods in TNR (ANOVA $F = 2.0$, $P = 0.097$ over all experiments), which was close to the theoretical value expected to be get with a random guess, 0.9.

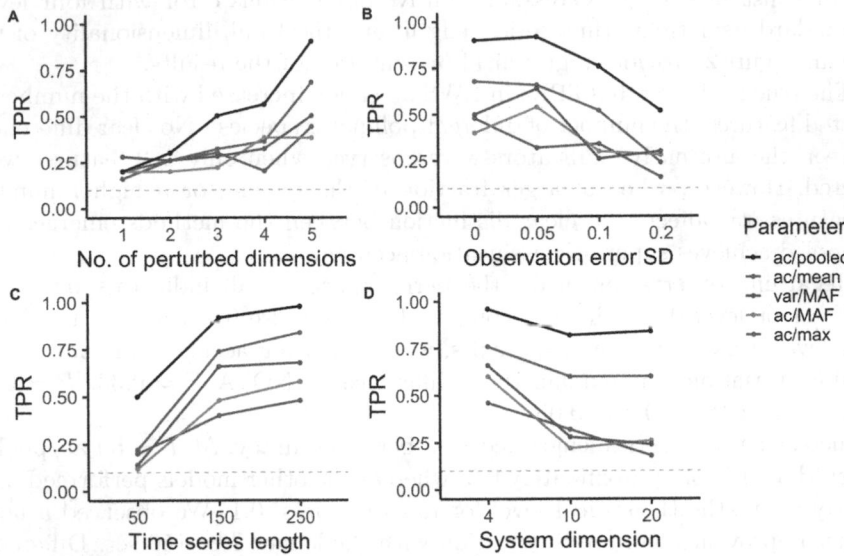

Fig. 2. EWS detection accuracy measured in true positive rate (TPR) depends on the data characteristics. **A** TPR increases as a function of perturbed dimensions. All replicates had 10 dimensions in total. **B** Increasing Gaussian additive observation error is associated with decreasing TPR. **C** Increasing time series lengths increase classification accuracy. The analyzed data sets were slightly shorter than indicated as only the part prior to the transition was used in analysis, and the exact number varied by simulation (49, 141 and 234 on average, respectively). **D** The total dimensionality of the system also influences the TPR, when half of the features are perturbed. In all panels the horizontal dashed line marks the theoretical level of a random classification, 0.1. At each x-axis value in each panel the TPR is based on 50 replicates of the simulation with randomly selected parameter and initial values.

3.2   Sensitivity Analysis

The detected EWS signal depends on chosen hyperparameters, and the values can cause spurious false positives or false negatives. Here, we investigated the effect of data detrending bandwidth and sliding window length, or Gaussian process length scale prior for the probabilistic model, on the EWS detection in representative time series.

We noticed that for ac/pooled and ac/mean the choice of these parameters did not notably influence the EWS detectability, and ac/pooled in fact produced posterior evidence exceeding the EWS detection limit in all cases. At the lowest levels of the detrending bandwidth the *P*-values for both methods decreased, see Fig. 3. For the other methods the experiment showed that an EWS was correctly identified only in a small set of hyperparameter combinations, indicating remarkable sensitivity to critical modeling choices.

In time series with constant conditions and no expected EWS the results were more uniform: all methods correctly identified a true negative with practically all hyperparameter combinations (results not shown).

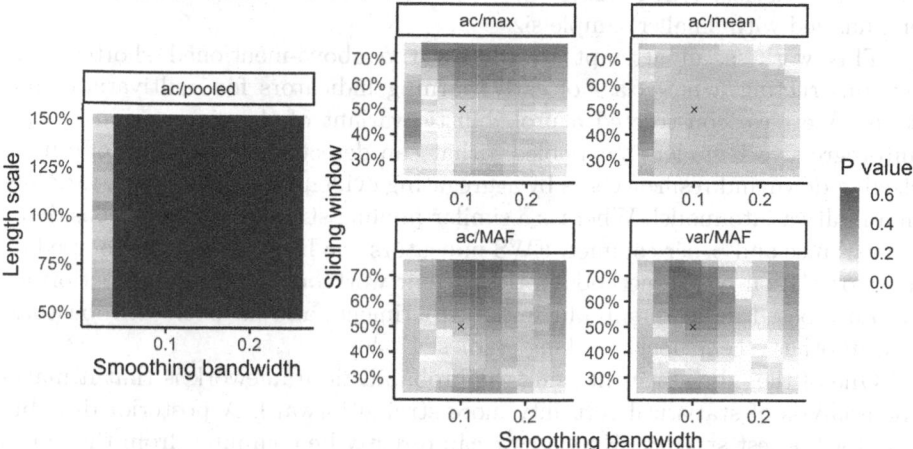

Fig. 3. Statistical evidence of the detected EWS depends on the hyperparameter combinations. The heatmap colors denote model evidence (posterior evidence and *P*-value for ac/pooled and all other models, respectively) for an increasing systemic autocorrelation at the hyperparameter values. The y-axis values represent proportion of the time series length. The black cross in the panels denotes the hyperparameter values we used in other experiments, Gaussian smoothing bandwidth of 0.1, and 100% and 50% of the times series length for length scale and sliding window, respectively.

4   Discussion

Early warnings have become an active area of research in the study of complex systems. Extensions of univariate indicators into the multivariate context provides opportunities to improve the sensitivity and accuracy of early warnings, as summarized in a recent review [40]. These earlier methods quantify (auto)correlation and variance in multivariate systems by optimized or averaged aggregate features and univariate projections. Related approaches have been also proposed based on neural networks [5], network analysis [26], and epidemic models [27]. These techniques typically rely on relatively large sample sizes and manual parameter adjustment, and lack explicit generative models for the data. These shortcomings form bottlenecks for practical application and interpretation. There is room for developing alternatives that are applicable to longitudinal data sets with limited sample sizes, provide explicit quantification of uncertainties, and avoid the need for manual parameter adjustment.

Our work is motivated by biomedical and ecological applications, where the number of available time points is typically low even in the best case scenarios. Therefore we have limited our experiments to relatively low sample sizes of up to 250 time points. It is noteworthy that the typical EWS methods in the literature generally rely on several hundreds or even thousands of time points, while these sample sizes are inaccessible in many applications. Hence, our experiments also provide a useful comparison for the alternative methods in the low sample size scenarios. Furthermore, the importance of analysing uncertainties are emphasized with smaller sample sizes.

This work is an attempt to address the above-mentioned shortcomings by constructing a new class of early warning indicators for multivariate time series. We have constructed a probabilistic variant of the time-varying vector autoregressive-1 model ("ac/pooled") that can detect early warnings of critical slowing down and resilience loss by aggregating evidence through latent variables in a multivariate model. Whereas a similar pooling strategy could be considered for variance and other common EWS indicators, we have exclusively focused on autocorrelation-based methods in the current work because these outperformed variance-based indicators in our initial experiments and have shown robustness also in other recent benchmarking studies [13, 40].

One of the advantages in using the probabilistic framework is that it makes the analysis of statistical certainty more straightforward. A posterior distribution for the test statistic, Kendall's τ can directly be computed from the model parameter ϕ_t. In contrast, in the non-probabilistic setting one needs to resort to indirect and time-consuming surrogate data analysis methods to generate an approximate sampling distribution for the test statistic. This can be relatively simple when a data generating model is known as in our simulation experiments. However, the data generating processes are often unknown in practice and need to be approximated in order to generate surrogate models. In univariate setting, for instance, the ARMA(p, q) model has been used to identify the optimal ARMA model parameters, in order to generate surrogates from this model [12]. In multivariate context the corresponding model is VARMA(p, q), but fitting this model is slow and potentially unreliable with higher dimension. Hence, the ability to directly estimate uncertainty in model parameters without such extra steps is beneficial. Another key advantage of the probabilistic framework is the ability to use prior distributions to regularize model fitting, and to incorporate available knowledge. This can be particularly useful when sample sizes are limited. By utilizing Gaussian process (GP) priors on ϕ_t we could restrict its posterior to differentiable functions, and by GP hyperparameter selection we could emphasize longer term trends in the target variable which are of most interest in EWS context. On the other hand the Matérn-3/2 covariance structure remains sufficiently flexible to detect relatively sudden changes as well [30, 37].

Benchmarking experiments showed systematic and robust improvements of the new model compared to available alternatives. As expected, the detection accuracy was in general better with larger perturbations. In all experiments, our proposed probabilistic multivariate indicator (ac/pooled) was systematically more sensitive than the other alternative autocorrelation based models and shorter time series were sufficient to achieve similar levels sensitivity than with alternative indicators. The highest TPR was achieved in systems with the smallest dimensionality. We speculate that such behavior could occur for instance when transients are more short-living with less species and mutualistic links that stabilize the system. All models performed equally well in terms of the true negative rate, close to their theoretical true negative rate corresponding to a random guess. No significant trends favouring any particular model in this regard were observed. Finally, our proposed method (ac/pooled) also outperformed the other methods in hyperparameter sensitivity analysis as it correctly detected the true EWS in a representative time series at all hyperparameter combinations.

The current work provides a proof-of-concept study on the potential of probabilistic multivariate early warning signals based on a single well-characterized ecological model that has been used also in other EWS studies [11,25,40]. Additional simulation models, and variations in data resolution, interaction structures, multiplicative noise, or large dimensionality, will help to assess the broader utility of the approach in practical scenarios [2,8].

Naturally, the ultimate aim is to apply the presented method on real data. While a large proportion of EWS literature has been motivated by questions in ecology, the methods are mostly agnostic to the application. In many fields, the bottleneck that has prevented wide adoption of EWS methods has been the scarcity of time series data with sufficient sample size. In biomedicine, however, developments in measurement technologies are making it increasingly feasible to collect comprehensive data sets. For instance, the human gut microbiome provides a potential target system where alternative stable states have been associated with health outcomes, such as in *C. difficile*-induced dysbiosis [4]. Other potential applications include deterioration of complex diseases [6] and predicting epidemic outbreaks [35].

The proposed method relies on a simple diagonal structure for the transition matrix with tied parameters. This aggregates information across the system and reduces the deterministic part of the dynamics into a single variable. The downside is that this will neglect interactions and does not inform us about the specific parts of the system that are affected and under risk. Future extensions could hence benefit from allowing off-diagonal terms in the transition matrix with a suitable regularization or sparsity inducing priors in order to increase model flexibility and capture important additional aspects of covariance within the system. A similar but more restricted approach would be to enhance automated feature selection by allowing the diagonal elements to vary independently, analogously to the maximal lag-1 autocorrelation in the non-probabilistic context. This could be regularized for instance with a composite GP prior, consisting of a common process, as in our model, in addition to separate GPs for the distinct elements which would be used to characterize additional, individual trajectories.

Furthermore, while quantifying uncertainties in parameter inference, our current method lacks an explicit model for observation error; adding this would allow the analysis of alternative error structures and potentially expand the scope of the method to different types of systems. Whereas a time-varying vector autoregressive-1 state space model has been studied in EWS context as a potential solution [18], this was only tested on 2-dimensional simulated data and convergence issues could arise in higher dimensions, and parameter estimation with state space models can be challenging even in the simplest cases [3]. Further extensions could consider variance, network structure, and other aspects of the system. For instance, principal component analysis has been used to detect changes in maximal variance and to identify features that are potentially most vulnerable [7,10]. Probabilistic PCA [38] could add sensitivity to these analyses by explicitly distinguishing between measurement error and random variations in the data. Incorporating other aspects of dynamics, such as estimated exit times [1] or memory properties [19], could provide further means to enhance EWS, and combining the analysis of longitudinal time series and aspects of multivariate survival analysis (see e.g. [31]) as prior information, could provide interesting avenues for future research.

The detection of early warning signals for critical transitions is a highly topical yet challenging task. Given the limitations typically encountered in applied scenarios, it is paramount that the available information can be utilized in the most optimal way, and uncertainties communicated effectively. Our proof-of-concept study presents a step towards this direction, providing an example on how probabilistic multivariate analysis can support the development of more sensitive, robust, and intuitive alternatives for the currently available early warnings signals in complex dynamical systems.

Acknowledgements. This work has been supported by Academy of Finland (decisions 295741, 330887) and by Turku university graduate school (UTUGS). The authors wish to acknowledge CSC - IT Center for Science, Finland, for computational resources. The authors declare no conflict of interest.

Code Availability. R source code for the experiments is available at 10.5281/zenodo. 6472720.

References

1. Arani, B.M.S., Carpenter, S.R., Lahti, L., van Nes, E.H., Scheffer, M.: Exit time as a measure of ecological resilience. Science **372**(6547), eaay4895 (2021). https://doi.org/10.1126/science.aay4895
2. Arkilanian, A.A., Clements, C.F., Ozgul, A., Baruah, G.: Effect of time series length and resolution on abundance- and trait-based early warning signals of population declines. Ecology **101**(7), e03040 (2020). https://doi.org/10.1002/ecy.3040
3. Auger-Méthé, M., et al.: State-space models' dirty little secrets: even simple linear Gaussian models can have estimation problems. Sci. Rep. **6**(1), 26677 (2016). https://doi.org/10.1038/srep26677

4. Belizário, J.E., Faintuch, J.: Microbiome and gut dysbiosis. In: Silvestre, R., Torrado, E. (eds.) Metabolic Interaction in Infection. ES, vol. 109, pp. 459–476. Springer, Cham (2018). https://doi.org/10.1007/978-3-319-74932-7_13
5. Bury, T.M., et al.: Deep learning for early warning signals of tipping points. Proc. Natl. Acad. Sci. **118**(39), e2106140118 (2021). https://doi.org/10.1073/pnas.2106140118
6. Chen, L., Liu, R., Liu, Z.P., Li, M., Aihara, K.: Detecting early-warning signals for sudden deterioration of complex diseases by dynamical network biomarkers. Sci. Rep. **2**, 342 (2012). https://doi.org/10.1038/srep00342
7. Chen, S., O'Dea, E., Drake, J., Epureanu, B.: Eigenvalues of the covariance matrix as early warning signals for critical transitions in ecological systems. Sci. Rep. **9**, 2572 (2019). https://doi.org/10.1038/s41598-019-38961-5
8. Clements, C.F., Drake, J.M., Griffiths, J.I., Ozgul, A.: Factors influencing the detectability of early warning signals of population collapse. Am. Nat. **186**(1), 50–58 (2015). https://doi.org/10.1086/681573
9. Clements, C.F., Ozgul, A.: Indicators of transitions in biological systems. Ecol. Lett. **21**(6), 905–919 (2018). https://doi.org/10.1111/ele.12948
10. Dakos, V.: Identifying best-indicator species for abrupt transitions in multispecies communities. Ecol. Ind. **94**, 494–502 (2018). https://doi.org/10.1016/j.ecolind.2017.10.024
11. Dakos, V., Bascompte, J.: Critical slowing down as early warning for the onset of collapse in mutualistic communities. Proc. Natl. Acad. Sci. **111**(49), 17546–17551 (2014). https://doi.org/10.1073/pnas.1406326111
12. Dakos, V., et al.: Methods for detecting early warnings of critical transitions in time series illustrated using simulated ecological data. PLoS ONE **7**(7), 1–20 (2012). https://doi.org/10.1371/journal.pone.0041010
13. Dakos, V., van Nes, E.H., D'Odorico, P., Scheffer, M.: Robustness of variance and autocorrelation as indicators of critical slowing down. Ecology **93**(2), 264–271 (2012). https://doi.org/10.1890/11-0889.1
14. Gelman, A., Carlin, J., Stern, H., Dunson, D., Vehtari, A., Rubin, D.: Bayesian Data Analysis, 3rd edn. Chapman and Hall/CRC (2013). https://doi.org/10.1201/b16018
15. Gelman, A., Rubin, D.B.: Inference from iterative simulation using multiple sequences. Stat. Sci. **7**(4), 457–472 (1992). https://doi.org/10.1214/ss/1177011136
16. Hastings, A., Wysham, D.B.: Regime shifts in ecological systems can occur with no warning. Ecol. Lett. **13**(4), 464–472 (2010). https://doi.org/10.1111/j.1461-0248.2010.01439.x
17. Held, H., Kleinen, T.: Detection of climate system bifurcations by degenerate fingerprinting. Geophys. Res. Lett. **312**, L23207 (2004). https://doi.org/10.1029/2004GL020972
18. Ives, A.R., Dakos, V.: Detecting dynamical changes in nonlinear time series using locally linear state-space models. Ecosphere **3**(6), 58 (2012). https://doi.org/10.1890/ES11-00347.1
19. Khalighi, M., Sommeria-Klein, G., Faust, K., Gonze, D., Lahti, L.: Quantifying the impact of ecological memory on the dynamics of interacting communities. PLOS Comput. Biol. (2022)
20. Kuss, M., Rasmussen, C.E.: Assessing approximate inference for binary Gaussian process classification. J. Mach. Learn. Res. **6**(Oct), 1679–1704 (2005)
21. Lahti, L., Salojärvi, J., Salonen, A., Scheffer, M., de Vos, W.M.: Tipping elements in the human intestinal ecosystem. Nat. Commun. **5**, 4344 (2014). https://doi.org/10.1038/ncomms5344

22. Laitinen, V., Dakos, V., Lahti, L.: Probabilistic early warning signals. Ecol. Evol. **11**(20), 14101–14114 (2021). https://doi.org/10.1002/ece3.8123
23. Lenton, T.M.: Tipping positive change. Philos. Trans. R. Soc. B: Biol. Sci. **375**(1794), 20190123 (2020). https://doi.org/10.1098/rstb.2019.0123
24. Lenton, T.M., et al.: Tipping elements in the earth's climate system. Proc. Natl. Acad. Sci. **105**(6), 1786–1793 (2008). https://doi.org/10.1073/pnas.0705414105
25. Lever, J.J., Nes, E., Scheffer, M., Bascompte, J.: The sudden collapse of pollinator communities. Ecol. Lett. **17**, 350–359 (2014). https://doi.org/10.1111/ele.12236
26. Liu, R., Chen, P., Aihara, K., Chen, L.: Identifying early-warning signals of critical transitions with strong noise by dynamical network markers. Sci. Rep. **5**(1), 17501 (2015). https://doi.org/10.1038/srep17501
27. Proverbio, D., Kemp, F., Magni, S., Gonçalves, J.: Performance of early warning signals for disease re-emergence: a case study on Covid-19 data. PLoS Comput. Biol. **18**(3), 1–22 (2022). https://doi.org/10.1371/journal.pcbi.1009958
28. Quax, R., Apolloni, A., Sloot, P.: The diminishing role of hubs in dynamical processes on complex networks. J. R. Soc. Interface **10**, 20130568 (2013)
29. Quax, R., Kandhai, D., Sloot, P.M.A.: Information dissipation as an early-warning signal for the Lehman brothers collapse in financial time series. Sci. Rep. **3**(1), 1898 (2013). https://doi.org/10.1038/srep01898
30. Rasmussen, C.E., Williams, C.K.I.: Gaussian Processes for Machine Learning. The MIT Press (2006). https://doi.org/10.7551/mitpress/3206.001.0001
31. Salosensaari, A., et al.: Taxonomic signatures of cause-specific mortality risk in human gut microbiome. Nat. Commun. **12**, 2671 (2021). https://doi.org/10.1038/s41467-021-22962-y
32. Scheffer, M.: Critical Transitions in Nature and Society. Princeton University Press, New Jersey (2009). https://doi.org/10.1515/9781400833276
33. Scheffer, M., et al.: Early-warning signals for critical transitions. Nature **461**(7260), 53–59 (2009)
34. Scheffer, M., Carpenter, S., Foley, J.A., Folke, C., Walker, B.: Catastrophic shifts in ecosystems. Nature **413**(6856), 591–596 (2001). https://doi.org/10.1038/35098000
35. Southall, E., Brett, T.S., Tildesley, M.J., Dyson, L.: Early warning signals of infectious disease transitions: a review. J. R. Soc. Interface **18**(182), 20210555 (2021). https://doi.org/10.1098/rsif.2021.0555
36. Stan Development Team: RStan: the R interface to Stan (2020). R package version 2.21.2
37. Stein, M.L.: Interpolation of Spatial Data: Some Theory for Kriging. Springer Series in Statistics, Springer, New York (1999). https://doi.org/10.1007/978-1-4612-1494-6
38. Tipping, M.E., Bishop, C.M.: Probabilistic principal component analysis. J. R. Stat. Soc.: Ser. B (Stat. Methodol.) **61**(3), 611–622 (1999). https://doi.org/10.1111/1467-9868.00196
39. Weinans, E., et al.: Finding the direction of lowest resilience in multivariate complex systems. J. R. Soc. Interface **16**, 20190629 (2019). https://doi.org/10.1098/rsif.2019.0629
40. Weinans, E., Quax, R., van Nes, E.H., van de Leemput, I.A.: Evaluating the performance of multivariate indicators of resilience loss. Sci. Rep. **11**(1), 9148 (2021). https://doi.org/10.1038/s41598-021-87839-y

Software

MobsPy: A Meta-species Language for Chemical Reaction Networks

Fabricio Cravo[1,2], Matthias Függer[2(✉)], Thomas Nowak[1(✉)],
and Gayathri Prakash[2]

[1] LISN, Université Paris-Saclay, CNRS, Gif-sur-Yvette, France
`thomas.nowak@lri.fr`
[2] LMF, ENS Paris-Saclay, Université Paris-Saclay, CNRS, Inria,
Gif-sur-Yvette, France
`mfuegger@lsv.fr`

Abstract. Chemical reaction networks are widely used to model biochemical systems. However, when the complexity of these systems increases, the chemical reaction networks are prone to errors in the initial modeling and subsequent updates of the model.

We present the Meta-species-oriented Biochemical Systems Language (MobsPy), a language designed to simplify the definition of chemical reaction networks in Python. MobsPy is built around the notion of meta-species, which are sets of species that can be multiplied to create higher-dimensional orthogonal characteristics spaces and inheritance of reactions. Reactions can modify these characteristics. For reactants, queries allow to select a subset from a meta-species and use them in a reaction. For products, queries specify the dimensions in which a modification occurs. We demonstrate the simplification capabilities of the MobsPy language at the hand of a running example and a circuit from literature. The MobsPy Python package includes functions to perform both deterministic and stochastic simulations, as well as easily configurable plotting. The MobsPy package is indexed in the Python Package Index and can thus be installed via `pip`.

Keywords: chemical reaction networks · modeling language · simulation

1 Introduction

Chemical reaction networks (CRNs) model the dynamics of biochemical systems by a set of species and reactions acting on them [8]. In particular, in synthetic biology, which studies the engineering of new behavior in biological entities [10], CRNs have proved useful to model genetic circuits. Examples are the toggle switch, the repressilator, and logic gates, among others [1, 3, 16]. While synthetic

Supported by the ANR project DREAMY (ANR-21-CE48-0003) and the CNRS project BACON.

I. Petre and A. Păun (Eds.): CMSB 2022, LNBI 13447, pp. 277–285, 2022.
https://doi.org/10.1007/978-3-031-15034-0_14

biology has primarily addressed single-cell behavior, recent work has studied the engineering of bacterial populations to reduce cell burden and allow for population-averaging of circuit responses [15,17].

However, the modeling of genetic circuits using CRNs is error-prone due to the complexity of the resulting model. Lopez, Muhlich, Bachman, and Sorger [13] studied several works from literature and found discrepancies when comparing their description and the provided models. There is thus a need for tools that simplify the modeling of complex biological circuits through CRNs, both for single-cell and population-level dynamics.

Several solutions have been proposed to facilitate the modeling and simulation of such systems. For instance, COPASI [11] and iBioSim [14] are simulators that can be either used via a graphical interface or by writing a model in SBML format [12]. They come with state-of-the-art stochastic and ordinary differential equation (ODE) solvers, but are not directly suited for automated prototyping due to design entry by a GUI. BasiCO [2] is a Python interface for using COPASI, thus allowing for automation. However, BasiCO does not support any language-based model simplifications: species and reactions are added one by one. BSim [9] is a geometric, agent-based simulator for population dynamics and uses ODEs for internal cellular dynamics.

In this work, we propose MobsPy, a language and Python framework that is based on the concept of meta-species and meta-reactions. Meta-species are sets of species and meta-reactions act upon meta-species. This allows for model simplification via the following features: (i) *Inheritance of reactions.* Meta-reactions construct reactions for all the species that inherit from the meta-species present in the meta-reaction. (ii) *The definition of species via independent characteristics-space structures.* Products of meta-species create a meta-species whose characteristics space is the Cartesian product of the characteristics space of its elements. (iii) *The possibility to query and transform the characteristics of a species.* Queries can be used to restrict meta-species to subsets of species. Transformations can be used to change specific characteristics within a meta-species' characteristics space. These features will be discussed in greater detail in Sect. 2, along with an example model.

The most closely related frameworks to MobsPy are Kappa [6] and iBio-Gen [7], both providing a rule-based language that can define multiple reactions with a single rule. A rule defines how individual species (typically molecules) combine and separate based on their state. While Kappa was designed to focus on complex formation in chemical pathways, MobsPy targets more general dynamics, where the reacting species do not necessarily form complexes upon interaction but may change states, produce, or consume species. Kappa also proposes the utilization of gadgets for keeping track of complex state changes, while MobsPy uses Python itself. In [5], an extension with general inheritance has been proposed to Kappa. To achieve this, they propose a distinction between two types of agents called concrete and generic agents, which is not required for inheritance in MobsPy. Also, Kappa's inheritance extension does not allow for multiple inheritances.

2   MobsPy Syntax and Simulator

We next discuss the main features of MobsPy along with a running example. Consider a system of two groups of trees (Fig. 1a): one group where trees grow and reproduce in a dense population and one in a sparse population. We assume that the leaves of a tree change color from green to yellow to brown in a cyclic fashion. Further, trees die, with young trees in the dense group dying at a higher rate due to competition for resources and space.

We start modeling the system by defining base-species with characteristics, and a reaction for how the base-species `Ager` changes its characteristics. We also assign a rate to this reaction using units via the unit registry u and the unit name, e.g., `u.year` for years.

```
1   Ager, Mortal, Colored, Location = BaseSpecies(4)
2   Colored.green, Colored.yellow, Colored.brown
3   Location.dense, Location.sparse
4   Ager.young >> Ager.old [0.1/ u.year]
```

A base species has a single set of characteristics that can either be defined explicitly using the dot notation as in code line 2 or implicitly as in code line 4. In the implicit case, new characteristics are automatically added, when they are used for the first time inside a reaction.

Meta-reactions have a CRN-like syntax of the form `reactants >> products [rate_spec]`. The `rate_spec` can be a (non-negative) real or a function whose parameters are the meta-species reactants, and that returns a real or a string. In the case of a (returned) real, the reaction rate follows mass-action kinetics, with the real being the rate constant. In the case of a returned string, one can define different kinetics in terms of concentrations of species/characteristics of the reactants. Meta-reactions can define multiple reactions in the CRN, as they act upon all the species from the meta-species. As an example, we give a list of all CRN reactions generated by the meta-reactions of the Tree model in the appendix.

To assign different death rates for old and young trees, we make use of a `rate_spec` in terms of a function that returns the respective rate-constant (see code line 5), and the special base-species `Zero` that contains no species.

```
5   Mortal >> Zero [lambda r1: 0.1/ u.year if r1.old else 0]
6   Tree = Ager*Colored*Mortal*Location
```

Multiplication (see code line 6) is used to combine base-species and meta-species into more complex meta-species. The product meta-species inherits all characteristics and reactions from its factors. In our example, `Tree` contains twelve species (see Fig. 1b). Further, the product meta-species inherits the meta-reactions of its factors. Therefore, `Tree` receives the aging reactions from `Ager` and the death reactions from `Mortal`. This allows to significantly simplify the model: only one death and aging meta-reaction need to be specified for all the model's species.

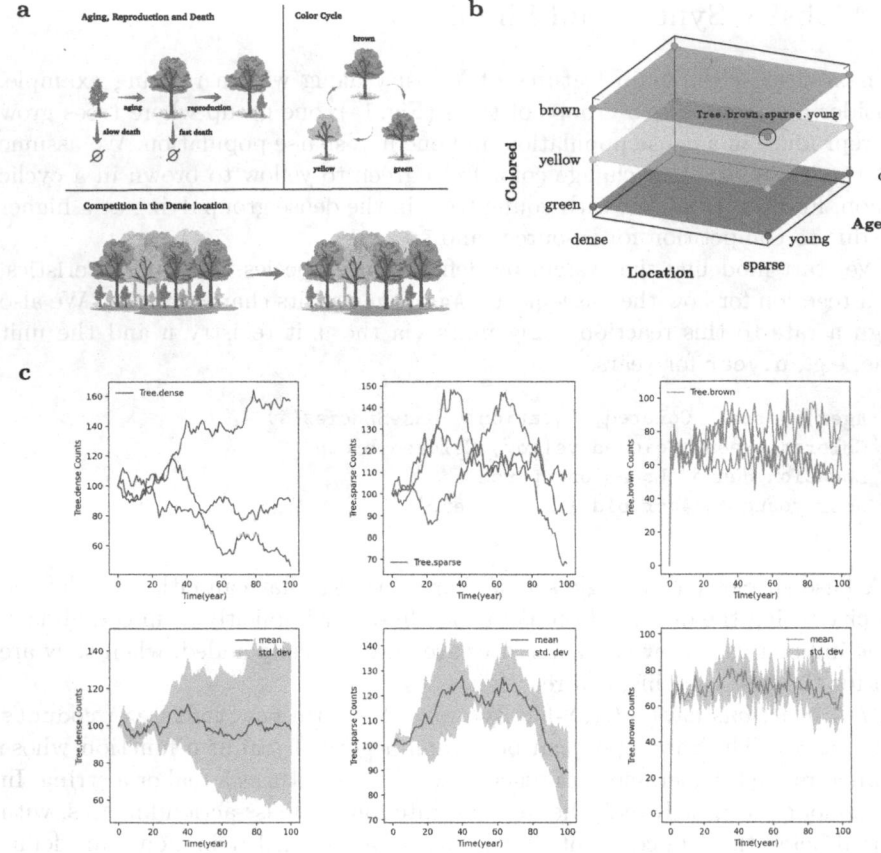

Fig. 1. (a) System dynamics: schematic representation of all the reactions in the Tree model, which are aging, reproduction, color cycle, and competition. **(b)** Meta-species `Tree` created by multiplication of the three base-species `Ager`, `Location`, and `Colored`. One of the twelve species generated, `Tree.brown.sparse.young`, has been labeled. **(c)** MobsPy default plots after simulating the Tree model ($n = 3$ stochastic runs). In the top row of the panel individual runs are shown. The bottom plots depicts mean and standard deviation.

Within reactants, the *dot operator* is used to query for species inside meta-species, making it possible to assign reactions to only the preferred sub-set. Queries can be composed arbitrarily and independently of order. It is further possible to query over a string value stored inside a variable, say `s`, using `species.c(s)`. In our example, we make use of queries to specify the color-cycle exhibited by the leaves:

```
 7  # color cycle
 8  colors = ['green','yellow','brown']
 9  for color, next_color in zip(colors, colors[1:] + colors[:1]):
10      Tree.c(color) >> Tree.c(next_color) [10/ u.year]
11  # competition
12  Tree.dense.old + Tree.dense.young >> Tree.dense.old
13      [1e-10*u.decimeter/ u.year]
14  # replication
15  Tree.old >> Tree + Tree.green.young [0.1/ u.year]
```

Unlike in reactions of a CRN, *order of products and reactants* matters in meta-reactions: In the competition meta-reaction (code line 12), `Tree.old` wins the competition against `Tree.young` and becomes `Tree.old`. MobsPy, by default, uses a round-robin order, where products cycle through the list of available reactants from the same meta-species. Alternatively, labels can be used to explicitly declare which reactant becomes which product.

The *initial count* of the default species within a meta-species S is set via $S(i)$, where i is either an integer count, or a real-valued concentration with respective units. Here, the default species is the species with the first characteristic added to each meta-species from which it inherits. For example, in the Tree model, `Tree(50)` assigns the count 50 to `Tree.green.dense.young`. To assign counts to other species, one can add more characteristics.

```
16  # initial conditions
17  Tree.dense(50),Tree.dense.old(50),Tree.sparse(50),Tree.sparse.old(50)
```

A simulation object of the species in the meta-species `Tree` is finally constructed by:

```
18  MySim = Simulation(Tree)
```

The simulation can then be parametrized with solver parameters (stochastic and ODE), plotting options, and data export options. As a backend, MobsPy generates SBML models [12] and runs simulations using BasiCO. Simulation results can be accessed directly in Python or exported as JSON files. Examples for default stochastic plots are shown in Fig. 1c. The plots can be easily configured via a Python dictionary, where a single parameter can be set for all figures, for individual figures, or for each curve in a hierarchical fashion. For a detailed description of the modeling, simulation, and plotting features of MobsPy, we refer the reader to the GitHub repository [4]. It contains several examples like a simple reaction A + B >> 2*C + D, a Hill function model of a genetic oscillator, a bacteria-phage system, a bacteria and phage random walk, a simple repressor, a toggle switch, genetic gates such as the NOR and AND gates, an example for the hierarchical plot structure and tutorial models.

In the next section, we demonstrate the modeling capabilities of MobsPy at the hand of a real-world synthetic design from literature [3].

3   Genetic Circuits with MobsPy: The CRISPRlator

To show that MobsPy is well-suited to model genetic circuits, we modeled the CRISPRlator, a CRISPRi oscillator proposed by Santos-Moreno, Tasiudi, Stelling, and Schaerli [16]. The circuit is a CRISPRi-based repressilator where each gRNA represses the next gRNA-encoding gene in a cyclic manner. The model is shown below, with reaction rates taken from Clamons and Murray [3].

```
1   Promoter, dCas, CasBinding = BaseSpecies(3)
2   Promoter.active, Promoter.inactive, CasBinding.no_cas, CasBinding.cas
3   DNAPro = New(Promoter)
4   gRNA = New(CasBinding)
5   gRNAs_list = ['g1', 'g2', 'g3']
6   Promoters_list = ['P1', 'P2', 'P3']
7
8   rev_rt = (1.8e-3/(u.nanomolar * u.second), 2.3e-2/ u.minute)
9   # Rev defines a reversible reaction
10  Rev[gRNA.no_cas + dCas >> gRNA.cas][rev_rt]
11  gRNA.no_cas >> Zero [0.0069/ u.second]
12  Promoter.active >> 2 * Promoter.active [2.3e-2/ u.minute]
13  Promoter >> Zero [2.3e-2/ u.minute]
14
15  for prom, grna in zip(Promoters_list, gRNAs_list):
16      act_rt = lambda dna: 5/ u.minute if dna.active else 0
17      DNAPro.c(prom) >> gRNA.no_cas.c(grna) + DNAPro.c(prom)[act_rt]
18
19  gRNA_rep_List = ['g3', 'g1', 'g2']
20  for prom, grna in zip(Promoters_list, gRNA_rep_List):
21      dna_rt1 = 1.2e-2 * u.liter/ (u.nanomoles * u.second)
22      dna_rt2 = 2.3e-2/ u.minute
23      DNAPro.active.c(prom) + gRNA.cas.c(grna) >> DNAPro.inactive.c(prom)[dna_rt1]
24      DNAPro.inactive.c(prom) >> 2 * DNAPro.active.c(prom) + gRNA.cas.c(grna)[dna_rt2]
25
26  Rev[Zero >> dCas][1/ u.minute, 2.3e-2/ u.minute]
27
28  DNAPro.active.P1(1), DNAPro.active.P2(1), DNAPro.active.P3(1)
29  gRNA.no_cas.g1(0), gRNA.no_cas.g2(40*u.nanomolar), gRNA.no_cas.g3(1*u.nanomolar)
30  dCas(43*u.nanomolar)
31
32  MySim = Simulation(gRNA | DNAPro | dCas)
33  MySim.volume = 1*u.femtoliter
```

The code to run ODE simulations over a range of 650 is shown in Fig. 2a. The resulting plot is depicted in Fig. 2b.

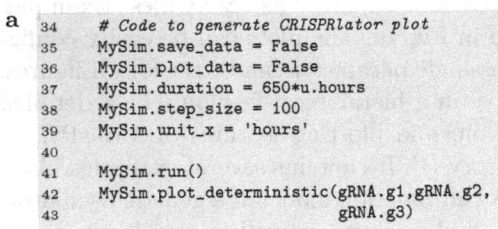

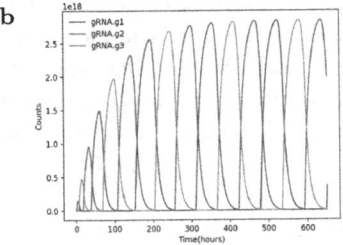

Fig. 2. (a) Code for simulating the CRISPRlator model and generating default plots. **(b)** CRISPRlator simulation results.

4 Conclusions

We presented MobsPy, a simple CRN programming language to simplify the modeling of complex biochemical reaction networks. The simulator creates SBML code and uses BasiCO/COPASI as a stochastic and ODE solver. We discussed MobsPy's features at the hand of a Tree toy example and a real-world genetic design, with succinct model descriptions instead of a direct specification as CRNs (12 species and 45 reactions for the Tree and 13 species and 32 reactions for the CRISPRi model). MobsPy is open-source (MIT license) and easy to install through `pip` or via the git repository.

A Comparison of Meta-reactions and Reactions for the Tree Model

In the following we list the MobsPy meta-reactions for the Tree model in Sect. 2 along with their corresponding CRN reactions.

```
Ager.young >> Ager.old
```

$$
\begin{cases}
\text{Tree . brown . dense . young} & \longrightarrow \text{Tree . brown . dense . old} \\
\text{Tree . brown . sparse . young} & \longrightarrow \text{Tree . brown . sparse . old} \\
\text{Tree . green . dense . young} & \longrightarrow \text{Tree . green . dense . old} \\
\text{Tree . green . sparse . young} & \longrightarrow \text{Tree . green . sparse . old} \\
\text{Tree . yellow . dense . young} & \longrightarrow \text{Tree . yellow . dense . old} \\
\text{Tree . yellow . sparse . young} & \longrightarrow \text{Tree . yellow . sparse . old}
\end{cases}
$$

```
Tree.old >> Tree + Tree . green . young
```

$$
\begin{cases}
\text{Tree . brown . dense . old} & \longrightarrow \text{Tree . brown . dense . old} + \text{Tree . green . dense . young} \\
\text{Tree . brown . sparse . old} & \longrightarrow \text{Tree . brown . sparse . old} + \text{Tree . green . sparse . young} \\
\text{Tree . green . dense . old} & \longrightarrow \text{Tree . green . dense . old} + \text{Tree . green . dense . young} \\
\text{Tree . green . sparse . old} & \longrightarrow \text{Tree . green . sparse . old} + \text{Tree . green . sparse . young} \\
\text{Tree . yellow . sparse . old} & \longrightarrow \text{Tree . yellow . sparse . old} + \text{Tree . green . sparse . young} \\
\text{Tree . yellow . dense . old} & \longrightarrow \text{Tree . yellow . dense . old} + \text{Tree . green . dense . young}
\end{cases}
$$

```
Tree.dense . old + Tree.dense . young >> Tree.dense . old
```

$$
\begin{cases}
\text{Tree . brown . dense . old} + \text{Tree . brown . dense . young} & \longrightarrow \text{Tree . brown . dense . old} \\
\text{Tree . brown . dense . old} + \text{Tree . green . dense . young} & \longrightarrow \text{Tree . brown . dense . old} \\
\text{Tree . brown . dense . old} + \text{Tree . yellow . dense . young} & \longrightarrow \text{Tree . brown . dense . old} \\
\text{Tree . green . dense . old} + \text{Tree . brown . dense . young} & \longrightarrow \text{Tree . green . dense . old} \\
\text{Tree . green . dense . old} + \text{Tree . green . dense . young} & \longrightarrow \text{Tree . green . dense . old} \\
\text{Tree . green . dense . old} + \text{Tree . yellow . dense . young} & \longrightarrow \text{Tree . green . dense . old} \\
\text{Tree . yellow . dense . old} + \text{Tree . brown . dense . young} & \longrightarrow \text{Tree . yellow . dense . old} \\
\text{Tree . yellow . dense . old} + \text{Tree . green . dense . young} & \longrightarrow \text{Tree . yellow . dense . old} \\
\text{Tree . yellow . dense . old} + \text{Tree . yellow . dense . young} & \longrightarrow \text{Tree . yellow . dense . old}
\end{cases}
$$

```
Mortal >> Zero
```

$$
\left\{
\begin{array}{l}
\text{Tree . brown . dense . old} \longrightarrow \\
\text{Tree . brown . dense . young} \longrightarrow \\
\text{Tree . brown . sparse . old} \longrightarrow \\
\text{Tree . brown . sparse . young} \longrightarrow \\
\text{Tree . green . dense . old} \longrightarrow \\
\text{Tree . green . dense . young} \longrightarrow \\
\text{Tree . green . sparse . old} \longrightarrow \\
\text{Tree . green . sparse . young} \longrightarrow \\
\text{Tree . yellow . dense . old} \longrightarrow \\
\text{Tree . yellow . dense . young} \longrightarrow \\
\text{Tree . yellow . sparse . old} \longrightarrow \\
\text{Tree . yellow . sparse . young} \longrightarrow
\end{array}
\right.
$$

```
Tree.c(color) >> Tree.c(next_color)
```

$$
\left\{
\begin{array}{l}
\text{Tree . brown . dense . old} \longrightarrow \text{Tree . green . dense . old} \\
\text{Tree . brown . dense . young} \longrightarrow \text{Tree . green . dense . young} \\
\text{Tree . brown . sparse . old} \longrightarrow \text{Tree . green . sparse . old} \\
\text{Tree . brown . sparse . young} \longrightarrow \text{Tree . green . sparse . young} \\
\text{Tree . yellow . dense . old} \longrightarrow \text{Tree . brown . dense . old} \\
\text{Tree . yellow . dense . young} \longrightarrow \text{Tree . brown . dense . young} \\
\text{Tree . yellow . sparse . old} \longrightarrow \text{Tree . brown . sparse . old} \\
\text{Tree . yellow . sparse . young} \longrightarrow \text{Tree . brown . sparse . young} \\
\text{Tree . green . dense . old} \longrightarrow \text{Tree . yellow . dense . old} \\
\text{Tree . green . dense . young} \longrightarrow \text{Tree . yellow . dense . young} \\
\text{Tree . green . sparse . old} \longrightarrow \text{Tree . yellow . sparse . old} \\
\text{Tree . green . sparse . young} \longrightarrow \text{Tree . yellow . sparse . young}
\end{array}
\right.
$$

References

1. Arkin, A., Ross, J.: Computational functions in biochemical reaction networks. Biophys. J. **67**(2), 560–578 (1994)
2. Bergmann, F.T.: BasiCO (2022). https://github.com/copasi/basico
3. Clamons, S.E., Murray, R.M.: Modeling dynamic transcriptional circuits with CRISPRi (2017). https://www.biorxiv.org/content/early/2017/11/27/22531
4. Cravo, F., Függer, M., Nowak, T., Prakash, G.: MobsPy (2022). https://github.com/ROBACON/mobspy
5. Danos, V., Feret, J., Fontana, W., Harmer, R., Krivine, J.: Rule-based modelling and model perturbation. In: Priami, C., Back, R.-J., Petre, I. (eds.) Transactions on Computational Systems Biology XI. LNCS, vol. 5750, pp. 116–137. Springer, Heidelberg (2009). https://doi.org/10.1007/978-3-642-04186-0_6
6. Danos, V., Laneve, C.: Formal molecular biology. Theoret. Comput. Sci. **325**(1), 69–110 (2004)
7. Faeder, J.R., Blinov, M.L., Hlavacek, W.S.: Rule-based modeling of biochemical systems with BioNetGen. In: Maly, I.V. (ed.) Systems Biology Methods in Molecular Biology (Methods and Protocols). Methods in Molecular Biology, vol. 500, pp. 113–167. Springer, New York (2009). https://doi.org/10.1007/978-1-59745-525-1_5

8. Gillespie, D.T.: Exact stochastic simulation of coupled chemical reactions. J. Phys. Chem. **81**(25), 2340–2361 (1977)
9. Gorochowski, T.E., et al.: BSim: an agent-based tool for modeling bacterial populations in systems and synthetic biology. PLoS One **7**(8), e42790 (2012)
10. Heinemann, M., Panke, S.: Synthetic biology-putting engineering into biology. Bioinformatics **22**(22), 2790–2799 (2006)
11. Hoops, S., et al.: COPASI-a COmplex PAthway SImulator. Bioinformatics **22**(24), 3067–3074 (2006)
12. Hucka, M., et al.: The systems biology markup language (SBML): a medium for representation and exchange of biochemical network models. Bioinformatics **19**(4), 524–531 (2003)
13. Lopez, C.F., Muhlich, J.L., Bachman, J.A., Sorger, P.K.: Programming biological models in Python using PySB. Mol. Syst. Biol. **9**, 646 (2013)
14. Myers, C.J., Barker, N., Jones, K., Kuwahara, H., Madsen, C., Nguyen, N.-P.D.: iBioSim: a tool for the analysis and design of genetic circuits. Bioinformatics **25**(21), 2848–2849 (2009)
15. Regot, S., et al.: Distributed biological computation with multicellular engineered networks. Nature **469**(7329), 207–211 (2011)
16. Santos-Moreno, J., Tasiudi, E., Stelling, J., Schaerli, Y.: Multistable and dynamic CRISPRi-based synthetic circuits. Nat. Commun. **11**(1), 1–8 (2020)
17. Tamsir, A., Tabor, J.J., Voigt, C.A.: Robust multicellular computing using genetically encoded NOR gates and chemical 'wires'. Nature **469**(7329), 212–215 (2011)

Automated Generation of Conditional Moment Equations for Stochastic Reaction Networks

Hanna Josephine Wiederanders[1,2], Anne-Lena Moor[1,2],
and Christoph Zechner[1,2,3(✉)]

[1] Center for Systems Biology Dresden, Dresden, Germany
[2] Max Planck Institute of Molecular Cell Biology and Genetics, Dresden, Germany
[3] Cluster of Excellence Physics of Life, TU Dresden, Dresden, Germany
`zechner@mpi-cbg.de`

Abstract. The dynamics of biochemical reaction networks require a stochastic description when copy number fluctuations become significant. Such description is provided by moment equations that capture the statistical properties of the involved molecular components such as their average abundance and variability. Certain applications require a special form of moment equations, where the statistics of some components are described conditionally on complete trajectories of other components. Typical examples include information theoretical analyses of biochemical networks, model reduction and subnetwork simulation, or statistical inference where time-varying molecular signals are inferred from counting observations. These conditional moment equations have so far been limited to relatively simple reaction systems as their manual derivation becomes difficult for systems involving many components and interactions. Here, we present a Python tool for the automated derivation of moment equations conditional on complete time trajectories for arbitrary user-defined reaction systems and showcase its utility in the context of subnetwork simulation. With this automated tool, conditional moment equations become applicable to a broad class of biochemical systems.

Keywords: Stochastic biochemical networks · Conditional moments · Moment generator

1 Introduction

The method of moments provides an effective way to analyze biochemical networks in the presence of noise. With this approach, a system of exact or approximate differential equations is derived from the chemical master equation (CME) [8,10,13] which capture how the statistics of the involved molecular components (e.g., means and variances) evolve with time. Moment equations have been applied successfully in the analysis of diverse intracellular systems, ranging from relatively simple gene networks [18] to complex circadian clock models involving many chemical species [6].

A special class of moment equations – termed *conditional* moment equations – have recently attracted interest in the context of stochastic biochemical

© The Author(s), under exclusive license to Springer Nature Switzerland AG 2022
I. Petre and A. Păun (Eds.): CMSB 2022, LNBI 13447, pp. 286–293, 2022.
https://doi.org/10.1007/978-3-031-15034-0_15

networks [4, 11]. Generally speaking, conditional moment equations describe the statistics of molecular components *conditionally* on certain other events. In [11], for instance, the authors derive equations for the moments of some molecular components conditionally on the abundance of certain other components. As an example, one could derive equations for the average mRNA and protein abundance in a gene network at a certain time given that the gene is transcriptionally active at that time. As was shown in this study, this idea can be exploited to greatly simplify the analysis of networks involving both lowly and highly abundant species.

Certain practical applications require moments of molecular species conditional on full trajectories as opposed to instantaneous values. This is the case, for instance, in the context of subnetwork simulation [1, 3, 4], which aims at directly simulating a subset of network components $\mathcal{A}$ embedded into a larger network comprising components $\mathcal{A} \cup \mathcal{B}$. As we have shown previously [4], the exact solution to this problem can be found by calculating a conditional expectation of the underlying propensity functions where the conditioning goes with respect to the full trajectory of components $\mathcal{A}$ between time zero and t. The resulting simulation algorithm, called selected-node stochastic simulation algorithm (snSSA), can lead to a substantial speedup when compared to standard SSA [9] because only a fraction of events needs to be simulated explicitly. The very same type of conditional moments is encountered also when performing information theoretical analyses of biochemical networks [5], or in certain inference problems, where dynamic molecular signals are reconstructed from counting observations [22].

Moments that are conditional on complete trajectories can be obtained from the theory of stochastic filtering [2]. The first step is to derive a stochastic differential equation that captures the time-evolution of the required conditional probability distribution. This differential equation can be understood as a counterpart to the CME for conditional Markov processes. Approximate moment equations can then obtained by employing standard moment-closure schemes [17, 18]. While the derivation of these conditional moments is in principle straightforward, it becomes tedious for complicated networks involving multiple reactants and interactions. Although several software packages for automatically deriving moment equations exist [7, 12, 14, 20], they do not apply to the generation of moments conditional on full trajectories. In this paper, we present a Python tool which allows the user to generate conditional moment equations for arbitrary mass-action reaction networks. In the following, we introduce the theoretical background of conditional moment equations, provide a brief description of the software tool and demonstrate its use in the context of subnetwork simulation.

2   Theory

Stochastic Reaction Networks. Consider a stochastic reaction network consisting of N distinct chemical species $Y_1, \ldots, Y_N$ and M reaction channels:

$$\sum_{i=1}^{N} s_{ij}^r Y_i \xrightarrow{c_j} \sum_{i=1}^{N} s_{ij}^p Y_i, \quad j = 1, \ldots, M. \tag{1}$$

We define stoichiometric change vectors $\nu_j = (s_{ij}^p - s_{ij}^r)_{i=1,\ldots,N}$ which collect the net molecular changes of each species associated with reaction j. Moreover, we denote the state of the system by $Y(t) = (Y_1(t), \ldots Y_N(t)) \in \mathbb{N}_0^N$ which tracks the copy numbers of all species at time t. Note that $Y(t)$ remains constant until a reaction of type j occurs which causes the instantaneous change $Y(t) \to Y(t) + \nu_j$. Finally, we associate to each reaction channel j a propensity function $h_j(Y(t))$, which determines how likely this reaction happens per unit time. Throughout this work, we restrict ourselves to mass-action kinetics such that $h_j(Y(t)) = c_j\, g_j(Y(t))$, with c_j as a stochastic rate constant and $g(\cdot)$ a polynomial defined by the law of mass-action [13]. The state $Y(t)$ can be described by a continuous-time Markov chain (CTMC), whose state probability distribution $P(y,t) = P(Y(t) = y)$ satisfies a CME [8,10,13]

$$\frac{\mathrm{d}}{\mathrm{d}t}P(y,t) = \sum_{j=1}^{M} h_j(y - \nu_j)P(y - \nu_j, t) - \sum_{j=1}^{M} h_j(y)P(y,t), \tag{2}$$

with initial condition $P(y,0) = P(Y(0) = y)$.

Stochastic Filtering and Conditional Probability Distribution. Our aim is to express the dynamics of certain species conditionally on the full trajectories of all other species. To this end, we split up the state vector as $Y(t) = (X(t), Z(t))$ where $Z(t)$ collects the copy numbers of the species whose dynamics we are interested in and $X(t)$ corresponds to the species that we want to condition on. In line with our earlier work [4], we refer to $X(t)$ and $Z(t)$ as *selected* and *latent* components, respectively. We introduce the conditional probability distribution $\pi(z,t) = p(Z(t) = z \mid X_0^t)$, where X_0^t denotes a full trajectory of $X(t)$ between time zero and t. While $\pi(z,t)$ is generally intractable, it is possible to derive a stochastic differential equation that describes how $\pi(z,t)$ evolves with time [4]. We refer to this equation as *filtering equation*, which in the considered scenario is given by

$$\mathrm{d}\pi(z,t) = \Bigg[\sum_{j \in \mathcal{R}_Z} \big(h_j(X(t), z - \nu_j^z)\pi(z - \nu_j^z, t) - h_j(X(t), z)\pi(z,t) \big)$$

$$- \sum_{k \in \mathcal{R}_C} \big(h_k(X(t), z) - \mathbb{E}[h_k(X(t), Z(t)) \mid X_0^t] \big)\pi(z,t) \Bigg] \mathrm{d}t \tag{3}$$

$$+ \sum_{k \in \mathcal{R}_C} \Bigg[\frac{h_k(X(t), z - \nu_k^z)}{\mathbb{E}[h_k(X(t), Z(t)) \mid X_0^t]}\pi(z - \nu_k^z, t) - \pi(z,t) \Bigg] \mathrm{d}R_k(t)$$

with ν_j^z as the part of the change vector ν_j that acts on $Z(t)$ and $\mathrm{d}R_k(t)$ as a point process with rate h_k, which is 1 exactly when reaction k fires and 0 otherwise. The set $\mathcal{R}_Z$ collects all reaction types that modify latent species but no selected species. The set $\mathcal{R}_C$ comprises all reactions that modify selected species but also involve latent species as reactant or product (or both). For further details, the reader may refer to our original work [4].

Conditional Moment Equations. We are interested in deriving differential equations for conditional moments defined as

$$\mathbb{E}[f(Z(t)) \mid X_0^t] = \sum_z f(z)\pi(z,t), \qquad (4)$$

where $f(z)$ is a monomial in z. This can be achieved by multiplying Eq. 3 with f and summing over all possible values of the latent components $Z(t)$. In general, moment equations of a certain order depend on moments of higher order, leading to an infinite hierarchy of moment equations. To get a finite-dimensional system, moment closure techniques can be employed, which replace higher order moments by functions of lower order moments by making assumptions about the underlying distribution [17,18].

Automated Moment Generation. To simplify the derivation of conditional moments, we developed an automated moment generator which is the main result of this work. The tool applies to any mass-action reaction network and arbitrary decomposition into selected and latent components. It is implemented in Python 3.7.9 and symbolically derives the expressions for the moment equations using the symbolic math package Sympy (v1.7.1) [16]. Once generated, the equations are saved to an output file and can be used for downstream applications. The source code of the generator, a tutorial, as well as code for the case study in Sect. 4 are available at https://github.com/zechnerlab/conditional-moment-generation. We will now briefly discuss the general usage and requirements of the software tool.

3 Usage

The tool requires as input the reactant and product stoichiometric matrices of the considered reaction network. Additionally, the user has to specify a set of *selected* species (all other species are assumed *latent*), the order up to which moment equations should be generated, and a moment closure scheme (Fig. 1a–b). In the present version of the tool, the normal [21] and lognormal [15,19] closures at 2nd and 3rd order are available. Optionally, names for the chemical species and reaction rate constants can be specified. If not provided, default names (A, B, C, ...; c_1, c_2, c_3, ...) are used, respectively. Using the command generateEquations(), a pipeline of operations is initiated (Fig. 1b). If an output file name is specified, all information about the reaction network, the filter equation, and the raw and closed conditional moment equations are saved to a .txt file and to a state file which captures the symbolic expressions for later use. The generated equations can be accessed either directly after generation (Fig. 1c) or by loading the saved file using the command loadAndLambdify(). Further information on commands and options can be found in the online tutorial.

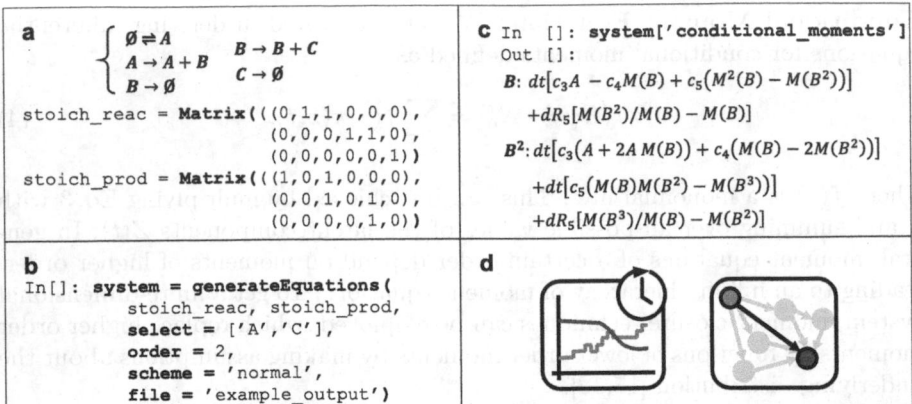

Fig. 1. Schematic workflow of the moment generator tool. (a) The user defines reactant and product stoichiometric matrices of the reaction network. (b) The command `generateEquations()` starts a pipeline to generate the moment equations. A list of selected species, the order up to which the equations are generated, a closure scheme and an output file name must be specified. (c) Generated conditional moment equations. (d) The output can be used for further applications such as subnetwork simulation [4] or the calculation of information transfer between molecular components [5].

4   Case Study

To demonstrate the use of the presented tool, we employ it in the context of subnetwork simulation. To this end, we consider a simple model of a bistable genetic switch that we studied also in our original work [4]:

$$\mathcal{R}_Z \begin{cases} \emptyset \underset{c_2}{\overset{c_1}{\rightleftharpoons}} R_i \\ R_i \xrightarrow{c_3} R_i + P_i \\ P_i \xrightarrow{c_4} \emptyset \\ G_i^* \xrightarrow{c_5} G_i^* + R_i \end{cases} \qquad \mathcal{R}_C \begin{cases} P_i + G_j^* \xrightarrow{c_6} P_i + G_j \\ \\ G_i \underset{c_8}{\overset{c_7}{\rightleftharpoons}} G_i^* \end{cases} \qquad \text{for } i,j = 1,2 \text{ and } i \neq j.$$

We denote by G_i^* and G_i the active and inactive promoters of two genes $i = 1, 2$, and their associated mRNA and protein molecules as R_i and P_i, respectively. The system incorporates positive feedback where the two proteins P_1 and P_2 mutually repress each other by enhancing the deactivation rate of the respective other promoter. Under certain parameter regimes, this model exhibits bistable behavior with trajectories switching between two meta-stable states [4]. Our goal is to use our conditional moment generator in combination with the selected-node stochastic simulation algorithm (snSSA) [4] to effectively simulate this system. The snSSA can be understood as a variant of Gillespie's stochastic simulation algorithm [9] where only a selected subset of the species is simulated explicitly. The dynamics of the remaining (i.e., latent) species are projected onto a set of conditional moment equations, which can be effectively integrated over time. In this way, the number of events that need to be simulated can be strongly reduced, often leading to a substantial speedup. For detailed information on how to use conditional moment equations for subnetwork simulation, the reader may refer to [4]. Here, we choose the active and inactive promoters G_i^* and G_i as selected species and use our moment generator to derive expressions for the 1st and 2nd conditional moments of the latent species R_i and P_i. With this choice, the reaction subsets $\mathcal{R}_C$ and $\mathcal{R}_Z$ are as indicated in the reaction scheme above. The generated moment equations were closed using the multivariate normal closure. The resulting equations were then used to run $n = 10000$ snSSA simulations of the network, which we compared to the results obtained from exact stochastic simulation. Our simulations reproduce the results from [4], where we found the probability distributions over the gene states to be very accurately approximated by the snSSA (Fig. 2a). While the distribution over the latent states is no longer simulated explicitly, it can be reconstructed from the snSSA simulations as a mixture distribution. Under the assumption that $\pi(z, t)$ is approximately Gaussian (inline with the employed closure function), we can reconstruct $P(Z(t) = z)$ by

$$P(Z(t) = z) = \mathbb{E}[P(Z(t) = z \mid X_0^t)] \approx \frac{1}{n} \sum_{i=1}^{n} \mathcal{N}(z \mid \mu^{(i)}, \Sigma^{(i)}), \qquad (5)$$

where $\mathcal{N}()$ is a Gaussian distribution with parameters $\mu^{(i)}$ and $\Sigma^{(i)}$ as the conditional means and covariances for a particular realization of X_0^t. Also for latent species, we found very good agreement with exact stochastic simulations (Fig. 2b), in line with our previous work [4].

For this reaction network and a final simulation time of $t = 3000$, the snSSA simulates on average only 19 reaction events while conventional SSA requires on average more than 2.8 million events, making snSSA about 460-times faster in this case. This extreme difference is due to the fact that the fast dynamics of the latent protein and mRNA species are handled through conditional moments and only the much slower promoter switching dynamics have to be simulated explicitly. In summary, this case study demonstrates how the presented tool can simplify the generation of conditional moment equations – here in the context of subnetwork simulation via the snSSA.

5   Discussion

We presented a Python tool to automatically generate conditional moment equations for arbitrary stochastic reaction networks. These moment equations are derived conditionally on full time trajectories of selected chemical species of the network, as needed for certain applications such as subnetwork simulation. Currently, the tool supports normal and lognormal moment closure approximations and is restricted to mass-action kinetics. For the future, we envision to incorporate further closure schemes and more complex, user-defined rate laws.

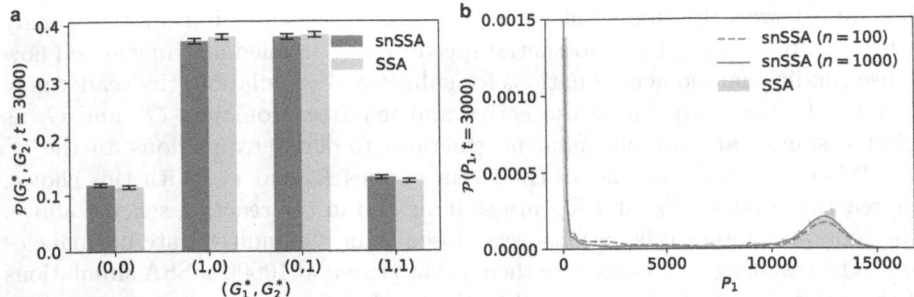

Fig. 2. snSSA and SSA display matching distributions for selected and latent species. (a) Frequencies $\pm$SE of the active promoters G_1^* and G_2^* at $t = 3000$ from $n = 10000$ snSSA and SSA simulations. (b) Probability density for the latent species P_1 estimated from $n = 100$ and $n = 1000$ snSSA simulations. For comparison, the relative frequencies of P_1 molecule numbers from $n = 10000$ SSA simulations are shown. Reaction rate constants: $\{c_1, c_2, c_3, c_4, c_5, c_6, c_7, c_8\} = \{0.01, 0.05, 5.0, 0.04, 10^{-6}, 5.0, 0.003, 10^{-6}\}$; initial conditions for SSA: $(G_i^*, G_i, P_i, R_i)_0 = (0, 1, 1, 1)$; initial conditions for snSSA: $(G_i^*, G_i)_0 = (0, 1)$, all 1st and 2nd order moments of P_i and R_i were set to 1.

Data and Code Availability. The code of the generator tool, a tutorial and the code implementing the case study are available at https://github.com/zechnerlab/conditional-moment-generation.

Acknowledgements. We would like to thank Tommaso Bianucci for testing the proposed moment generator tool. This work has been supported by core funding from the Max Planck Institute of Molecular Cell Biology and Genetics.

References

1. Albert, J.: A hybrid of the chemical master equation and the Gillespie algorithm for efficient stochastic simulations of sub-networks. PLoS ONE **11**(3), e0149909 (2016). https://doi.org/10.1371/journal.pone.0149909
2. Bain, A., Crisan, D.: Fundamentals of Stochastic Filtering. Springer, New York (2009). https://doi.org/10.1007/978-0-387-76896-0
3. Bronstein, L., Koeppl, H.: Marginal process framework: a model reduction tool for Markov jump processes. Phys. Rev. E **97**(6), 062147 (2018). https://doi.org/10.1103/PhysRevE.97.062147
4. Duso, L., Zechner, C.: Selected-node stochastic simulation algorithm. J. Chem. Phys. **148**(16), 164108 (2018). https://doi.org/10.1063/1.5021242
5. Duso, L., Zechner, C.: Path mutual information for a class of biochemical reaction networks. In: 2019 IEEE 58th Conference on Decision and Control (CDC), pp. 6610–6615. IEEE (2019). https://doi.org/10.1109/CDC40024.2019.9029316
6. Engblom, S.: Computing the moments of high dimensional solutions of the master equation. Appl. Math. Comput. **180**(2), 498–515 (2006). https://doi.org/10.1016/j.amc.2005.12.032

7. Fan, S., et al.: MEANS: python package for moment expansion approximation, inference and simulation. Bioinformatics **32**(18), 2863–2865 (2016). https://doi.org/10.1093/bioinformatics/btw229
8. Gardiner, C.: Stochastic Methods, Springer Series in Synergetics, vol. 4. Springer, Berlin (2009). https://link.springer.com/book/9783540707127
9. Gillespie, D.T.: Exact stochastic simulation of coupled chemical reactions. J. Phys. Chem. **81**(25), 2340–2361 (1977). https://doi.org/10.1021/j100540a008
10. Gillespie, D.T.: A rigorous derivation of the chemical master equation. Physica A **188**(1–3), 404–425 (1992). https://doi.org/10.1016/0378-4371(92)90283-V
11. Hasenauer, J., Wolf, V., Kazeroonian, A., Theis, F.J.: Method of conditional moments (MCM) for the chemical master equation. J. Math. Biol. **69**(3), 687–735 (2013). https://doi.org/10.1007/s00285-013-0711-5
12. Hespanha, J.P.: `StochDynTools` – a MATLAB toolbox to compute moment dynamics for stochastic networks of bio-chemical reactions, May 2007. https://web.ece.ucsb.edu/~hespanha/software/stochdyntool.html
13. van Kampen, N.G.: Stochastic Processes in Physics and Chemistry. Elsevier, 3rd (edn.) (2007)
14. Kazeroonian, A., Fröhlich, F., Raue, A., Theis, F.J., Hasenauer, J.: CERENA: ChEmical REaction Network Analyzer-a toolbox for the simulation and analysis of stochastic chemical kinetics. PLoS ONE **11**(1), e0146732 (2016). https://doi.org/10.1371/journal.pone.0146732
15. Keeling, M.J.: Multiplicative moments and measures of persistence in ecology. J. Theor. Biol. **205**(2), 269–281 (2000). https://doi.org/10.1006/jtbi.2000.2066
16. Meurer, A., et al.: SymPy: symbolic computing in Python. PeerJ Comput. Sci. **3**, e103 (2017). https://doi.org/10.7717/peerj-cs.103
17. Schnoerr, D., Sanguinetti, G., Grima, R.: Approximation and inference methods for stochastic biochemical kinetics-a tutorial review. J. Phys. Math. Theor. **50**(9), 093001 (2017). https://doi.org/10.1088/1751-8121/aa54d9
18. Singh, A., Hespanha, J.P.: Approximate moment dynamics for chemically reacting systems. IEEE Trans. Autom. Control **56**(2), 414–418 (2010). https://doi.org/10.1109/TAC.2010.2088631
19. Singh, A., Hespanha, J.P.: Lognormal moment closures for biochemical reactions. In: Proceedings of the 45th IEEE Conference on Decision and Control, pp. 2063–2068. IEEE (2006). https://doi.org/10.1109/cdc.2006.376994
20. Sukys, A., Grima, R.: MomentClosure.jl: automated moment closure approximations in Julia. Bioinformatics **38**(1), 289–290 (2022). https://doi.org/10.1093/bioinformatics/btab469
21. Whittle, P.: On the use of the normal approximation in the treatment of stochastic processes. J. Roy. Stat. Soc.: Ser. B (Methodol.) **19**(2), 268–281 (1957)
22. Zechner, C., Seelig, G., Rullan, M., Khammash, M.: Molecular circuits for dynamic noise filtering. Proc. Natl. Acad. Sci. **113**(17), 4729–4734 (2016). https://doi.org/10.1073/pnas.1517109113

An Extension of ERODE to Reduce Boolean Networks By Backward Boolean Equivalence

Georgios Argyris[1] , Alberto Lluch Lafuente[1] , Mirco Tribastone[2] ,
Max Tschaikowski[3] , and Andrea Vandin[4,1] (✉)

[1] DTU Technical University of Denmark, Kongens Lyngby, Denmark
andrea.vandin@santannapisa.it
[2] IMT School for Advanced Studies Lucca, Lucca, Italy
[3] University of Aalborg, Aalborg, Denmark
[4] Sant'Anna School for Advanced Studies, Pisa, Italy

Abstract. Boolean Networks (BN) are established tools for modelling biological systems. However, their analysis is hindered by the state space explosion: the exponentially many states on the variables of a BN. We present an extension of the tool for model reduction ERODE with support for BNs and their reduction with a recent method called Backward Boolean Equivalence (BBE). BBE identifies maximal sets of variables that retain the same value whenever initialized equally. ERODE has been also extended to support importing and exporting between different formats and model repositories, enhancing interoperability with other tools.

Keywords: Boolean Network · Backward Equivalence · Reduction

1 Introduction

Boolean Networks (BNs) [8] are established models for biological systems which have gained a lot of interest due to their simplicity; they consist of Boolean variables which denote active/inactive genes, high/low concentration of substances, etc. The variables are updated according to functions which are encoded into logical rules as we display in the left part of Fig. 1. This BN was published in [6], and models neurogenesis: the process by which nervous system cells, the neurons, are produced by neural stem cells.

A major hurdle in analyzing large BNs is the state space explosion, i.e., the presence of exponentially many *BN states*, the different configurations of (de)activation values of each variable, with respect to the number of BN variables. For example, Fig. 2 shows the state space of the BN of Fig. 1; the BN has 6 variables, leading to 2^6 states. For the tractability of large BNs, several reduction techniques have been proposed (e.g., [1,15,17]). One of the most popular reduction methods is based

Partially supported by the DFF project REDUCTO 9040-00224B, the Poul Due Jensen Grant 883901, the Villum Investigator Grant S4OS, and the PRIN project SEDUCE 2017TWRCNB.

I. Petre and A. Păun (Eds.): CMSB 2022, LNBI 13447, pp. 294–301, 2022.
https://doi.org/10.1007/978-3-031-15034-0_16

$$Her6(t+1) = \neg miR9(t) \wedge \neg N(t)$$
$$HuC(t+1) = \neg miR9(t) \wedge \neg P(t)$$
$$N(t+1) = HuC(t)$$
$$P(t+1) = Her6(t) \vee Zic5(t)$$
$$Zic5(t+1) = \neg miR9(t) \wedge \neg N(t)$$
$$miR9(t+1) = \neg Her6(t) \wedge \neg N(t)$$

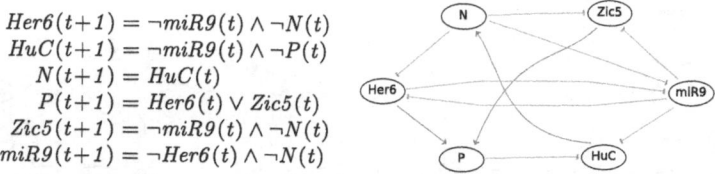

Fig. 1. A BN from [6]. (Left) the variables and update functions. (Right) an abstract graphical representation known as *interaction graph* where the nodes correspond to the variables, and the green/red arrows denote positive/negative effect to the activation value of the target variable, resp. (Color figure online)

on fast-slow decomposition, studied in [15,17] and implemented in GINsim [5]. Here, we present an extension of ERODE [3], a tool for modelling, analysis, and reduction of biological models that implements a complementary method of reduction for BNs called *Boolean Backward Equivalence* (BBE) [1]. Originally, ERODE was developed to support chemical reaction networks, and systems of ordinary differential equations [3]. Here, we present an extension to support BNs and the new importing and exporting capabilities between three different formats: a native format of ERODE to describe BNs (*.ode*), the *.bnet* format [10], and the *SBML-qual* format [4]. Notably, the latter is a standard for modelling biological systems[1]. These formalisms allow to interface with popular online BN model repositories like BioModelsDB [9] and the GinSim repository [11], as well as tools for BN analysis like those fostered by the COLOMOTO initiative [14].

2   Preliminaries

Model. A BN model is a pair (X, F) with X being a set of variables, and F a set of update functions. In the model of Fig. 1, the set of variables and the set of update functions are: $X = \{Her6, HuC, N, P, Zic5, miR9\}$ and $F = \{f_{Her6}, f_{HuC}, f_N, f_P, f_{Zic5}, f_{miR9}\}$ with, e.g., $f_{Her6} = \neg miR9 \wedge \neg N$.

BBE Partition. The crucial aspect of BBE is the notion of BBE partition (or BBE equivalence), which is a partition of the BN variables that satisfies the following criterion:

> *if the variables within each block have same activation value, they will retain the same value in all subsequent steps.*

An example of partition is $P^1 = \{\{Her6, HuC, N, P, Zic5, miR9\}\}$, which consists of one unique block. Another partition is $P^2 = \{\{Her6, Zic5, P\}, \{HuC\}, \{N\}, \{miR9\}\}$, which consists of four blocks. The partitions P^1, P^2 are not BBE partitions. Instead, $P^3 = \{\{Her6, Zic5\}, \{P\}, \{HuC\}, \{N\}, \{miR9\}\}$ is a BBE partition.

---

[1] The artifact can be downloaded from www.erode.eu/examples.html with further guidelines to replicate the experiments in this document.

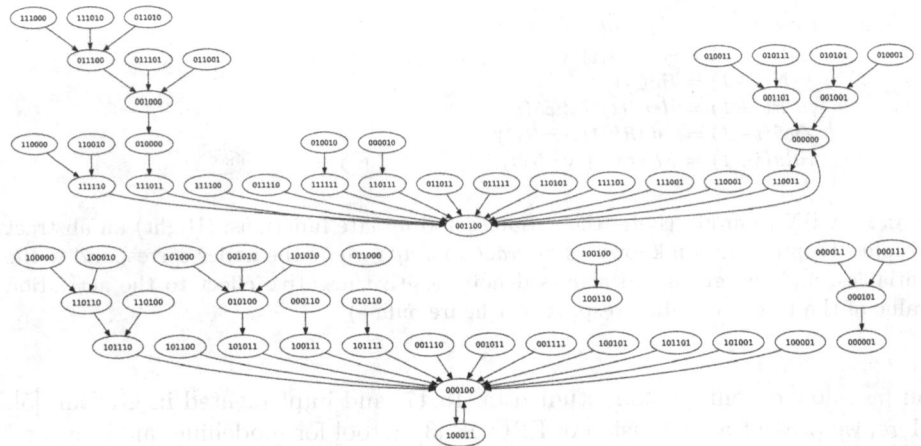

Fig. 2. The state transition graph (STG) of the BN of Fig. 1. An STG encodes the state space (nodes) and the dynamics (transitions) of a BN. The STG consists of 4 disconnected components. Each node contains digits denoting the activation values of each variable in that particular state. The transitions are obtained by synchronously applying the update functions in Fig. 1 to the activation values of the source state.

The BBE reduction method requires the user to specify an initial partition. Following a partition refinement approach [16], BBE proceeds by iteratively splitting the blocks of such partition until a BBE partition is obtained. The maximal BBE reduction of a BN can be obtained by using *trivial* initial partitions with one block only like P^1. By using P^2 as initial partition we get P^3.

Given a BBE partition, we can create a BBE-reduced BN containing one variable per partition block. We have shown in [1] that this can be used to study selected part of the original dynamics.

3   ERODE

Figure 3 provides a screenshot of ERODE. The middle panel provides the BN of Fig. 1 in ERODE format. The variables shall be declared in a block `begin init ... end init`. We illustrate by comment (`//`), how one could specify initial conditions for some of the variables (set to false by default).

The initial partition for the partition refinement algorithm can be specified in a block `begin partition ... end partition`. In the example of Fig. 3 we declare P^2.

Finally, we declare the update functions for each of the variables in a block `begin update functions ... end update functions`.

After BN definition which is encoded in the previous three blocks, we can provide either reduction or exporting commands. For example, BBE reduction is obtained with command `reduceBBE` which requires 3 parameters. The first, `fileWhereToStorePartition`, names the .txt file to store the blocks of

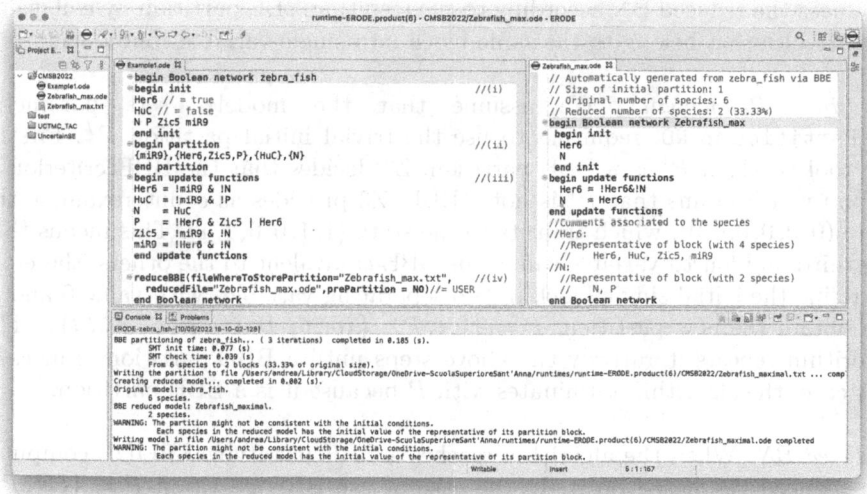

Fig. 3. ERODE GUI: (left) the project explorer; (middle) the BN from Fig. 1 in ERODE format; (right) The BBE reduction of the BN; (bottom) A console with log information. In (right) we see how to specify (i) the variables, (ii) a partition of the variables, (iii) the update functions, and (iv) the reduction commands.

the obtained BBE partition. The second parameter, `reducedFile`, names the ERODE file wherein the reduced BN is stored. We display this .ode file in the right window of Fig. 3. The parameter `prePartition` can take three values: `USER` to declare as initial partition the one specified above, `NO` for plugging the trivial partition wherein all variables belong to one block (e.g., P^1), or `IC` to define an initial partition according to the initial conditions specified by the user (one block for all variables initialized to true, and one for those initialized to false).

Before discussing the importing/exporting commands in Sect. 5, we provide the steps, implemented by ERODE, in our running example.

4   An Illustration of BBE Reduction

In this section, we exemplify how ERODE performs BBE reduction in the BN of Fig. 1. The reduction is summarized in three steps: *(i)* the first step is done by the modeller who provides the BN and an initial partition of the BN variables. ERODE implements the other two steps as follows: *(ii)* it splits the blocks P_i of the initial partition until a partition P that satisfies the BBE criterion is reached. The splitting is done by an iterative partition refinement algorithm [16]. In each iteration, the Z3 SAT solver [7], integrated in ERODE, checks the validity of the SAT-encoding of the BBE criterion (see [1]); if valid, the current partition is a BBE; if invalid, Z3 returns a counter-example according to which the splitting is performed. Once we get to a partition for which the formula is valid, *(iii)* ERODE produces the reduced BN according to the resulting BBE partition by collapsing all variables that belong to the same block into single variable components.

Partition Refinement. We assume that the modeler sets parameter prePartition to NO, requiring to use the trivial initial partition P^1. Firstly, the tool checks if P^1 is a BBE partition. Z3 decides that the BBE criterion is invalid which means that P^1 is not a BBE. Z3 provides as counterexample the state $(0,0,0,0,0,0)$, which transits to the state $(1,1,0,0,1,1)$. This means that the third and fourth variable cannot be BBE-equivalent to the others, therefore we refine the initial single block in two separating variables with value 0 and 1. We obtain the new partition: $P = \{\{N, P\}, \{Her6, HuC, Zic5, miR9\}\}$. The algorithm repeats iteratively the above steps until a BBE partition is met. In this case, the algorithm terminates with P because it is a BBE partition.

Reduced BN. When the algorithm reaches a BBE partition, ERODE computes the reduced BN based on it. In the case of the BBE partition $P = \{\{N, P\}, \{Her6, HuC, Zic5, miR9\}\}$, the variables N, P are collapsed into one component $x_{\{N,P\}}$, and the variables $Her6, HuC, Zic5, miR9$ into the variable component $x_{\{Her6,HuC,Zic5,miR9\}}$. The update function of the variable $x_{\{N,P\}}$ is given by selecting the update function of one variable (either N or P), and replacing each occurrence of an original variable with the new one corresponding to its block (i.e., N, and P, are replaced by $x_{\{N,P\}}$, and the others by $x_{\{Her6,HuC,Zic5,miR9\}}$). It can be shown that any update function of the variables in a block can be chosen without affecting the dynamics of the obtained reduced model. In this example we select the variables with the simplest function. We obtain:

$$x_{\{N,P\}}(t+1) = x_{\{Her6,HuC,Zic5,miR9\}}(t)$$
$$x_{\{Her6,HuC,Zic5,miR9\}}(t+1) = \neg x_{\{Her6,HuC,Zic5,miR9\}}(t) \wedge \neg x_{\{N,P\}}(t)$$

We display this BN in the right panel of Fig. 3, where variable $x_{\{N,P\}}$ is denoted by N, and the variable $x_{\{Her6,HuC,Zic5,miR9\}}$ by $Her6$.

Application of BBE Reduction for Model Analysis. Several tasks in model analysis are intractable due to the high dimensionality of BNs e.g., the generation of the STG, and the computation of attractors. Attractors are sets of states towards which the BN tends to evolve and remain. They are usually associated with important behaviours of the underlying system: for instance, different attractors correspond to different cell types in cell differentiation processes [2]. In [1], we present cases wherein BBE reduction can enable of facilitate these tasks.

5   Importing and Exporting Capabilities

Input and Output Variables. The variables of a BN can
be divided in 3 categories [13]: *inputs* which denote
external stimuli, *outputs* which model readout/response
of the modelled system, and internal variables. These
categories can be easily observed in the interaction graph
of a BN (e.g., Fig. 1 and Fig. 4). Inputs have no incoming edges, outputs have no outgoing edges, and internal variables have both incoming and outgoing edges.
ERODE features automatic identification of these three
categories basing on the update functions. Input variables can be identified in the BN model as variables that
are regulated only by themselves or have a stable update
function, i.e. the update function of an input variable x

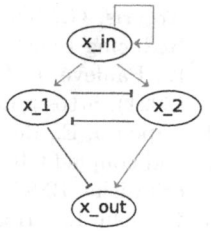

Fig. 4. The interaction
graph of a BN with 1
input (x_{in}) and 1 output (x_{out}).

has the form: x, 1, or 0. Instead, output variables do not appear in the update
functions of other variables.

Importing a Model. ERODE has importing capabilities from the SBML-qual
format [4], which is a standard format for biological models, and the .bnet format.
Importing can be done with the following commands:

```
importSBMLQualFolder(folderIn="BNs_sbml",folderOut="BNs_ode")
importBNetFolder(folderIn="BNs_bnet",folderOut="BNs_ode")
```

which load all models from folder `folderIn`, and store the ERODE versions
in folder `folderOut`. The commands have an additional optional parameter `guessPrepartition`, triggering the generation of corresponding `partition`
block. If set to `outputs`, the outputs are split in singleton blocks, while the
others belong to another single block. Similarly for `inputs`. We can also specify `outputsOneBlock` or `inputsOneBlock` in which cases we put all outputs (or
inputs) in the same block.

Exporting a Model. ERODE can export BNs, e.g. reduced ones, in the
above mentioned formats. This is done using commands `exportBoolNet` or
`exportSBMLQual`.

6   Conclusion

We extended ERODE to reduce Boolean Networks (BN) with Boolean Backward Equivalence (BBE) which collapses variables such that if initialized equally,
retain the same value in all subsequent steps. The scalability and the efficiency
the tool has been illustrated in [1] wherein we apply our method to the whole
GINsim repository. As future work, ERODE will be extended with further reduction techniques for BNs, complementary to BBE that we presented here. In our
future work, we also aim to incorporate ERODE in COLOMOTO notebook [12]
to further promote interoperability.

References

1. Argyris, G., Lluch Lafuente, A., Tribastone, M., Tschaikowski, M., Vandin, A.: Reducing Boolean networks with backward Boolean equivalence. In: Cinquemani, E., Paulevé, L. (eds.) CMSB 2021. LNCS, vol. 12881, pp. 1–18. Springer, Cham (2021). https://doi.org/10.1007/978-3-030-85633-5_1
2. Azpeitia, E., Benítez, M., Vega, I., Villarreal, C., Alvarez-Buylla, E.R.: Single-cell and coupled GRN models of cell patterning in the Arabidopsis thaliana root stem cell niche. BMC Syst. Biol. 4(1), 1–19 (2010)
3. Cardelli, L., Tribastone, M., Tschaikowski, M., Vandin, A.: ERODE: a tool for the evaluation and reduction of ordinary differential equations. In: Legay, A., Margaria, T. (eds.) TACAS 2017. LNCS, vol. 10206, pp. 310–328. Springer, Heidelberg (2017). https://doi.org/10.1007/978-3-662-54580-5_19
4. Chaouiya, C., et al.: SBML qualitative models: a model representation format and infrastructure to foster interactions between qualitative modelling formalisms and tools. BMC Syst. Biol. 7(1), 1–15 (2013)
5. Chaouiya, C., Naldi, A., Thieffry, D.: Logical modelling of gene regulatory networks with GINsim. In: van Helden, J., Toussaint, A., Thieffry, D. (eds.) Bacterial Molecular Networks. Methods in Molecular Biology, vol. 804, pp. 463–479. Springer, New York (2012). https://doi.org/10.1007/978-1-61779-361-5_23
6. Coolen, M., Thieffry, D., Drivenes, Ø., Becker, T.S., Bally-Cuif, L.: miR-9 controls the timing of neurogenesis through the direct inhibition of antagonistic factors. Dev. Cell 22(5), 1052–1064 (2012)
7. de Moura, L., Bjørner, N.: Z3: an efficient SMT solver. In: Ramakrishnan, C.R., Rehof, J. (eds.) TACAS 2008. LNCS, vol. 4963, pp. 337–340. Springer, Heidelberg (2008). https://doi.org/10.1007/978-3-540-78800-3_24
8. Kauffman, S.: Metabolic stability and epigenesis in randomly constructed genetic nets. J. Theor. Biol. 22(3), 437–467 (1969)
9. Malik-Sheriff, R.S., et al.: BioModels–15 years of sharing computational models in life science. Nucleic Acids Res. 48(D1), D407–D415 (2020). https://doi.org/10.1093/nar/gkz1055
10. Müssel, C., Hopfensitz, M., Kestler, H.A.: Boolnet: an R package for generation, reconstruction and analysis of Boolean networks. Bioinformatics 26(10), 1378–1380 (2010)
11. Naldi, A., Berenguier, D., Fauré, A., Lopez, F., Thieffry, D., Chaouiya, C.: Logical modelling of regulatory networks with GINsim 2.3. Biosystems 97(2), 134–139 (2009). https://doi.org/10.1016/j.biosystems.2009.04.008, https://www.sciencedirect.com/science/article/pii/S0303264709000665
12. Naldi, A., et al.: The CoLoMoTo interactive notebook: accessible and reproducible computational analyses for qualitative biological networks. Front. Physiol. 9, 680 (2018)
13. Naldi, A., Monteiro, P.T., Chaouiya, C.: Efficient handling of large signalling-regulatory networks by focusing on their core control. In: Gilbert, D., Heiner, M. (eds.) CMSB 2012. LNCS, pp. 288–306. Springer, Heidelberg (2012). https://doi.org/10.1007/978-3-642-33636-2_17
14. Naldi, A., et al.: Cooperative development of logical modelling standards and tools with CoLoMoTo. Bioinformatics 31(7), 1154–1159 (2015)

15. Naldi, A., Remy, E., Thieffry, D., Chaouiya, C.: Dynamically consistent reduction of logical regulatory graphs. Theoret. Comput. Sci. **412**(21), 2207–2218 (2011)
16. Paige, R., Tarjan, R.E.: Three partition refinement algorithms. SIAM J. Comput. **16**(6), 973–989 (1987)
17. Veliz-Cuba, A.: Reduction of Boolean network models. J. Theor. Biol. **289**, 167–172 (2011)

eBCSgen 2.0: Modelling and Analysis of Regulated Rule-Based Systems

Matej Troják[✉], David Šafránek, Branislav Brozmann, and Luboš Brim

Systems Biology Laboratory, Masaryk University, Brno, Czech Republic
trojak@mail.muni.cz

Abstract. eBCSgen is a software tool for developing and analysing models written in Biochemical Space Language (BCSL). BCSL is a rule-based language designed for the description of biological systems with rewriting rules in the form of behavioural patterns. This tool paper describes a new version of the tool, implementing the support for regulations, a mechanism suitable for reducing the branching behaviour of concurrent systems. Additionally, the presented version provides export to SBML, and support for CTL model checking. The paper artefact is available via https://doi.org/10.5281/zenodo.6644973.

1 Introduction

Rule-based modelling is a well-established approach in system biology. It makes a natural extension to the mechanistic reaction-based approach used in chemistry by introducing abstraction of detailed properties of molecules, forming the behavioural *patterns*. These are then used in rewriting *rules* that allow avoiding the combinatorial explosion that occurs when underlying molecules are specified explicitly in reactions, thus allowing to express complex models compactly. There are multiple existing rule-based modelling languages [4,7,9,12,13,15,16,22,24]. To improve the usage and the interoperability among individual languages, the rule-based approach is supported by SBML [11] in terms of the package multi [23].

While the mechanistic approach is beneficial for understanding the modelled system, some processes or properties are not easy to describe in such detail. They can often be expressed quantitatively, but such information is not always easy or even possible to determine. Still, they are crucial to capturing the behaviour of biological phenomena. In [20], we introduced *regulation* mechanisms in the context of multiset rewriting systems, and in [21], we formalised them as an extension of a rule-based languages representative Biochemical Space Language (BCSL) [22]. In general, regulations influence conditions when a rewriting rule can be applied. The regulations allow compactly modelling the additional knowledge about the system, otherwise too laborious or even impossible to express. We support five types of regulations – *regular* given by regular language over

This work has been supported by the Czech Science Foundation grant 22-10845S.

I. Petre and A. Păun (Eds.): CMSB 2022, LNBI 13447, pp. 302–309, 2022.
https://doi.org/10.1007/978-3-031-15034-0_17

rules, *ordered* given by strict partial order on rules, *programmed* given by successors for each rule, *conditional* given by prohibited context for each rule, and *concurrent-free* given by resolving concurrent behaviour of multiple rules.

In [18], we introduced eBCSgen, a software tool to provide a development environment and analyse models written in BCSL. The tool is integrated into Galaxy [2], a web-based platform for data-intensive biomedical research. It provides a convenient way to use eBCSgen due to the extensive popularity of Galaxy in the biology-oriented community. Among other features, the tool provides an interactive model editor for the development of models, simulations, and analysis of the model with respect to PCTL [10] properties.

In this paper, we present the second version of eBCSgen. Compared to the original version described in [18], the new version supports models with regulations, a non-probabilistic variant of CTL model checking, and model export to SBML [11] standard. The regulations are directly integrated into the core of the tool, which required redesigning the architecture. Consequently, it enabled to perform simulations and analyses even on models with a defined regulation. We employ an explicit CTL model checking [5,6] to support analysis of the branching behaviour on a qualitative level (in addition to quantitative PCTL analysis [19]), which can be directly influenced by the effects of regulations. To join the community effort on interoperability among other rule-based languages, we implemented support for model export to SBML [11] standard using the package multi [23].

eBCSgen is available as a standalone Python package [17] distributed using conda package manager [1] in the bioconda channel [8]. Moreover, it can be accessed online within our Galaxy instance[1]. It is accompanied with a tutorial[2] on how to use the tool in the Galaxy environment and a detailed explanation of regulations on a running example.

2 Regulated Biochemical Space Language

In this chapter, we briefly summarise the main features of BCSL using an illustrative yet simplistic example. Then we intuitively explain the effects of regulations in general and describe individual classes of regulation approaches introduced in [20] on multiset rewriting systems (MRS). The regulations are formally established within BCSL in a technical report in [21], where MRS are used as a low-level formal basis for BCSL. To grasp regulations in BCSL in more detail, we recommend studying the language itself [22] and then going through chapter four of tutorial (See footnote 2), providing a comprehensive description of individual regulations on illustrative examples.

In Fig. 1, there is an example of BCSL model. It consists of a definition of three rewriting *rules*, each tagged by a unique label (r1_S, r1_T, and r2). The rules describe possible modifications of a molecule P. This molecule has two domains, S and T, with an internal state (i for inactive and a for active).

---

[1] https://biodivine-vm.fi.muni.cz/galaxy/.
[2] https://biodivine.fi.muni.cz/galaxy/eBCSgen/tutorial.

Additionally, the molecule has assigned a physical compartment, denoted by its name after a double colon. The model is initialised, containing a single molecule of an inactive form of P.

```
#! rules
r1_S ~ P(S{i})::cell ⇒ P(S{a})::cell
r1_T ~ P(T{i})::cell ⇒ P(T{a})::cell
r2   ~ P()::cell      ⇒ P()::out

#! inits
1 P(S{i},T{i})::cell
```

Fig. 1. Example of BCSL model. It consists of a set of rules and an initial state, containing a single molecule of inactive form of P.

The rule r1_S can change the internal state of domain S from inactive to active. Similarly, the rule r1_T can do the same with the domain T. Please note such a modification can be done regardless of the internal state of the other domain. This is the main feature of the rule-based approach, which allows us to omit context which does not play any role in the modification. We often say that such a rule describes a *pattern* instead of a particular biochemical interaction. Finally, the rule r2 can export the molecule P outside the cell by changing its compartment to value out.

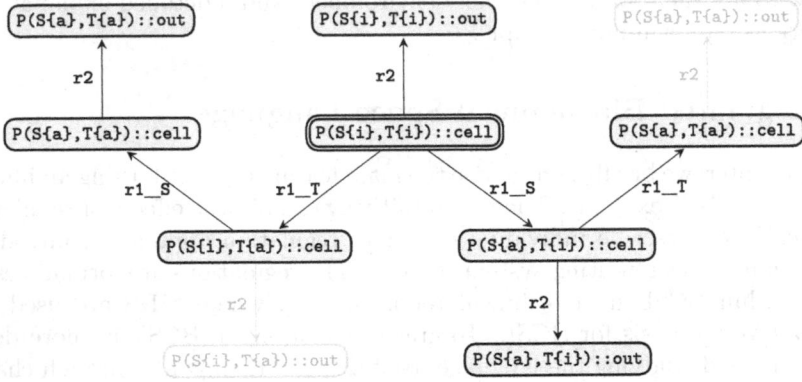

Fig. 2. Set of runs compactly represented by a transition system, starting in the double circled initial. In this case, the states always contain a single molecule P in a particular form. The label on an edge represents the rule responsible for a particular transition. The grey states and transitions are absent due to the effects of the regulation.

The semantics of BCSL model is based on *rewriting* process. A state, a (multi)set of molecules, can be rewritten by a rule to another state. There must be enough molecules present in the state that fit the pattern described by the rule. By chaining such rewriting, a *run* is formed. Transitive rewriting of the initial state by the rules forms a set of all possible runs, usually represented compactly by a transition system. The transition system of the model from Fig. 1 is available in Fig. 2 (including greyed-out states and transitions).

Regulated rewriting poses additional conditions on the rewriting process. A particular mechanism depends on an individual regulation approach, but it usually depends on the history of rewriting in the current (the system is not memoryless) or concurrent (the parallelism of biological systems) run. In general, the regulation restricts the set of runs, rendering some of the runs prohibited in the regulated model. In the following, we briefly present established regulation approaches.

Regular regulation is defined as a regular language over the rules. For practical reasons, it can be specified using a regular expression. Having a transition system, each run can be labelled by a sequence of used rules in the respective order. For the transition system of the regulated model then holds that every run must have its label from the defined regular language. This regulation is specifically useful in cases when we want to explicitly target particular runs that correspond to the desired behaviour of the model.

Ordered regulation is given by a partial order over the rules. It can be specified as a set of pairs. Focusing on each subsequent pair of rules in a run, it is required that the latter rule is greater than the former one. Every run in the transition system of the regulated model has to respect this order, otherwise it is eliminated. This regulation can be applied in cases when we want to ensure that some rules are not used immediately after each other in a sequence.

Programmed regulation specifies a set of successor rules for every rule. It is given by a function assigning an allowed set of successors to every rule. Each rule sequence can only be extended by the successor of the last rule. In terms of the regulated transition system, this eliminates runs that do not respect the successor function.

We provide an example to support this regulation. Let us assume the successor function for model from Fig. 1 given by r1_S → {r1_T, r2} and r1_T → {r1_S}. It allows specified successors for rules r1_S and r1_T, while the successors for rule r2 are not given, which is considered as an empty set by default. The effect of the regulation is illustrated in Fig. 2 by greyed-out states and transitions. In particular, the rule r2 is not allowed to be used after the rule r1_T, resulting in the elimination of two states.

Conditional regulation is a natural extension to the rewriting mechanism where instead of requiring the presence of some molecules in the state, it requires their absence. The regulation is specified by a function assigning prohibited context to every rule. Every transition of every run in the transition system of the regulated model is enabled only if the prohibited context is not present in the corresponding state. Conditional regulation is beneficial when we need to eliminate transitions under precise conditions while allowing the respective rule to be applied in all the other cases.

Concurrent-free regulation is targeting a natural phenomenon of concurrency in biological processes. It allows controlling concurrent processes by choosing the prioritised ones. The regulation is defined by an enumeration of pairs of rules, where the first member of the pair has priority over the second one. In the regulated transition system, branches of runs with a non-prioritised rule while a prioritised one is also enabled are disposed of.

3   Implementation

eBCSgen is implemented as a command-line Python package distributed using conda package manager [1] in the bioconda channel [8]. To improve its usability outside of computer science, it is wrapped into a series of tools for Galaxy [2], a web-based scientific analysis platform with three primary features – accessibility, reproducibility, and communication. It allows connecting individual tools into comprehensive pipelines and distribution of results among other scientists by sharing computation histories and datasets.

The previous version of the tool already supports a variety of analysis techniques and features. The Galaxy interface of eBCSgen offers interactive *model editor*, which can be used to create and edit BCSL models with automatic syntax highlighting and real-time code validation. Models with quantitative attributes (rule rates) exhibiting probabilistic behaviour can be analysed with respect to PCTL [10] properties. Such models can also be simulated, either using a *stochastic* (modified version of the standard Gillespie algorithm) or deterministic (inferring ODEs from generated reaction network) approach. Finally, there are several *static analysis* techniques to improve the scalability issues of exhaustive computational methods.

The major contribution to the new version of eBCSgen is the support of regulated models. This novelty influenced the tool on several levels. We extended the syntax of the BCSL and introduced these changes to the interactive editor. The presence of regulation in a model directly impacts the rewriting process of individual rules. This required changing the core of eBCSgen to support checking these extra conditions. Since the regulations typically require some sense of the history of the run, this also had to be included in the rewriting mechanism. To be more precise, individual regulations differ in their requirements on the depth of the history. Conditional and concurrent-free regulations do not need history. They can be validated directly based on local information (by inspecting the contents of the current state and checking other enabled rules, respectively). Ordered and programmed regulations require one-step history – to determine whether a rule can be used, information about the previously used rule is needed. Finally, regular regulation needs full history of used rules to determine whether the run belongs to the specified regular language.

Besides regulations, we also implemented support for CTL model checking. The main motivation behind this step was to support CTL analysis of qualitative models and to be able to analyse the behaviour of such models with regulations. Indeed, the inherent feature of regulations is to reduce the branching of the transition system, making CTL model checking an ideal tool for validation of its effects.

We implemented this feature using a python package `pyModelChecking` [5] and created a Galaxy wrapper.

Finally, to join the community effort on interoperability among other rule-based languages, we implemented support for model export to SBML standard using the package `multi`. The package extends the SBML Level 3 core with the *type* concept, and therefore reaction rules may contain species that can be patterns and be in multiple locations in reaction rules. It allows the SBML standard for encoding rule-based models using their native concepts for describing reactions instead of having to apply the rules and unfold the networks prior to encoding in an SBML format. This allows us to save BCSL models in a standard format, which enables their analysis beyond the scope of eBCSgen. We implemented the export using a package `libSBML` [3].

4 Evaluation

We demonstrated regulations on several models from biological domain[3]. In this section, we show usage of regulations on a well-studied fragment of `MAPK/ERK` signalling pathway [14], responsible for signal transduction from a receptor on the surface of the cell to the DNA in the nucleus of the cell. `MAPK` cascade contains three interconnected cycles of `MAPK`, `MAPK` kinase (`MAPKK`), and `MAPKK` kinase (`MAPKKK`). To capture the inhibitory effect of active `MAPKKK` on the activity of `MAPK`, we have applied conditional regulation to the model. It ensures that rule `r3k_fw` (responsible for the activation of `MAPKKK`) is not enabled when the active form of `MAPK` is present.

This regulation, in consequence, causes that when the deactivated form of `MAPKKK` and the activated form of `MAPK` coexist, then the activation of `MAPKKK` is not possible. To enable it again, `MAPK` first needs to be deactivated. In terms of biological effects, the active form of `MAPK` inhibits the activation of `MAPKKK`.

$$\text{EF [MAPK(R1\{a\},R2\{a\})::cyt} > 0 \ \wedge \ \text{MAPKKK(R\{i\})::cyt} > 0 \ \wedge \ \text{EX [P(S\{a\},T\{a\})::out} > 0]] \quad (1)$$

To support this claim about the behaviour of the regulated model, we perform CTL model checking. The property formulated above can be expressed as a CTL formula in Eq. 1. The analysis concludes that while this formula is indeed `true` in the unregulated system, its truth value changes to `false` in the regulated system, confirming the desired behaviour effects were achieved. CTL analysis of both unregulated and regulated models is available as a Galaxy history[4].

5 Conclusion

We presented the tool eBCSgen with a focus on new features introduced in version 2.0. The new version supports the development and analysis of models written in BCSL with regulations. These mechanisms can be used to ensure

---

[3] https://biodivine.fi.muni.cz/galaxy/eBCSgen/case-studies/cmsb-2022-regulations.
[4] https://biodivine.fi.muni.cz/galaxy/eBCSgen/case-studies/cmsb-2022.

the correct sequence of execution of the individual processes and their mutual effects. We showed how the regulations could be used in a short case study and demonstrated their effects using CTL model checking, another newly introduced feature to eBCSgen. Finally, the tool supports SBML export as a key step toward standardisation of the rule-based representation.

References

1. Anaconda software distribution (2020). https://docs.anaconda.com/
2. Afgan, E., et al.: The Galaxy platform for accessible, reproducible and collaborative biomedical analyses: 2018 update. Nucleic Acids Res. **46**(W1), 537–544 (2018)
3. Bornstein, B.J., Keating, S.M., Jouraku, A., Hucka, M.: libSBML: an API library for SBML. Bioinformatics **24**(6), 880–881 (2008)
4. Calzone, L., Fages, F., Soliman, S.: BIOCHAM: an environment for modeling biological systems and formalizing experimental knowledge. Bioinformatics **22**(14), 1805–1807 (2006)
5. Casagrande, A.: pyModelChecking (2022). https://pypi.org/project/pyModelChecking
6. Clarke, E.M.: Model checking. In: Ramesh, S., Sivakumar, G. (eds.) FSTTCS 1997. LNCS, vol. 1346, pp. 54–56. Springer, Heidelberg (1997). https://doi.org/10.1007/BFb0058022
7. Danos, V., Laneve, C.: Formal molecular biology. Theoret. Comput. Sci. **325**, 69–110 (2004)
8. Grüning, B., et al.: Bioconda: sustainable and comprehensive software distribution for the life sciences. Nat. Methods **15**(7), 475–476 (2018)
9. Harris, L.A., et al.: BioNetGen 2.2: advances in rule-based modeling. Bioinformatics **32**, 3366–3368 (2016)
10. Hasson, H., Jonsson, B.: A logic for reasoning about time and probability. FAOC **6**, 512–535 (1994)
11. Hucka, M., et al.: The systems biology markup language (SBML): a medium for representation and exchange of biochemical network models. Bioinformatics **19**, 524–531 (2003)
12. Lopez, C.F., Muhlich, J.L., Bachman, J.A., Sorger, P.K.: Programming biological models in python using PySB. Mol. Syst. Biol. **9**, 646 (2013)
13. Maus, C., Rybacki, S., Uhrmacher, A.M.: Rule-based multi-level modeling of cell biological systems. BMC Syst. Biol. **5**(1), 1–20 (2011). https://doi.org/10.1186/1752-0509-5-166
14. Pearson, G., et al.: Mitogen-activated protein (MAP) kinase pathways: regulation and physiological functions. Endocr. Rev. **22**(2), 153–183 (2001)
15. Pedersen, M., Phillips, A., Plotkin, G.D.: A high-level language for rule-based modelling. Plos One **10**(6), 1–26 (2015)
16. Romers, J.C., Krantz, M.: rxncon 2.0: a language for executable molecular systems biology. bioRxiv (2017)
17. Troják, M.: eBCSgen: a bioconda package (2022). https://anaconda.org/bioconda/ebcsgen
18. Troják, M., Šafránek, D., Mertová, L., Brim, L.: eBCSgen: a software tool for biochemical space language. In: Abate, A., Petrov, T., Wolf, V. (eds.) CMSB 2020. LNCS, vol. 12314, pp. 356–361. Springer, Cham (2020). https://doi.org/10.1007/978-3-030-60327-4_20

19. Troják, M., Šafránek, D., Mertová, L., Brim, L.: Parameter synthesis and robustness analysis of rule-based models. In: Lee, R., Jha, S., Mavridou, A., Giannakopoulou, D. (eds.) NFM 2020. LNCS, vol. 12229, pp. 41–59. Springer, Cham (2020). https://doi.org/10.1007/978-3-030-55754-6_3
20. Troják, M., Pastva, S., Šafránek, D., Brim, L.: Regulated multiset rewriting systems (2021). https://arxiv.org/abs/2111.13036
21. Troják, M., Šafránek, D., Brim, L.: Biochemical space language in relation to multiset rewriting systems (2022). https://arxiv.org/abs/2201.08817
22. Troják, M., Šafránek, D., Mertová, L., Brim, L.: Executable biochemical space for specification and analysis of biochemical systems. PLoS ONE **15**(9), 1–24 (2020)
23. Zhang, F., Meier-Schellersheim, M.: Multistate, multicomponent and multicompartment species package for SBML level 3. COMBINE Specifications (2017)
24. Zimmer, R.H., Millar, A.J., Plotkin, G.D., Zardilis, A.: Chromar, a language of parametrised objects. Theor. Comput. Sci. (2017)

Author Index

Printed in the United States
by Baker & Taylor Publisher Services